T0222417

"This book is a major contribution to the history of modern skepticism. It will also be of considerable interest to the history and philosophy of science more generally. Martin Gardner comes back to life in the form of sparkling, never-before-seen correspondence with one of the most enigmatic figures of skepticism, sociologist and gadfly (and eventual critic) Marcello Truzzi. The lively and surprisingly substantive exchanges are filled with fascinating insights into the intense intellectual debates and arguments they and others had over how to identify crank science and scientists and how best — or even whether — to counter pseudoscience and occultism The letters animate the divergent perspectives and personalities of key figures who pioneered the effort. Editor Dana Richards (George Mason University) expertly guides us through the nuanced issues."

— KENDRICK FRAZIER
Editor, *Skeptical Inquirer*

"The skeptical movement today has gone global and occupies a central place in popular culture, but in the 1970s it was nothing more than an idea in letters exchanged between a handful of people concerned about the growth of pseudoscience and irrationalism. First among equals as founders of this movement were Martin Gardner and Marcello Truzzi, whose correspondence — so thoughtfully edited in this splendid volume — covers some of the most important subjects in science, philosophy, and culture.

A page turner that belongs on the bookshelves of every thinking reader."

— MICHAEL SHERMER
Publisher, *Skeptic* magazine; columnist, *Scientific American*,
and Presidential Fellow, Chapman University

Dear Martin Dear Marcello

Gardner and Truzzi on Skepticism

Edited by

Dana Richards
George Mason University, USA

World Scientific

NEW JERSEY · LONDON · SINGAPORE · BEIJING · SHANGHAI · HONG KONG · TAIPEI · CHENNAI · TOKYO

Published by

World Scientific Publishing Co. Pte. Ltd.

5 Toh Tuck Link, Singapore 596224

USA office: 27 Warren Street, Suite 401-402, Hackensack, NJ 07601

UK office: 57 Shelton Street, Covent Garden, London WC2H 9HE

Library of Congress Cataloging-in-Publication Data
Names: Gardner, Martin, 1914–2010, correspondent. |
    Truzzi, Marcello, correspondent. | Richards, Dana, editor.
Title: Dear Martin, dear Marcello : Gardner and Truzzi on skepticism /
    edited by Dana Richards (George Mason University, USA).
Description: 1st edition. | New Jersey : World Scientific Publishing Co. Pte. Ltd, 2017. |
    Includes index.
Identifiers: LCCN 2017002301| ISBN 9789813203693 (hardcover : alk. paper) |
    ISBN 9789813203709 (pbk. : alk. paper)
Subjects: LCSH: Parapsychology--Miscellanea. | Occultism--Miscellanea. |
    Pseudoscience--Miscellanea. | Skepticism--Miscellanea. | Gardner, Martin,
    1914–2010--Correspondence. | Truzzi, Marcello--Correspondence.
Classification: LCC BF1040 .G37 2017 | DDC 130.92/2--dc23
LC record available at https://lccn.loc.gov/2017002301

British Library Cataloguing-in-Publication Data
A catalogue record for this book is available from the British Library.

Printed in Singapore

Contents

Introduction

Marcello Truzzi was a sociologist of science. Martin Gardner was a philosopher of science. Both were prolific writers. Both were strong voices in the fight against pseudoscience. However their legacy was one of public disagreements. Both were diligent letter writers and their correspondence has filled many scrapbooks. In the future all of their correspondence should be published. In this book we restrict our attention to their mutual correspondence to fully explain their public stances.

We will begin by describing the two men. Martin Gardner's life story is better known, so we will only emphasize the facts that allow parallels to be drawn with Marcello Truzzi's biography. We will then explain the core topic of their correspondence, along with some background about the people and organizations the reader may need help with.

Marcello Truzzi

Marcello Truzzi was born in Copenhagen in 1935 into an interesting Italian-Russian family. There had been many generations in circus life. His grandfather was an impressario in Moscow and other cities and owned a circus. His father was a famous juggler, whose legacy is strong today. By 1940 his father was in America as the center-ring juggler for the Barnum and Bailey Circus. Marcello learned to juggle, worked as a clown and had other jobs off-stage. Circus life certainly informed Marcello's view on sociology. He learned the public and private aspects of performance and deception.

He also became fascinated by magic, occasionally publishing tricks and effects in magic journals. He was active in national magic organizations (such as the International Brotherhood of Magicians) and local "rings," where he talked about the occult, in addition to traditional magic. He was a member of the Order of Merlin. In the 1980s he became quite active in the Psychic Entertainers Association, a group of magicians who specialized in mentalist routines.

He majored in sociology at Florida State University, and then studied law at the University of Florida. However he switched back to sociology, getting his Masters' degree. His doctorate in sociology was from Cornell. He taught at Cornell, the University of Southern Florida, the University of Michigan, and New College (in Sarasota) before coming to Eastern Michigan University (Ypsilanti) to serve as chairman.

His interest in science was not immediate. His earliest publications (starting in 1966) were typical of his profession. However he quickly started to specialize in the sociology of the circus, a topic he returned to often in his published work:

> "The American Circus as a Source of Folklore: An Introduction," 1966
> "The Decline of the American Circus: The Shrinkage of an Institution," 1968
> "Folksongs of the American Circus," 1968
> "Lilliputians in Gulliver's Land: The Social Role of the Dwarf," 1968
> "Carnival, Road Shows and Freaks," 1972
> "Towards an Ethnography of the Carnival Social System," 1972
> "Introduction: Circuses, Carnivals and Fairs," 1972
> "Notes Towards a History of Juggling," 1974
> "Circus and Side Shows," 1979
> "On Keeping Things Up in the Air," 1979

His interest in pseudoscience was not directly from this background.

Cornell had a fine collection of literature on the occult. He was fascinated by the counter-culture sociological aspects of the growing interest in the occult. In 1969 he offered an adult-education class "Witchcraft, Black Magic, and Modern Occultisms," which drew 225 registrants. Thereafter he wrote "Definitions and Dimensions of the Occult: Towards a Sociological Perspective" (1971) and "Towards a Sociology of the Occult: Notes on Modern Witchcraft" (1972). This led to his pivotal article "The Occult Revival as Popular Culture: Some Random Observations on the Old and Nouveau Witch" (*Sociological Quarterly*, 1972).

The "occult revival" was the motivation for Marcello's interest in pseudoscience. In that article he gives an array of numbers documenting the phenomenal growth of the public's interest in all things occult; for example, two-fold yearly increases in the number of books. His article was largely descriptive and taxonomic. He identified four "foci": astrology, witchcraft/satanism, parapsychology, and Eastern religious thought. He only discusses the first two, since parapsychology was "not undergoing any significant increase in popularity," and Eastern mysticism peaked with the beat-poets in the 1950s. (A fifth miscellaneous group includes latter-day prophets, strange monsters, and UFOs, among others. He ignores these since they "have small scope and influence or are in an actual state of decline.")

In his discussion of astrology he does not use an "open-minded" tone. He describes it as a "pseudo-science" and refers the reader to "reviews of the scientific repudiation of astrology." (He favorably stressed one such repudiator, Gauquelin, who later became a contentious figure.) "To [most astrology-believers], the heart of the matter, is not that they use science as the basis for their acceptance of astrology as truth but that science must catch up with their truth." He concludes astrology, for the majority of believers, is a leisure-time activity and not a search for spiritual truth.

On the subject of witchcraft and satanism, which he is careful to separate, he has more to say. Presumably that is because in the previous years he had been researching and discussing witchcraft. He concedes there are hard-core elements but most involvement has a sociological basis (sex cults, reaction to organized religion, etc.). He concludes that most of the occult focus can be viewed as "pop religion," a "playful contempt for what many once viewed seriously." In fact he is optimistic: "The more we eliminate these old fears and myths, the more we develop a naturalistic rationalism, a scientific view of the universe."

After 1972 Marcello must have noticed that pseudoscience (parapsychology, UFOs, etc.) was not in a "state of decline." Even if he wrote very little on these topics, we can tell by his correspondence he is adding to his library (which was extensive) and that his interests were broadening.

He edited school humor magazines at Florida State University and the University of Florida. *The Subterranean Sociology Newsletter* was Marcello's first serious effort as an editor/publisher. Started at the University of South Florida (Tampa) in 1967, it was published irregularly for a decade, consisting of updates and playful news in the field (because "I was ... disenchanted with pomposity"). In 1973 he started another low-key newsletter called *Explorations* concerned with the occult and pseudo-science. It was soon retitled *The Zetetic*. When CSICOP was formed (see below), their official organ was also called *The Zetetic*. Since Marcello was the editor of the new magazine, he ceased publishing his own newsletter with that title. Soon he parted ways with CSICOP and as a result the CSICOP's publication was thereafter called *The Skeptical Inquirer* and Marcello started a smaller journal *The Zetetic Scholar* with a very different editorial board.

Marcello died in 2003.

Martin Gardner

Martin Gardner was born in Tulsa, Oklahoma, in 1914, to an upper-middle-class family. His father, with a doctorate in geology, became a successful independent oil man. Martin was always interested in science. His mother was a staunch Methodist, and his father was a pantheist. The arc of his religious beliefs varied from Protestant fundamentalism to "philosophical theism." What is relevant for us is that this engendered an early skeptical outlook; initially he believed, and then rejected, the fundamentalists' young earth geology.

Martin developed an early interest in magic, publishing literally hundreds of tricks, from 1930 until 2010. While he loved to meet with fellow magicians he was loath to perform in public. He considered magic as exciting the scientific mind of the audience, in the sense that trying to figure out a trick is like trying to understand Nature.

Hoping to major in science, Martin attended the University of Chicago. However he changed to the philosophy of science and graduated in 1936, with honors. He took a desultory series of graduate courses, both before and after the war, including some at the Chicago Theological Seminary. The most influential professor he had was Rudolph Carnap. Gardner's personal library was enormous, including an extraordinary number of works on the philosophy of science. He dwelt on this subject far more than he wrote about it. His interest was a personal quest, not an academic exercise.

In 1950 Martin wrote "The Hermit Scientist" for *Antioch Review* about UFOs, Velikovsky, and Dianetics. These topics were just a few of the many pseudo-scientific fads at the time; a 1950s-era "occult revival." These were expanded in the book *In the Name of Science* (1952) that enjoyed a good critical reception (but sales were weak until the Dover paperback reprint *Fads and Fallacies*, 1957). From testimonials we know that many, if not most, of today's skeptics were influenced by this book. After this, Martin moved on to other subjects, returning to pseudo-science because of the same occult revival (circa 1970) that attracted Marcello's attention. (Martin had no interest in the unscientific aspects of the occult, such as satanism.)

As more data flowed to Gardner he wrote more, and his influence grew even larger. He had a column in the *Skeptical Inquirer* for decades. In time he changed the name of the column to "Notes of a Fringe-Watcher" to explicitly indicate that *some* of his subjects were not necessarily crazy, "beyond the pale," but were merely on the fringe of scientific orthodoxy. His style was scientific but journalistic. He rarely did any original research but reported on others' work.

Martin died in 2010.

Marcello and Martin

Gardner and Truzzi were similar in many ways. They both had a background in the magic fraternity. This gave them intimate knowledge of how people can be tricked and come to believe things that are not supported by the evidence. They both had long careers as writers. Both wrote articles backed by enormous personal libraries; they rarely expressed opinions that were not supported in the literature. They knew what they were talking about. They both applauded good science and deplored bad science.

They also shared another trait: they were not scientists. Martin made his living writing on many topics, some frivolous some serious. He wrote about mathematics for *Scientific American* for more than a quarter of a century, and about pseudo-science for longer. Literally before Marcello was born, Martin had published articles on science and skepticism in the University of Chicago newspaper. And, as stated earlier, the philosophy of science was near the center of his intellectual life for 75 years.

Marcello was trained in sociology. While it is always contentious to ask if sociology (or economics or historiography or ...) is a true science, it is clear that Marcello's subfield was divorced from, say, the working astronomer or chemist. It is also clear that even though he was not a working scientist he worked diligently to understand the methods of science, deeper than most working scientists. He spent the last 25 years of his life trying to bring attention to the areas of the sociology of science that he felt were not properly addressed.

Cranks

The "demarcation problem," how to tell good science from bad science, is their true difference. It is a core problem in the philosophy of science. They agree about fundamentals, but disagree about practical aspects. Martin felt that practical aspects of *doing* science favor ignoring and/or debunking pseudoscientists. Marcello felt that practical aspects of *adjudicating* science trump the exigencies of doing science. He often explicitly used legalistic language which, he acknowledged, reflected his brief time as a law student. At a colloquium in 1989 he stated:

> As a sociologist of science I remained outside of the controversies surrounding unconventional claims in science. My commitment is to the judicial process within the scientific community rather than the resolution of specific debates.

He wrote to Martin saying:

> You guys can act as lawyers for the defense of orthodoxy while the opposition advocates their own side. The general scientific community can act as judge, and I hope to act as a kind of *amicus curiae*, a friend of the court who recognizes the rules of evidence and the adjudication procedure and tries to help the process work more efficiently and fairly.

In a review of one of Martin's books, Marcello wrote:

> Like the little girl with the curl, when Gardner is good, he is very, very good (there may be no critic better), but when he is bad he is horrid. ...Gardner uses many tricks of the good lawyer arguing a case, and we jurors may not always agree with him, but he is a fine advocate and always worth reading if only for his excellence as a science writer. But it is imperative that the reader remember that Gardner is a self-confessed advocate and not—as I am afraid many scientists see him—as the gatekeeper to truth in science.

Marcello always contended that Martin (and CSICOP, etc.) act as lawyers, more interested in winning a case than abiding by the rules of conduct in science.

What are the rules of conduct? Probably very few scientists envision them in a legalistic framework. Marcello felt everyone should be given a fair hearing. Martin felt that this, taken to its logical extremes, is impractical and foolish. Marcello remained unclear on what criteria would permit someone to be excluded.

The word "crank" appears hundreds of times in their correspondence. It is the touchstone for the practical application of the demarcation problem. Martin repeatedly identifies people he feels are cranks, explains why they are cranks, and pleads with Marcello to ignore them, if not disavow them. Marcello repeatedly says that anyone who acts like a scientist (uses data, etc.) must not be dismissed; what matters is that they obey the rules—being clearly wrong matters little.

While Martin was well-schooled in every nuance of the philosophy of science, he had no truck with ideas that hindered the enterprise of science. Philosophy may illuminate, or obfuscate, but he felt it was meant to help scientists better understand what they do. Marcello, with a different background, felt the sociology of science would help scientists better understand what they *should* do.

The Correspondence

Gardner has a global reputation as letter writer. Many of his files are now accessible (at Stanford University and at the Committee for Skeptical Inquiry in Amherst, NY) and it can be seen that even people with no academic standing get long insightful letters in response to requests. His inbox was always full. The correspondence was so extensive, over more than six decades, that he only retained letters that he thought he might have use for in some future book or article. Luckily for this volume, Truzzi and Ray Hyman saved them more systematically. ("I recently put our letters into a large notebook and was amazed to see that there must be 600 pages of correspondence between us over these last 10 years.") It is also clear that Truzzi was a reliable correspondent. Many of his letters to other people have been transferred to the Eastern Michigan University Libraries.

In a letter to Marcello (July 23, 1979) Martin wrote:

> I write now to ask if you would mind if I sent copies of our exchange to Ken Frazier with the suggestion that he consider publishing them as a letter exchange? If so, we should each have the privilege of a bit of polishing, after which we could exchange and okay our final copies. In my case I am thinking mainly of such trivial things as adding first names here and there (Scott Rogo, Stanley Jaks, etc), changing Baker Streeters to Baker Street Irregulars, and in general making the letter more readable. If you prefer not to have our exchange published, do you have any objection to my copying your letter for people like Ray and Persi who would be especially interested? Maybe you sent a copy to Ray already.

Marcello responded (July 30, 1979) that he liked the idea but was unwilling to let the *Skeptical Inquirer* publish them:

> Always feel perfectly free to let any of our mutual friends see my letters to you unless I specifically say something about confidentiality. So certainly do send copies to Persi and Ray (along with your own letter with the questions, presumably) if you want to. ... As to publishing the exchange(s) in *Skeptical Inquirer*, I appreciate your intentions but must decline at this time. There are many reasons. The most obvious include words of derogation about Rogo and similar things not really meant for all eyes. Far more important, however, is the lack of context for the exchange which was, after all, written for you in light of all our past correspondence. People like Ray and Persi know my positions about as well (or better) than you do, but the

typical *Skeptical Inquirer* reader would not have that context. ... Perhaps, if you think the matter worth pursuing, we could find some more neutral territory for publication. I could understand your possible reluctance to publish in *The Zetetic Scholar* for reasons that to a degree parallel mine for *Skeptical Inquirer*, but there might be some external forum which both of us might find more comfortable. I would be quite flattered to publish any exchange with you, so I am reluctant to give you a flat "no" on your suggestion.

So this volume is that correspondence, published in "neutral territory." The remainder of this Introduction gives "context" the reader may appreciate. As both authors are deceased, we have resisted the urge to polish the letters. They are, except for minor grammatical errors, all unedited. As for giving people's full names, our index supplies those.

The reader will find many abbreviations in the correspondence, most of which are disambiguated in the preceding text. Some will be known to the reader who has some familiarity with the history of the skeptical movement. For example, "P and T" refers to Harold Puthoff and Russell Targ, two researchers at SRI (Stanford Research Institute, not affiliated with the University) who imagined they knew how to do research in parapsychology, but actually dealt with "psychics" who were more clever than they were. Their experiments with Uri Geller were discussed at length by Gardner in articles. Uri Geller was an Israeli magician who convinced many people he had psychic powers; Randi and Gardner wrote about how the tricks were done. In the letters "V" stands for Immanuel Velikovsky, who promoted a fanciful combination of history and astronomy in his books, starting with *Worlds in Collision*. The abbreviation "G" usually stands for Gauquelin, who is discussed below.

CSICOP

The history of CSICOP, the Committee for the Scientific Investigation of Claims of the Paranormal, has been told many times. Here we will add some background that may be new to some. The only intent is to set the stage for Truzzi's views of the organization. In recent years the group has streamlined the name to CSI, Committee for Skeptical Inquiry. (Perhaps because they got tired of people referring to them as "psi-cops.")

Martin Gardner, Ray Hyman, and James Randi started planning a group that would combat the occult revival. Ray Hyman is a professor (now emeritus) of Psychology at the University of Oregon. In the early 1970s he was giving lecture/demonstrations around the country on belief systems and how people are fooled. He had a background in magic and deception and briefly worked as a psychic doing "cold readings." This led to friendships with Gardner and Randi, and contact with Truzzi. James "The Amazing" Randi is a colorful character, starting out in magic as an escape artist and by the 1970s had developed an interest in debunking, much as his role-model Houdini had done. Unlike Hyman he favored a strategy of debunking to change people's minds. His efforts yielded him a MacArthur Genius Award, which he used to start JREF, the James Randi Educational Foundation. Much has been written about his life, including the feature-length movie *An Honest Liar*.

Let us begin with a letter from Hyman to Randi (August 14, 1973):

> I like your idea for setting up a nonprofit organization with a title such as Sanity In Research. Obviously much remains to be done in setting out its objectives, scope, administrative machinery, financing, and the like before anyone can say just how much of a commitment he wants to make towards it. I'll give you my immediate reactions to your proposal now. Later I will relay my further thoughts if any.
>
> First I think it is important that SIR not duplicate nor seem to be duplicating the function of existing organizations. There already exists the Society for Psychical Research and a Parapsychological Foundation. We would have to make it clear how SIR differs from these groups and why there is a need for the role that SIR will play.
>
> Second I think it is extremely important that SIR offers something POSITIVE rather than project an image of being a debunking organization. The names of Randi, Martin Gardner and Ray Hyman on a letterhead, SIR will have difficulty in conveying an image of an organization that is trying to offer something positive. I think that these three names will immediately connote that SIR belongs to the hard-headed skeptical extremist wing. This is a defect, I feel, and one that we should try to overcome if we want to reach individuals who not already confirmed skeptics, and we have to avoid coming on as radical debunkers.
>
> One way to accomplish this end, I think, would be to emphasize that our goal is not to debunk the supernatural nor to prove that extra-sensory is impossible. Rather our goal is to help individuals and organizations evaluate evidence. There exist many simple tools and precautions that are frequently overlooked by

people who encounter psychics or seemingly supernatural happenings. Also psychologists have discovered many quirks of perception, thinking, memory and emotion that can lead people astray when they try to account for bizarre events. One goal SIR could have to be to make sure that these tools and knowledge of relevant psychology are made more widely available.

Hyman returned to the subject in letter to Gardner (November 2, 1973):

> I have not thought much more about Randi's idea for the SIR foundation. But I do think that the attempt to supervise parapsychological research in other laboratories or the head-on collisions with individual psychics will not be a productive enterprise. On the other hand, I think that a foundation could play a very important educational role. It could generate materials to supply to grade schools, the media, and others to show that there is more than one way to view psychic claims. It could also supply speakers and serve as a clearinghouse for questions that are raised. Another function would be to monitor the media and ask for "equal time" to reply to programs like "In Search of Ancient Astronauts" or the *Psychology Today* article on Krippner.

Another letter from Hyman to Gardner (December 4, 1973):

> The creation of an organization to provide the other side of the occult story sounds better and better to me. My vision of it and its functions is now something like this. The basic approach would be "soft sell." We would take care to build an image of credibility, integrity, and fairness. We would not carry out crusades, vendettas, accept or offer challenges, or otherwise indulge in publicity stunts.

> Our goal would be to let individuals and organizations know there is another side to the stories the media keeps telling about them about UFO's, Uri Geller, psychic surgery, astrological forecasts, and the like. We could provide information kits, press releases, bibliographies, film, TV tapes and other materials to the media, public schools, colleges and other organizations.

> Probably it would be best to begin on a modest basis at first. The organization would need a full-time director and some clerical staff at the outset. It might eventually build its own library.

Writing to Gardner, Hyman is now planning for the renamed group (December 10, 1974):

You cannot fail to have noticed that no prospectus on our Committee for Constructive Skepticism has appeared. The usual excuses of other duties and no time apply in part. But the main problem is that I am still struggling with various alternatives and issues of justification. When I work out tentative budgets for launching an operation that covers several of the functions we have in my mind, I always come up with something on the order of $250,000 per year or more. This bothers me because I hate to get involved in something so huge without some prior experience. Also I think it will frighten away potential backers. So I want to work out a way of starting on a very modest scale and then building up from there. Say an operation that can start with a full-time secretary and an office. This might be an undertaking that would be in the $10,000 to $20,000 per annum range. Maybe a newsletter as the first venture; then with data on reactions and with experience at one level we could gradually add on other activities and enlarge the venture.

Within the year the group had a new name RSEP, Resources for the Scientific Evaluation of the Paranormal. Hyman's letter had the salutation "Dear fellow RSEP members" and was addressed to Randi, Gardner, and now Truzzi was included (September 19, 1975):

I have accumulated a haphazard assortment of comments inspired, in the main, by uncorrected proofs of Randi's book [*Flim Flam*]. They also relate to Martin's pessimism about the effectiveness of RSEP, to my recent interest in Sherlock Holmes and Arthur Conan Doyle and what remains to be done now that Randi's case against Uri is completed.

That pessimism was expressed by Gardner in a letter to Hyman just weeks earlier (September 9, 1975):

I continue in my pessimistic mood with respect to convincing anybody. Maybe pessimistic isn't the right word since I am personally more amused than I am depressed by it. I agree with Lewis Carroll, who once recorded in his diary that he had concluded it is impossible to "convince anybody of anything." (Cf. a famous remark by John Dewey to the effect that nobody is ever persuaded that a point of view is wrong, they just outgrow it.) I am now of the opinion that trying to combat the occult wave with rational argument is as hopeless as trying to persuade a devout Catholic or Billy Graham follower that Jesus was not God in human flesh, or trying to persuade a dedicated Communist something is wrong with the communist ideology. My mother was a great admirer of Richard Nixon, and stoutly defended him until she discovered that he used four letter words in private, then suddenly she became disenchanted. Marxism, orthodox Freudianism, orthodox Christianity, Judaism and Moslemism, and now "occultism" are self-sealing systems. Nothing you or I can say can have the slightest effect on someone like Eisenbud, or Charles Panati, or Judy Skutch, or the average person who is into astrology. The only way their mindset can be altered is by a long tedious process of education, covering many years in contact with a different environment and a set of influences — and even *that* may have little effect.

In short I am arguing in defense of H. L. Mencken's attitude towards the irrationalism of his day. Forget about trying to persuade any true believer. Enjoy the spectacle, poke fun at the boobs, get some laughs from the people on your side, and hope that what you say may have an effect on the very young who have not yet formed strong opinions. I note that the current issue of *The Humanist* is devoted to anti-astrology articles, and the signed statement by eminent scientists saying that astrology is nonsense. I'll wager that not one astrology buff alters his opinion as a result of reading the publicity this issue is getting, or would alter his opinion if he read the articles. The social causes of the astrology revival or too deep-seated to be countered by any kind of argument. There may be remarkable exceptions, of course, I am speaking about people in general. A few months ago I had dinner with a young man, a friend of a friend, who is devoting a great deal of time to astrology. It was a friendly conversation, and I raised all the arguments I could, proposed tests he could make and so on. My final impression was that I had been writing on water.

That anti-astrology issue of *The Humanist*, mentioned above, is the next chapter in the story. It was the result of work by Paul Kurtz, the editor of that journal, who was a philosopher and one of the vocal proponents of "secular humanism" (which had had religious overtones). He was not an advocate for science as much as he was an opponent of irrationalism. The publicity that issue attracted spurred Kurtz to organize a session in April, 1976, at the State University of New York, Buffalo, to discuss science and the paranormal. This was done as part of a previously scheduled humanist meeting. As a result of organizing this meeting Kurtz contacted Truzzi and became aware of the group that Randi had proposed and the plans of

Hyman, Gardner, and later Truzzi.

As can be gleaned from the correspondence above, Hyman, Gardner and Randi had worked out the details but each was unable, both by temperament and skills, to run such a group. Kurtz however was a natural organizer. Further he was an editor of a major magazine and had many contacts. In addition, Truzzi had the temperament and experience to edit a magazine. CSICOP (the fourth name used) was formed soon thereafter, and their journal/magazine was published by September, 1976. It was called *The Zetetic* (from the Greek for "inquirer"), the name Truzzi was already using for a newsletter.

We should stop and recall the confluence of backgrounds. Ray Hyman was interested in the educational role almost exclusively. James Randi was originally interested in debunking, though with his book *Flim Flam* he was branching out to the educational role. Martin Gardner was increasingly drawn back to a subject he had written about earlier, even though he was pessimistic. Gardner was not a debunker (at least not actively) and rarely attacked anyone, but preferred to hold things up in an unflattering light. Marcello Truzzi was newer to the game and preferred to pass on information with less judgment than Gardner. Paul Kurtz was a humanist, crusading to stamp out irrationalism.

CSICOP was, however, never intended to be a scientific organization that performed experiments or carried out field studies. It was an educational organization. It may have lacked a single vision of how to educate the public but the union of their approaches was potent and the organization thrived. (Gardner, while a founder of CSICOP, rarely cared to be involved in the administration of it. He only attended a handful of the early meetings, preferring to stay at home.)

Truzzi and CSICOP

In two issues of *Fate* magazine, in 1979, Truzzi gave an interview that detailed his experiences with CSICOP. The interviewers did not like CSICOP and the process was one-sided; even so, by all accounts Truzzi's facts were correct and his attitude was conciliatory. These facts in the interview are necessary to understand the correspondence in this volume.

In a letter to Randi (September 10, 1975) Truzzi wrote:

> Martin writes that he has become somewhat disenchanted as has, to a lesser degree, Ray. I am still hot on the organization, but I must admit that I am disappointed thus far. I had hoped that Ray and Martin would be able to find funding but none

> so far. In the meantime, I am approaching several journal pub-
> lishers about doing a serious journal in this area. I hope to get
> all of you and others like Sagan, Klass, Kusche, etc. on the
> editorial board. This may be our best bet. Will let you know
> the results as I get them. Any suggestions? I have gotten good
> initial responses.

This shows his enthusiasm before the hoped-for RSEP became CSICOP.

He contributed to the early discussions and was clear in his own mind about what CSICOP should become. He felt the Executive Council should be merely administrative and the Fellows "would be the people in control, in the sense that they would be responsible for the money, the by-laws, the journal and so on." He also felt that non-Fellow members could be from the very communities that they were investigating, since "these people were specialists, it would be ridiculous to pretend they didn't know anything." Paul Kurtz, who was the de facto chairman, asked Truzzi to be co-chairman when things became official; he agreed. After the group was announced at the conference in Buffalo, Truzzi drafted a set of by-laws.

In New York City, at the first meeting of the Executive Council (Summer 1976), his idea of a more inclusive membership was voted down. This was principally due to the strong opinions of Martin Gardner, Randi and Dennis Rawlins (who is discussed below). Truzzi "was disappointed but [he] accepted the decision." At this meeting Truzzi was given the responsibility of editing the new journal. For the first issue he asked Ron Westrum, a colleague in his own department, to write an article questioning the wisdom of Paul Kurtz's *Humanist* anti-astrology issue. Truzzi had been vocally opposed to the heavy-handed tone of the Manifesto in that issue, so it is not a stretch to imagine that he had talked to Westrum and asked Westrum to write the article since it would voice Truzzi's own opinion (and presumably Westrum's as well). Paul Kurtz was not happy, but allowed it as long as it was followed by a rebuttal written by himself.

Before an August 1977 news conference to announce the second issue of *The Zetetic*, Paul Kurtz issued a press release that Truzzi thought was too strident. For example, it suggested the rise of occultism in Nazi Germany implied occultism leads to fascism. Truzzi complained inasmuch as he was co-chairman and was not consulted. He asked the Executive Council for a vote of confidence. Truzzi did not get it so he resigned as co-chairman.

At this meeting it was suggested that the journal should be "hard-hitting." Truzzi took this to mean that it was to be less scholarly and more suitable for a newstand, emphasizing debunking. Worse, Truzzi took it to mean that his vision of a balanced treatment, including articles from the unorthodox communities was to be abandoned. He proposed that instead they have two publications, one popular and one scholarly. His ideas were voted down so he resigned as editor. (He was replaced by Kendrick Frazier, from *Science News*, who Truzzi respected.) He remained on the Council.

He immediately announced that he would start his own journal, independent from CSICOP, called *The Zetetic Scholar*. The idea was roundly supported. When he asked Kurtz for the CSICOP mailing list, Kurtz was reluctant. Kurtz asked all the other Council members to vote on the request; they voted no. Since Truzzi was still on the Council he was upset at what he regarded as back-room politics. Only then, did he appreciate that the by-laws allowed the Council to do anything they wanted without oversight by the Fellows. He soon resigned not only from the Council but from CSICOP entirely.

Some other of Truzzi's opinions should be mentioned. He thought "debunking is a perfectly honorable practice.... There's fraud, there's chicanery, there's a lot of bad science." His point was "while there can be lousy parapsychology, there can also be good astrobiology." He concedes both he and CSICOP "started with the assumption — shared, I think, by thoughtful proponents of the paranormal — that 90 percent of the occult explosion is intellectual and scientific garbage." However, he feels some studies of unorthodox claims are examples of a "protoscience" (his neologism). His only requirement for this label was that the researchers "aspire to be scientific."

Should CSICOP do research? Truzzi did not think so, and neither did Ray Hyman. In fact CSICOP has tried to be clear that research done by its members is not done under its auspices. Randi was active and in the public eye, and confused people by reminding them that he was a member CSICOP. (In fact, Randi eventually had to resign to protect CSICOP from court cases by his detractors.) Another important example of research that people thought was sponsored by CSICOP, the Gauquelins, occurred in the first year of the Committee.

Gauquelin and the Mars Effect

This case appears often in the correspondence in this volume. Michel Gauquelin is Truzzi's strongest example of a protoscientist. We will try to explain the issues involved briefly, but there have been scores of articles written about this controversy, often very selective and biased. It is impossible to reconcile them all but we will try.

Michel Gauquelin was a French psychologist and was interested in astrology as a young man. He wrote *L'influence des astres* in 1955, to report that his statistical studies failed to support traditional astrology, to his disappointment. However, among his voluminous data he was able to tease out what he felt were statistically significant correlations with the position of the planets at the time of birth (the hour, not the date). Of these he only promoted one correlation as he felt it was the strongest. His thesis was that if the position of Mars was in one of two unrelated 2-hour sectors of the sky then the person was more likely to be a sports champion.

No one outside of the astrological community took it seriously. However he insisted that as a scientist his study must either be refuted or taken seriously. The astrological community (even though they had no real interest in planetary influences) latched onto this and insisted they had the smoking gun that proved human influences from the heavens. The study has been repeatedly attacked by proposing potential flaws. Gauquelin and his wife (who was his research partner) rarely berated their detractors. Instead they patiently showed that each proposed flaw was not present. Of these potential flaws we are concerned with one proposed by the statistician Marvin Zelen.

Recall that the origin of CSICOP revolved around the anti-astrology issue of the *Humanist* (September 1975). It contained questions about Gauquelin by the Belgian Committee for the Scientific Investigation of Alleged Paranormal Phenomena (hereafter Comitè Para) which had 20 years earlier confirmed Gauquelin's calculations but distrusted his statistical assumptions. Gauquelin was given space to respond in the January 1976 issue. Kurtz, in the same issue, had asked statistician Marvin Zelen to follow up with some specific statistical test. So there was the "Zelen test," which asked for contemporary data (same time and same place) for non-athletes. It is important to emphasize that this predated CSICOP and was never later sponsored by them, as some believe. Gauquelin did the test, using many birth records, and it was shown to not invalidate the Mars Effect hypothesis.

At this point Kurtz was loath to simply report the results, because he felt that his efforts to get the new CSICOP up and running would be hurt. So he organized a post hoc rebuttal of the implementation of the Zelen test in an article by himself, Zelen and the astronomer George Abell. Dennis Rawlins was not a co-author but contributed to the post-analysis (see below). (They *did* indicate that the Mars Effect was only supported with data from one district (arrondissement) of Paris, but the data sets were small.) This was handled poorly, and people alleged that Kurtz and company were suppressing the truth (even though the *Humanist* published Gauquelin's rebuttal of the Zelen Test).

When each theory of how Gauquelin erred was disproven, the believers felt Gauquelin's case was stronger. Of course, it just meant the detractors had guessed wrong. What was needed was a new test. It was done, using American star athletes, and did not support the Mars Effect. Now it was Gauquelin who quibbled. The Comitè Para also did another test of a 1000 modern French athletes and it was negative too. Soon thereafter he committed suicide and so now there is no reputable defender of the Mars Effect.

This story has no ending. Today no orthodox researcher is much troubled by the Mars Effect even though it will be repeated endlessly by astrologers. The American test revealed some previously unsuspected flaws with his data. It was always contentious how "sports champion" was defined, allowing post hoc massaging of the data, but that was not the real problem. The main issue, as shown by Geoffrey Dean, was that Gauquelin relied on data from the century before 1950. This was a time when the hour of birth was self-reported; after 1950 it was more commonly and reliably reported by doctors. Dean showed that a modest 3% error rate in the birth hour would eliminate all statistical evidence for the Mars Effect.

What is important for the reader of the correspondence below is that Truzzi could state that Gauquelin was acting scientifically and responding to all questions from his detractors. Even so, Gauquelin never conceded that *a priori* it was more likely a statistical anomaly than an actual anomaly. He principally relied on the same data sample for 40 years, which was more dogged than scientific. It should be added that Truzzi never actually believed the Mars Effect was any direct cause-and-effect; only that Gauquelin was a protoscientist.

Dennis Rawlins

Dennis Rawlins' background is reconstructed here from his own statements. He said his background was in physics. Starting in the 1960s he took an active interest in both astronomy and history, which involved his spending a great deal of time going through the logbooks of astronomers. He seemed to stress the discovery of mistakes and missed opportunities. This was a type of debunking but not fervent. In 1973 he published *Peary at the North Pole: Fact and Fiction?* which stressed, again, the intersection of astronomical calculations and debunking. In fact an announcement by Rawlins, covered in the *Washington Post* in 1989, of supposed miscalculations by Peary caused a furor after it was shown Rawlins did not know what he was talking about. Kevin McManus, in the *Baltimore Magazine* said Rawlins was a "obnoxious, glory-starved show-boater... caught with his academic pants down" and was someone "who has practically made a career... of trashing other people's pet theories." By 1998 he was renewing his efforts to attack arctic explorers, this time concentrating on Cook.

Russell W. Gibbons of the Cook Society says Rawlins was "nursing a 25-year-old grudge against a pro-Cook writer whose publisher chose the same release date as Rawlins' [book], he has over the years combined scholarly papers with pure vitriol against Cook and in this latest tirade ... drips in venom." These quotes are meant to illustrate that others regard Rawlins the same way that CSICOP itself came to.

Apparently Rawlins had written to Gardner indicating his interest in debunking. In 1974 Gardner told Rawlins about some fraud in J. B. Rhine's ESP laboratory at Duke University, which Rawlins then passed on to a Baltimore newspaper. This indicates how Rawlins became involved in correspondence around the time CSICOP was being formed. He was not a founder of the group but was on the first Executive Council. He was, by all accounts, a strong proponent of debunking, stronger than Randi and Gardner. Truzzi has stated that Rawlins was instrumental in eliminating his vision of what CSICOP should be, leading to Truzzi's resignation from the council.

The reason for mentioning Rawlins is his transition from being a member of CSICOP to being its most vocal opponent. He set forward his reasons in a lengthy article called "sTARBABY" that appeared in *Fate* (October 1981). In the article he explains at length how, in his opinion, CSICOP revealed itself to be biased and involved in a cover-up. Unfortunately, while many of the facts were correct, he relied on biased language to color every event. Further he put quotes around statements made over the phone years earlier. And he omitted facts, i.e., selection bias. Philip Klass, also on the Executive Council, consulted archival letters and wrote a robust rebuttal, called "CRYBABY." *Fate* refused to publish it, but it has been circulated on the internet.

What was the issue? It involved the Zelen test of Gauquelin. Kurtz and Abell were, of course, dismayed that the Zelen test did not successfully challenge the Mars Effect. They did not do anything to block the outcome from being published in the *Humanist*. However, since the test was performed by Gauquelin himself they considered ways that the execution might be flawed. They did their own analyses (by separate geographical areas) that were not part of Zelen's original test. Rawlins, because of his background in checking astronomical calculations, did most of the new calculations. Ideas that Rawlins had proposed about another related test (but also negative) were ignored. Abell, an astronomer, was not sure Rawlins, "an amateur," should be the one doing the calculations; however Abell was not willing to do it himself. A paper raising questions about the execution of the Zelen test did not include Rawlins as a co-author.

Rawlins went wild. He took offense at being ignored, questioned and not given credit. Rawlins was universally described as "thin-skinned." The actions of the others might have been handled better. The hatchet piece "sTARBABY, " which should have been about a few individuals not treating him as a professional, became an indictment of CSICOP, as a dark political organization perhaps worse than the people they investigated. The article has been pointed to by all the varied critics of CSICOP, justifying their dismissal of it. Recall, CSICOP was not responsible for the Zelen test, the response to it, or the activities of Kurtz, Zelen and Abell.

In a letter to John Fuller, not a friend of CSICOP, Gardner was quite frank (June 18, 1983):

> From the outset I thought it was a mistake for anyone connected with *The Humanist* or with the newly formed CSICOP to spend time trying to "refute" Gauquelin. The result was as I feared. Mistakes were made in the refutation, which have been freely admitted by Abell. ... Allowing Rawlins to have his say in the *Skeptical Inquirer*, vituperative, angry, and libelous, is I think almost unique in the annals of literature about such things. ... When our committee was first formed we had no inkling of Rawlins' inability to get along with anyone with the slightest disagreements with him. No sooner had the committee been formed, than Rawlins began an intensive campaign to get rid of Marcello. He would keep me on the phone interminably telling me what an awful person Marcello was, and how he would damage our committee if we didn't fire him as editor. I strongly disagreed with this tactic, though I did agree that Marcello's idea of our magazine was quite different from what the rest of us wanted it to be. ... I think I was the last on the committee to realize how disturbed the man is.

Rawlins continued, for many years, a letter-writing campaign against CSICOP, trying to find dirt behind every door. Since the reaction was so out of proportion to the offense many observers, such as Gardner, came to believe that Rawlins might be mentally unbalanced. The paradox of this is Truzzi's involvement. Throughout this episode Rawlins never doubted Gauquelin was wrong and needed to be debunked. Truzzi did not agree with Rawlins' approach to debunking. Further Truzzi, at least initially, was conciliatory to CSICOP, stating they served an important service, just not in a manner he desired. However, Truzzi took great interest in Rawlins' poison pen letters, when he should have been very wary of anything Rawlins said. This became a source of much friction between Gardner and Truzzi.

Truzzi and Hofstadter

Before we get to the correspondence that is the point of this volume, we would like to give a letter from Truzzi, explaining in his own words the issues mentioned above (all capital letters in the original). The Center for Scientific Anomalies Research, CSAR, was created by Truzzi as an alternative to CSICOP. The letter (January, 31, 1982) was to Douglas Hofstadter, who had recently replaced Martin Gardner as a columnist for *Scientific American*. Hofstadter's handwritten response follows (used by permission), with a short note by Gardner.

Dear Mr. Hofstadter:

I was greatly disturbed and disappointed to read your column in the February issue of *SCIENTIFIC AMERICAN* because of its serious distortions about the character of the "schism" in CSICOP and the position and history of the *ZETETIC SCHOLAR*. Your article conveys the clear impression that *ZETETIC SCHOLAR* is somehow more sympathetic to pseudoscience, more "relativist," and "unjudgemental." That is completely untrue. I can not help but wonder if you have ever seen *ZETETIC SCHOLAR* or have information about it from anyone but Paul Kurtz or Martin Gardner. In case you do not, I enclose a copy of a typical issue (in this case, #6 which contains an article by Ray Hyman dealing with "pathological science" that I think would benefit you to read) and some other materials on ZS and the new Center for Scientific Anomalies Research.

I think you completely missed the issue between CSICOP and CSAR, The term "skeptic" has become unfortunately equated with disbelief rather than its proper meaning of *nonbelief*. That is, skepticism means the raising of doubts and the urging in inquiry. ZS very much stands for doubt and inquiry. The term "zetetic" was originally used by me in my newsletter *THE ZETETIC* and that name was taken for the CSICOP journal when I joined CSICOP as its co-chairman and the editor of its journal. I very much agree with C. S. Peirce that the principal obligation of a philosopher/scientist must be to do nothing that might block inquiry. I view much of CSICOP activity as obstructing inquiry because it has prejudged many areas of inquiry by labeling them pseudoscientific prior to serious inquiry. In other words, it is not judgement that I wish to avoid—quite the contrary—but it is prejudgement. CSICOP claims (in its goal statement—which I helped author and which I very much agree with) to not prejudge prior to inquiry, but that is demonstrably not the case in many instances.

The major problem is that CSICOP, in its fervor to debunk, has tended to lump the nonsense of *THE NATIONAL ENQUIRER* with the serious scientific research programs of what I call protosciences (that is, serious but maverick scientists trying to play by the rules of science and get their claims properly tested and examined). By scoffing at all claims of the paranormal (which should *not* be confused with claims of the supernatural; paranormal facts are those which are quite within nature but await adequate explanation by science). CSICOP inhibits (through mockery) serious work on anomalies. The fact is that the protoscientific community (made up of groups like the Parapsychological Association, the Cryptozoological Society, or the Center for UFO Studies) rejects and abhors the sensationalist accounts of *THE NATIONAL ENQUIRER* and similar junk publications as much as does CSICOP.

ZETETIC SCHOLAR tries to bring together protoscientific proponents (those willing to abide by the rules and evidence of science) and responsible critics (those willing to similarly accept normal scientific rules of discourse and not reverting to ad hominem and similar tactics) into rational dialogue. Please note that I use the word dialogue rather than debate, for the purpose is not to "win" or "defeat" an opponent. The purpose is to advance science.

My position is not a relativist one. I believe science does progress and is cumulative. But I do believe that skepticism must extend to all claims including orthodox ones. Thus, before I condemn fortune tellers as doing social evil, I think the effects of their use needs to be compared to the orthodox practitioners, psychiatrists, and clinical psychologists. The simple fact is that much nonsense goes on within science that is at least as pseudoscientific as anything going on in what we usually term pseudosciences.

Science is a method and not a substantive set of conclusions about the world. I am a staunch supporter of the scientific method and am philosophically completely committed to it. As others have pointed out, science is descriptive and not really prescriptive. Science can tell us what we might expect in the future based on the past, but it can not tell us that something can never be in the future. I do not believe in most paranormal claims, but I refuse to close the door on discussion of them. The simple fact is that I think I have more confidence in science than, say, Martin Gardner (with whom I corresponded for many years on these matters). For example, Martin resigned as a consulting editor for ZS when he was told that I planned to publish a "stimulus" article asking for a reconsideration of the views of Velikovsky. Martin was invited to comment, as were many critics of Velikovsky. But Martin felt that even considering Velikovsky seriously in ZS gave him undeserved legitimacy, so Martin resigned. I happen to think Velikovsky is dead wrong, but I also think that he has not been given due process by his critics. I have confidence that honest discourse will reveal the errors and virtues (if any) in any esoteric scientific claim. I see nothing to be afraid of. I have full confidence in science as a self-correcting system.

Some on CSICOP, like Martin, do not.

Luckily, many CSICOP Fellows agree with me and not Martin re the above. You seem to be unaware that Ray Hyman, George Abell, and (as of a month ago) James Randi are members of the Executive Council of CSICOP who have *also* taken positions on the Editorial Board of ZS (Ray Hyman is even my Associate Editor). Other Fellows of CSICOP are also on my editorial board and are also Senior Consultants for CSAR. So, your view of a "schism" is rather a distorted picture.

Finally, I wonder if you are even aware of the storm of controversy that has been going on surrounding the Mars Effect experiments conducted by CSICOP, as recently revealed by Dennis Rawlins, an astronomer and ex-member of the Executive Council. Rawlins was expelled from CSICOP for whistleblowing on the gross incompetence of CSICOP in handling these tests. The full story of these strange machinations by Kurtz and others is only emerging now and much of this controversy will be documented in the upcoming issue of *ZETETIC SCHOLAR*. This controversy has already resulted in several resignations from CSICOP including those of Daniel Cohen, Richard Kammann, and William Nolen. (I also wonder if you are aware that my resignation from CSICOP was followed by others over the same issue, e.g. Persi Diaconis and T.X. Barber's departure from CSICOP.)

I do not wish to inundate you with materials, but I enclose issues of ZS and I also enclose an interview with me published in *Fate* which gives the details of my departure from CSICOP.

Let me be clear. I do not wish to see CSICOP go out of business. It plays an important advocacy role. But I think it is imperative that CSICOP be publicly recognized as the advocacy voice for majority (or orthodox) science and not as a truly scientifically objective body willing to listen to all sides. CSICOP can be a wonderful lawyer for orthodoxy, completely honest and legitimate; but we must not confuse it with the role of judge and jury. I think your article makes that confusion.

CSICOP tries to present itself as *the* voice of rationality and reason. It is one voice among many, and I think we do a gross injustice to others seeking objective truth when we ignore them. The demonstrable fact is that far more UFO hoaxes have been discovered and debunked by the major pro-UFO organizations than has been done by CSICOP or its members; the major scandals in parapsychology have been revealed to us by fellow parapsychologists who blew the whistle; and it is outrageous to condemn the "policemen" in these groups for the "criminals" they have exposed.

THE NATIONAL ENQUIRER publishes garbage, and I am glad that *THE SKEPTICAL INQUIRER* debunks such stuff. But *THE NATIONAL ENQUIRER* represents no serious protoscientific effort anywhere. Debunking trashy publications like *THE NATIONAL ENQUIRER* may be good public relations for science to be doing, but it does not constitute doing science (adding to scientific knowledge). CSICOP claims to be promoting inquiry when it actually places priority on debunking. I am all for debunking, especially when responsibly done (which is not always the case, I am afraid), but ZS does not place debunking as its *major* priority. We need both *ZETETIC SCHOLAR* and *THE SKEPTICAL INQUIRER*, but the latter should not be confused with something it is not: an objective journal promoting inquiry.

   ZETETIC SCHOLAR is a "peanuts" operation compared to *THE SKEPTICAL INQUIRER*. TSI has about 7500 subscribers (prior to your story which should add about 5000 more). Its subscriptions therefore produce over $1,00,000 per year. ZS has subscribers numbering only about 300, but I am happy to say that most of these are serious researchers. Perhaps I should also note that the *JOURNAL OF THE AMERICAN SOCIETY FOR PSYCHICAL RESEARCH* only has about 4000 subscribers. So CSICOP is hardly a David struggling against a Goliath in any serious scientific sense. The fact is that *THE NATIONAL ENQUIRER* is not taken seriously by scientists any more than its Hollywood stories are taken seriously by those in show business. The question of any harm done by TNE is an *empirical* one. As scientists, we should try to learn what that influence/harm really is. CSICOP has offered no evidence of significant harm being done to science by TNE. But I assure you that it can be demonstrated that harm is being done by CSICOP to those wishing to do serious research into anomalies. The current Mars Effect controversy is the best evidence of that.

   I have written you at such length because I think you are sincerely looking for the truth about things. I greatly admire your book and columns, and I hear good things about you via friends like Persi Diaconis who have met you. I hope therefore that this letter and the documents I enclose will help correct your view of ZS. I also hope that you might contact Ray Hyman for corroborative information.

   I don't really expect you to do anything about this matter, but I hope we might remain in communication in the future. Certainly, if you are again interested in matters relating to scientific anomalies, please feel free to call upon me since I am in touch with quite a network of specialized scholars, especially now that CSAR has been established. And, of course, if I can clarify anything for you re this letter, just let me know.

   Sincerely,
   Marcello Truzzi, Ph.D.

P.S. I see I did forget one important matter: Thank you for giving the address of *ZETETIC SCHOLAR*.

In the final analysis, I think your comment that "science and pseudo-science coexist happily" in *ZETETIC SCHOLAR* demonstrates your misunderstanding about ZS. ZS is only interested in claims aspiring to validation as scientific ones. Labelling the protoscientific statements in ZS as pseudoscientific demonstrates a prejudgement.

If I am correct in thinking that Martin Gardner may be your source of information about ZS, for much in your comments his misperceptions stated to me in his letters, I urge you to ask Ray Hyman for any clarification on this matter since Ray received most of the letters (now about 600 pages), between me and Gardner over the last 8 years. He will confirm, I think, that Gardner misconstructs much about both my views and motives.

I am also enclosing a copy of ZS #7 because it has a reply by Ray Hyman to the article on pathological science which you might also find useful.

Hofstadter's response was written March 16, 1982.

Prof. Truzzi,

Thank you for your letter of Jan. 31 and additional material. I have thought quite a bit about the issues you raise, and the difference in tone, outlook, purpose, vision, etc. between *The Zetetic Scholar* and *The Skeptical Inquirer*.

I find myself more sympathetic than you are to the cause of out-and-out debunking. I am impatient with, and in fact rather hostile towards, the immense amount of nonsense that gets a lot of undue credit because of human irrationality. It is like not dealing with someone very unpleasant in a group of people because you've been trained to be very tolerant and polite. But eventually there comes a point when somebody gets up and lets the unpleasant person "have it"—verbally or physically or whatever—maybe just escorts them out—and everyone then is relieved to be rid of the nuisance, even though they themselves didn't have the courage to do it.

Admittedly, it's just an analogy, but to me, Velikovsky is just such a person. And there are loads more. I simply don't feel they should be accorded so much respect. One shouldn't lean over backwards to be polite to genuinely offensive parties. Carlos Casteñada—a fraud. And many other examples.

I also happen to feel that much of parapsychology has been afforded too much credibility. I feel that ESP (etc.) are incompatible with science for *very fundamental reasons*. In other words, I feel that they are so unlikely to be the case that people who spend their time investigating them really do not understand science well. And so I am impatient with them. Instead of welcoming them into scientific organizations, I would like to see them kicked out.

Now this doesn't mean that I feel that debating about the *reasons* I find ESP (etc.) incompatible with science at a very deep level is worthless. Quite the contrary: coming to understand *how* to sift the true from the false is exceedingly subtle and important. But that doesn't mean that all pretenders to truth should be accorded respect.

It's a terribly complex issue. None of us sees the full truth on it. I am sorry if I did you a disservice by describing your magazine as I did. However, I'm sure to comment much more on these issues as time goes on, and perhaps then I can state in print some more ideas that will be in some sense truer about your magazine that what I wrote.

I have nothing against your magazine in principle, except that I find its open-mindedness *so* open that it gets boring, long-winded, and extremely wishy-washy. Sometimes it reminds me of the senators and representatives who, during Watergate, seemed endlessly dense, either unable or unwilling to get the simple point: that Nixon was guilty, on many counts. And that was it. It was very simple. But they kept resisting the obvious, Ehrlichman and Haldeman kept on in their tricky smooth ways, making a seeming case for the White House. Oh—let's not forget Nixon's various lawyers—Buzhardt, Wilson, et al.

It was really quite an appalling spectacle to see people hold off, for as long as they possibly could, from seeing the obvious. Can you imagine if a journal had been put out, during the Watergate era, called "The Open Mind," which, at every single argument or counter-argument by Nixon or his opponents, contained a million delicate points about logic and so forth? It would have *exceedingly* nit-picky and would really have obscured the main point, which was simply that by a certain moment, Nixon was *obviously* guilty. After that moment *no* amount of legalistic quibbling should have been able to forestall his removal (notice how this harks back to my earlier analogy about "Throw da bum out!")—and yet Nixon did manage to obscure the obvious for many months, thanks to fuzzy-minded people who somehow couldn't "snap" into seeing something that was very black-and-white. They *insisted* on seeing it in endless shades of gray. And in a way I think that's what you're up to, in your magazine, a lot of the time: seeing shades of gray when it's black-and-white,

There is a legitimate, indeed, *very* deep question, as to *when* that moment of "obviousness," that moment of "snapping" or "clicking" comes about. Certainly, I'd be the first to say that that's as deep a question as one can ask. But that's a question about the nature of truth, evidence, perception, categories, and so forth and so on. It is *not* a question about parapsychology or Velikovsky and so on. If your magazine were a magazine about the nature of objectivity I'd have no quarrel with it. I'd love to see such a magazine.

But it's really largely a magazine that helps lend credibility to a lot of pseudoscientists.

Not to say that everyone who writes for it is a pseudoscientist! Not at all! But my view is that there is such a thing as being *too* open-minded. I am *not* open-minded about 13 being unlucky, about the earth being flat, about whether Hitler is alive today, about whether Jesus is the son of God (and born of virgin), and so on and so forth. I am also not open-minded about people who claim to have squared the circle, proven relativity (special) wrong, etc. etc. I am also not open-minded with respect to the paranormal, which includes many things. That is how I am. And I think that it is wrong to be open-minded about those things, just as I think it is wrong to be open-minded about whether or not the Nazis killed 6,000,000 Jews in WWII.

I *am* open-minded, to some extent, about questions of ape-language, dolphin-language and so on. I haven't reached any final, firm conclusion there. But I don't see that being debated in ZS (or in SI).

My viewpoint is that SI is doing service to the masses of the country, albeit indirectly, by writing articles that have flair and dash and whose purpose it is to combat the huge waves of nonsense that we are forced to swim in all the time. Of course most will never read SI themselves, but many teachers will, and will be much better equipped thereby to refute kids who come up and tell them about recognitive dreams or bent keys or magically fixed watches or ... you name it.

It's very valuable to have debunking books and magazines around (such as *Fads and Fallacies* and SI) because they can be read by large numbers of people.

On the other hand, in my opinion, ZS is a little too heavy and wishy-washy and pedantic and, mainly, *refuses to see the obvious*. I can't help it, nor can I understand it, that many people cannot see the obvious. Watergate was a painful case of that sort, where senators et al. made fools of themselves.

Well, I feel SI is playing the role of the chief prosecutor in some sense, of the paranormal, and ZS is a member of the jury who refuses, absolutely refuses to make up his/her mind, until more evidence is in. And after more, more, more, more, more, more, more evidence is in and this character still refuses to go one way or another, then one gets impatient.

So, *not* in brief, there you have a sketch of my reactions. I'm sure we could have a much longer discussion of these matters and perhaps come to some points of agreement but on others we'd have to agree to disagree.

In any case, I've enjoyed trying to express to you, here, some of my feelings. In fact, they are considerably more complex than this letter would allow me to say. But that'll have to remain for another time.

Thanks again for writing.

Sincerely,
Douglas Hofstadter

On August 19, 1983, Gardner read Hofstadter's letter and sent this note.

Dear Doug:

Ken Frazier let me see your exchange of letters with Marcello, and I want to say that I think your letter to Marcello is a small masterpiece. If I ever write an article about Marcello's philosophy of science (unlikely), I will want to quote from it, though of course not without letting you see the copy and getting permission.

This calls for no reply—just wanted to add to Ken's endorsement of every thing you said to Marcello.

best,
Martin

Chapter 1

The Road to CSICOP

The beginning of the correspondence is quite heavy with letters to Marcello from Martin. This is because Martin did not retain either part of the correspondence. His volume of letter writing was always large and he never kept letters that he would not use again. On the other hand, Marcello did retain Martin's letters but not his own. (Apparently that was his practice in all his correspondence at the time.)

Before long we have both sides of the conversation. However, throughout this book, there will be letters responding to another letter that was not retained. We have resisted the urge to say what the missing parts were; the reader may be more qualified to guess than we are.

These letters show the natural progression of ideas in the pre-CSICOP thinking. In these letters Marcello takes positions that seem more strident or, at least, less conciliatory than he voiced after leaving CSICOP. For example, he wants to embark on a book project that would embarrass people, he later called protoscientists. Also in a letter to Randi (September 10, 1975) he wrote "I particularly liked your tone in discussing the duped scientists. You were far kinder than I would have been."

It shows that plans for the new improved *Zetetic* were underway before Kurtz got involved. In fact, Paul Kurtz is rarely mentioned at all. The events surrounding Marcello's departure are not the center of any letter. Martin was never directly steering the Committee and Marcello knew that.

Two notes. Occasionally a letter from a third party is included. This is because that letter was filed with this correspondence and enhances it. Second, we have indicated for each letter the letterhead that was used, since it indirectly indicates "where they were coming from."

[Euclid Avenue]   19 May 1970
Dear Mr. Truzzi

Yes, I saw and clipped that *NY Times* piece, was intrigued by it, and would indeed be delighted to see a copy of your newsletter. I'm sure you've thought of making a book anthology out of it—as McConnell did with selections from his *Worm Runner's Digest.*

Our interests certainly overlap enormously—including circuses and carnivals. I was a friend of William Gresham (*Nightmare Alley*, etc), and numerous ex-carnies. One of the great regrets of my life is that I never joined a carnival for a while in my youth. An old friend in my home town of Tulsa, OK, Roger Montandon used to publish *The Juggler's Bulletin.* I myself once got as far as 4 balls—or rather 4 rolled-up socks, which I used for practice after washing them (in my bachelor days).

The article in *Psyche* was just a cut and edited version of my chapter on Fort in *Fads and Fallacies*; they obtained rights from Dover without telling Dover anything about the nature of the forthcoming periodical. My only pseudo-science in recent years was an article on D.O.P (dermo-optical perception)—about those Russian ladies who read with their fingers. It appeared in *Science*, February 11, 1966, and has been reprinted in a Scott Foresman paperback anthology, *Research in Psychology*, edited by B. L. Kintz and J. L. Bruning.

I'm generally a poor correspondent—but I'll do my best, and I certainly look forward to your book on occultism and would enjoy seeing your paper. I have been too busy with other things to keep up with the fantastic upsurge in occultism, astrology, (even Scientology!), though I clip articles of interest when I happen on them. (I'm sure you saw the good exposé of Ted Serios in *Popular Photography*, October 1967.) (I have an amusing letter from Rhine in which he praises Eisenbud as a competent scientist, and he says he is reserving judgment on Serios until more research has been done!)

All best,
Martin Gardner

[Euclid Avenue]   28 May 1970
Dear Mr. Truzzi

This calls for no response—just a quick note of thanks for the delightful material you sent—all of which I read with utmost interest.

Only two items come to mind worth mentioning:

In 1939 a mimeographed book appeared from Pyramid Publishers (Box 116, Edgewood P.O., Providence RI) called *Hurry, Hurry, Hurry!* (42 pages) subtitled "A Handbook of the Modern Carnival Midway," by "Doc" and "The Professor." Alas, I no longer have a copy and it's probably hard to locate. I mention it because it is not well-known and because it is the best reference I have seen on carnival game gaffs and it is accurate. (There is a detailed explanation for the loose floorboard technique of controlling a vertical wheel.)

The reference in your newsletter to Mencken's piece on Veblen (which I have in my files) prompts me to say I once wrote a curious short story (*Esquire*, April 1947) about Veblen, of whom I have been a lifelong admirer. It is called "The Conspicuous Turtle" and was about a Prof. at the Univ. of Chicago who taught Veblen's economic views by day and at night was a jewel thief who stole the most outrageous examples of conspicuous waste. Copies of this, too, have slipped out of my hands. I recall that I ended with a horrible pun, "The Veb and the Rocks."

I mentioned you a few days ago to William Kaufmann, editor of W. H. Freeman and Co. (now owned by *Scientific American*), as a possible source for a book of excerpts from the newsletter (or some other book). He seemed genuinely intrigued and took your name, so you may hear from him. If you have no publisher for your book on the occult revival, he might be interested. It is a small firm, but has excellent distribution in college bookstores.

Best,
Martin Gardner

P.S. Your paper on the occult boom is excellent—I learned a lot from it.

[Euclid Avenue]   6 June 1970
Dear Mr. Truzzi

Greatly enjoyed the newsletter you sent and was pleased also to get Sanderson's piece. I know of his society but do not take his journal. We have exchanged some letters on this and that, but I cannot count myself among his admirers—he is a charming rogue, not very knowledgeable in the sciences—and it's hard to say how much of what he has written he really takes seriously.

I did not know of Braithwaite's book, which sounds worth looking into when I get a chance. I checked a folder on carnivals and pass along the following references for whatever they may be worth:

Time, August 30, 1956, p. 38+ (on sociologist Krassowski, who joins a carny every summer and seems to have made a special study of them.)

Time, September 29, 1958, p. 41+ (long article on carnies.)

NYT Book Review, March 13, 1960, no page recorded (Gresham's review of the novel *And Where it Stops Nobody Knows*, by David Mark, Doubleday.

NYT Book Review, February 19, 1956, p. 5 (Gresham's long review of Herbert Gold's novel *The Man Who Was Not with It*, Atlantic-Little Brown.)

Life, September 13, 1948 (long article by Gresham, "The World of Mirth".)

NYT Magazine, May 18, 1952 ("Carny Biz—Bigger Than Ever," by Gilbert Millstein)

You know, of course, of the recent Trident Press book on Carnivals, which I have not yet seen, though I have heard the author on the Long John Nebel radio talk show here. (Long John himself is an ex-carny. He reviewed the book for *NYT Book Review*, but I failed to record the date.)

Frederic Brown's *Madball* (Fawcett paperbacks) is an amusing carnival novel. Clayton Rawson's *Headless Lady* is an earlier mystery about carny life.

Gresham's ex-wife, Joy Davidson, by the way, married C. S. Lewis, the Anglican Church apologist. (His book titled *Surprised by Joy*, puns on the event. It's a wild story—too long to tell now. When Gresham saw the book, his comment was: "It should have been called *Overwhelmed by Joy*.)

Will look forward to your carnival paper. This demands no reply.

All Best,
Martin Gardner

[Postcard]   8 June 1970
Dear Mr. Truzzi:

It occurs to me that Dennis Flanagan, editor of *Scientific American* [address withheld] might see in your talk on the occult revival the basis of an article. Anyway, if interested, you might send him a copy with a note saying I asked you to do so. (If I give him mine, I may never get it back!). He's a grad of the Univ. of Michigan, by the way.

Best,
Martin Gardner

[Euclid Avenue]   26 June 1970
Dear Marcello

I did indeed enjoy the *Newsletter* you sent. The only reason I don't seek
a subscription is that much of the humor is inside humor among sociologists,
so I miss much of it—but what I *do* understand is top grade.

I'm a Sherlockian only in my admiration for the Holmes saga and I
knew the late Bill Baring-Gould who did the mammoth *Annotated Sherlock
Holmes* for Clarkson Potter Inc (2 vols, $25.00). I've attended one Baker
Street Irregular Annual NYC Dinner as a guest, but I am not a member.
(I sneaked some Sherlockian material into the footnotes of my *Annotated
Casey at the Bat*—e.g. one footnote explains that Mudville, for a short
time before it vanished, changed its name to Moorville (in Kansas). This
clears up the mystery of Moorville mentioned in "The Adventure of the
Three Garridebs." (I said nothing about the story in my footnote, but the
Irregulars spotted it and reported it in their publication.)

The *Humbug Book* is a great idea. I do keep some refs on anti-Xmas
items and here is what I have:

1. *Time*—Dec 23, 1966, p. 44. Full page story headed "A Black Christmas." It reports on various anti-Xmas articles that year in December issues
of *Esquire, McCall's, Holiday, Red Book, Reader's Digest,* and other magazines.

2. *NY Times*—Dec 25, 1966. Russell Baker's essay, on editorial page,
is anti-Xmas—very funny. (another Baker essay on Xmas on December 14,
1967.)

3. *NY Times*—Dec 19, 1966, p. 1. Article headed "Office Parties?
Humbug! Santa? Needs Analysis!" Refers to article in *The American
Sociologist*, by Warren O. Hagstrom, on "What is the Meaning of Santa
Claus?", Nov/66 issue. (Hagstrom then at the U of Wisconsin.)

4. *NYT Magazine*—Dec 17, 1967. "Singing those Christmas Holiday
Blues," article by Edwin Diamond.

5. *Time*—Dec 10, 1965. Long picture article (several pages) on "The
Great Festival". Not specially anti-Xmas, but an interesting round-up of
pros and cons.

6. *NY Times*—Dec 15, 1968. News story headed "Youth in Sweden
Stage Protest: Against Xmas" about three anti-Xmas youth organizations
in Sweden that put up anti-Xmas posters in department store windows,
etc., and their reasons for opposition to Xmas.

7. There must be many anti-Xmas parodies of "Night before Xmas."
I heard one read one night, by a radio personality here, Jean Shepherd
(station WOR), which he said was by H. I. Phillips, but he did not give the
source. It is about a tired sales girl visited in her dingy apartment by Saint

Nick, ending with him calling "Merry Xmas!" as he goes up the air shaft and she yells back "Sez you!". A correspondent, Dorman Luke [address withheld] collects parodies of Moore's classic and he tells me he has more than 30. Don't know if he has the Phillips one or not, but you might induce him to xerox for you any anti-Xmas parodies in the lot.

I have no easy access to a photocopier (I used to have one, but it broke down and I tossed it out last year), but I can let you borrow any of the above items if you like. They are hardly suitable as items for the book, but they might be of use to you in doing your introduction.

Chesterton, in *Tremendous Trifles*, has a charming essay, "A Shop of Ghosts," that has as its theme the apparent perpetual dying of Santa Claus. If you want to close the book on a positive note, this could be it. I don't own any of the many pro-Xmas collections, so I don't know if this essay by GK was caught by any of the editors or not.

All Best,
Martin

[Euclid Avenue]   9 July 1970
Dear Marcello

No reply to this called for—I am sometimes a compulsive answerer—even worse than waiting 6 months—but I do want to thank you for the delightful book on caldron recipes, and your amusing inscription (and the marvelous jacket photo!).

I knew of the comic book buffs (through friends who used to write comic books for a living) and had heard of the French Crepitator, but not of the *book* about him. I, too, am not yet convinced he is not imaginary, but several people assure me he is not. The Agony Column Book was new to me also, and I shall look it up and mention it to Dover.

Yes, Holmes' inductions *were* pretty awful and even most of the plots are not very good. But the strong sense of reality Doyle achieved, not only for Holmes and Watson, but also for Victorian London, is amazing. I wish I knew how he did it!

Re: Those curious in-groups. I subscribe to two periodicals on origami, to the *Baum Bugle* (devoted to L. Frank Baum and Oz; the Oz fans have national and regional conventions), to *Kalki* (about James Branch Cabell), *The Wellsian* (British society of H G Wells admirers), *Jabberwocky* (British Lewis Carroll group) and *Word Ways* (a quarterly on recreational linguistics). Magic became so specialized that about 15 years ago a periodical devoted only to magic with *thimbles* was being published!

I hope you will consider Freeman and Co as a publisher. They are owned by *Scientific American* (which assumes good advertising in S.A.) and their distribution to college bookstores is better than the trade book publishers. I have just switched to them (from Simon and Schuster) as the publisher for my *Scientific American* column collections.

All Best,
Martin

[Euclid Avenue] 4 August 1970
Dear Marcello

Your carnival paper is certainly excellent in *all* respects, and thank you very much for the advance look at it. I was particularly glad to get an explanation of the z-language, the details of which I did not know.

The issue of *American Behavioral Scientist* devoted to Velikovsky, was later expanded into a hardcover book, *The Velikovsky Affair*, edited by Alfred de Grazia (University Books, 1966). It is still available, I think, in remainder stores at a greatly reduced price (original price, $5.95). I have made no attempt to keep up with Velikovsky articles, pro or con, and don't know of any important anti-V material in response to the *American Behavioral Scientist* blasts. There is a chapter on V in Daniel Cohen's *Myths of the Space Age* (Dodd, Mead, 1967), but it is mostly a reprinting of a *Science Digest* (which he edited—perhaps still does) article.

Just back from a week in Vermont, with a big backlog of mail and work to hand, so I'll stop for now.

Cordially,
Martin

[Postcard] 5 February 1971
Dear Marcello:

Thoroughly enjoyed your Sherlock piece (I hadn't realized he had so much to say about scientific method and theories!) Have passed it along to my old friend, John Shaw, now of Santa Fe. He spoke in NYC last month, at the annual dinner (picketed in 1970 by Women's Lib!) on *pornography* in the Canon (unintended word play, etc.). I'll retrieve the text when your book is published.

All best and thanks,
Martin

[Euclid Avenue]   27 April 1972
Dear Marcello

Many thanks for putting me on the list to receive *Explorations*. The forthcoming Norman Cousins magazine, *World*, has asked me to do a long review of Koestler's *Roots of Coincidence* (to be published here soon as a companion volume to his book on Kammerer), and I will try to remember to send you a copy. It is an incredibly naive book, though a cut above the crap in such books as *Psychic Discoveries Behind the Iron Curtain*. The interest in Russian parapsychology is amusing. I hope you saw Krippner's piece on the Russians in *Saturday Review* of last March 18. (He's the dream research man at Maimades Med Center NYC.) It features a photograph of Ninel Kulagina making a ping-pong ball float in the air (Krippner, of course, has to call it a "plastic sphere," which reminded me of B. Russell's famous footnote chiding Kohler for repeatedly referring to a banana as the "objective" in his monkey experiments.)

But I write mainly to mention that Dr. Ray Hyman, a psychologist at the University of Oregon, should be on your list. He is a fine amateur magician, an old friend, and co-author of a fine book on dowsing published by the U. of Chicago. Last year (maybe two years ago) he gave a course on occultism, with special lectures by astrologers, yogis, mind-readers, etc. He may be writing a book about it. If he kept a carbon of a long, detailed letter he sent me about the course, how his students reacted, and so on, you might want to ask him for a copy. He set some ingenious traps. For example, a girl graduate student who believed in astrology was asked to prepare character readings based on birthdates and submit them to a group of students to see if they could pick out their own reading with better than chance odds. She was jubilant when the results indicated high success. Hyman then suggested she check with the students to distinguish between those who were familiar with astrology and those who were not. It turned out, of course, that the successes were only with the former, whereas the latter guessed according to chance. He reports that the girl actually changed her mind as a result of all this, which strikes me as almost as remarkable as making a plastic sphere float. Anyway, you two ought to know each other.

Best,
Martin

[Postcard]   9 May 1972
Dear Marcello:

Could you spare me a second copy of your offprint on "The Occult Revival?" Reason: I gave my copy to Hal Bowser, the editor of the forthcoming journal, *World* (published by Norman Cousins, former publisher of *Saturday Review*). He has a lively interest (as a sceptic) in the occult book, so possibly you will hear from him.

Best,
Martin

[Euclid Avenue]   21 July 1972
Dear Marcello

I had hoped to include a copy of my Koestler review but *World* has not sent me any copies of the Aug. 1 issue, which I am told contains it; they are not on sale in the area; and we are taking off tomorrow for a two-week driving trip through Maine. I had to cut 90 lines from the galleys to make room for a cartoon and photo that the editors added at the last moment!

It seems impossible to hang on to copies of your paper on the Nouveau Witch. I am emboldened to ask for still another on the grounds that my last copy found, its way to *Time*, and is responsible for the quotes in their cover story on Occultism. A *Time* editor, Steve Kanfer, is now a neighbor. He borrowed my copy of your paper to pass along to the writer of the cover story. (He also took the first issue of the occult newsletter, so I'd like a copy of that also if any are still around). The probability of getting either item back is close to zero. You may hear from Kanfer, by the way. He is a marvelous writer—*NYT Magazine* pieces recently on Khalil Gibran, on ping-pong, etc., a recent *Time* cover piece on Woody Allen, and so on.

If you ever get around to a satirical study of Oz you must try to obtain a full run of the *Baum Bugle*, the publication of the Oz society. It is crammed with material you could use, and which is not available anywhere else.

Best,
Martin

[Euclid Avenue]   26 November 1973
Dear Marcello

Here's one dollar for issue just received of *Explorations*. Very informative issue. I was pleased to see your statement on the importance of
knowing the methods of mentalism before trying to evaluate someone like
Geller.

I've been keeping close tab on Geller, and have a fairly complete knowledge of his history and how he does everything he does. (For the most part,
his methods are so crude that Kreskin would be ashamed to use them.)
Would you be interested in my doing a piece on him for your Journal? If
so, it would be best to hold off writing it until the last possible moment
because Geller's star is still rising, and there may be much more to come in
the months immediately ahead. He has thrown out several hints of wanting
to move into psychic healing (where the big money is!). I enclose a recent
exchange of letters in the *NY Review of Books*, which you may have missed.
Coming up (next issue, I am told) is a review I did of *The Preachers*, in
which I get a chance to take another mild poke at Spraggett. Both Book of
the Month Club and Book Find Club have taken Harper's latest monstros-
ity: *The Secret Life of Plants*. You won't believe it, but last week I got a
serious request from Library of Science asking me to evaluate the book for
them (for a fee), and if I liked it, would I write their advertising brochure?
I returned a blistering evaluation, and it will be interesting to see if they
offer the book anyway. Bob Scott, who wrote to me, pointed out that the
previous book by Tompkins, *Secrets of the Great Pyramid* (which they did
take), had been one of their all-time best sellers. Scott was obviously hoping I would report favorably on the book; he even reminded me about how
orthodox geological opinion had shifted with respect to the continental drift
theory of Wegener!

Science News is now on the occult bandwagon. Recent big cover story
on Kirlian photography (Thelma Moss identified only as an "experimental
psychologist") and a couple of issues back, another cover story on ESP with
high praise of Geller as a genuine psychic. It's almost as funny as Watergate
and the Nixon tapes.

                                        Best,
                                        Martin

P.S. Did you catch the *Time* story (Oct 8, p. 104) on the little pyramid
that sharpens razor blades etc.? For Eric McLuhan's work on this (he's son

of Marshall) see Air Canada's magazine *En Route* (Feb/73, p.206) (cover story). I tipped off *Time* about it.

[Euclid Avenue]   4 December 1973
Dear Marcello

I'm a 40 minute ride up the Hudson, by train from Grand Central. If you don't mind the trip, and would like to come out any afternoon, or for lunch or dinner, would love to have you. I mention this possibility because you might want to check things in my library or files, and I own a good copying machine so I can run off copies of anything on paper, including book pages. The house is only a few minutes from the RR station, where I would pick you up, and trains leave about every hour. Also, I can easily get to Manhattan for lunch, unless I'm tied up with some prior commitment. I don't like to go to the city for dinners, partly because we keep early hours of retiring and rising, and getting back at night is somewhat complicated.

It's hard to believe, but the respected Library of Science actually sent me a copy of *The Secret Life of Plants* to ask me if I thought they should offer it as a main selection, and if so, would I write their brochure! It's the biggest collection of crap yet, and a selection by both Book of the Month, and Book Find Club. Almost every crank in my old *Fads and Fallacies* is there and treated as a great scientific genius.

Farrar and Straus will soon issue US edition of Dr. Christopher Evans' *Cults of Unreason*, a book I highly recommend. He should be on your mailing list—a top English psychologist. [address withheld] He runs a popular radio talk show, and would probably be delighted to interview you if you're ever in London. His book is mostly about Scientology (best treatment of it yet), but also covers other distinguished pseudo-scientific cults.

Yes, Clayton Rawson was a very good friend whom I saw often. (He organized the Witch Doctor's Club, which used to meet irregularly, and of which I was a member. Those with PhD or medical degrees were addressed as "Doctor doctor..." Very funny set of club rules, that I have filed somewhere.) His son, Hugh Rawson, is an editor at T. Y. Crowell.

No reply needed.

Best,
Martin

P.S. That Nixon bumper sticker spoonerism is great—hadn't heard of it before.

[Euclid Avenue]   24 January 1974
Dear Bob [Crowell]

I write with a heavy heart. As you know, Crowell recently offered me
a contract for a book on which I have been working. My editor would be
Hugh Rawson, whom I have known for many years, whom I respect greatly,
and whose father, Clayton Rawson, was an old and dear friend.

I find that I cannot, in clear conscience, do this book for Crowell. The
reason: Crowell is publishing in June a book by John Fuller, *Arigo: Surgeon
of the Rusty Knife.*

Let me say at once that, as a libertarian, I am resolutely opposed to
any type of government control over the free press. I would defend to my
death your right to publish this book. But as a libertarian I, too, feel free
to register a protest, and to do it in the only way in which it might be
effective.

The occult revolution in America is now reaching awesome heights of
insanity, heights paralleled only by the wave of occultism and pseudoscience
that swept Nazi Germany. Most of the hundreds of worthless books in the
occult field, pouring weekly from the presses, are amusing and relatively
harmless: books on astrology, witchcraft, demon possession, prophecy; spiritualism, and so on. But books on psychic healing are another matter.
Fuller's book, unlike his flying saucer volumes, will, in my opinion, cause a
great deal of harm. Thousands of sick souls, caught up in the psychic wave,
are now turning from their doctor's and spending life savings on psychic
healers. Arigo, the subject of Fuller's book, was a notorious quack. To
publish a book that presents him seriously is to mislead thousands. It will
result in needless tragedies.

I am ashamed to have a book of mine on the list of a company that
yields this readily to the temptation to make a quick killing on the occult
market. I hope that my protest will leave some small impress on those
at Crowell who are pleased with the prospect of promoting Fuller's sordid
work.

                                    Sincerely,
                                    Martin
CC: Hugh Rawson, misc. friends

[Postcard]   25 January 1974
Dear Marcello:

Did *not* know of the *New Horizon* articles on the pyramid—many thanks! Just in time. *Time* is preparing a big cover story on contemporary psychics, in which I am involved. Feb/74 *Penthouse* (article on Kirlian photography and a nice photo of a rose on p. 55). Feb/74 *Esquire* (article on a psychic healer). *Horizon* last issue (article on Geller).

                            Best,
                            Martin

[Euclid Avenue]   6 February 1974
Dear Marcello

  Thanks for yours of January 28. I hadn't heard anything about that Jeane Dixon trick of giving contradictory predictions, but I wouldn't be too surprised.

  Time's cover story still in preparation, but no date set yet for it. Stefan Kanfer is writing it, as I may have told you before. Perhaps you saw his marvelous Sherlock Holmes pastiche in *Time* a few weeks ago (Holmes solves the mystery of the gap in the tapes.)

  Things are happening so fast on the psychic scene that it's impossible to do more than hit a few highlights. Geller is now doing the Ted Serios bit. Big headlines in London papers when they published photos of Geller taken with the cap covering camera lens! Puharich has written a book, *Uri*, to be published in March by Doubleday. I just broke a contract with Crowell over their forthcoming book on Arigo, with introduction by Puharich. I enclose a copy of my letter. You can print it if you like. Christopher Evans' fine book on Scientology and other things, *Cults of Unreason*, is now off the press in US edition (Farrar, Strauss). Ray Hyman will be in New York next month to consult with Randi, me, Leon Jaroff of *Time*, etc, about forming some sort of organization to expose psychic frauds, but I doubt if much will come of it. Open Court will soon issue volume 1 of a mammoth study called *The Occult Underground*—vol. 1 going only to outbreak of W. War I. It's by a Scotsman named James Webb. I'm reviewing the book (along with Evans' book) for *NY Review*. If you know anything about Webb, would appreciate your shooting back what you know. The jacket blurb says he's a member of US Soc. for Psychic Research, and has contributed to two of the occult encyclopedias on sale. Can't tell from vol. 1 how much of the nonsense he buys. It's a patchy history, dull writing, but he catches a lot of interesting data and puts it in one place.

Best,
Martin

[New College, Sarasota]   14 February 1974
Dear Martin

Afraid I can not supply you with anything about James Webb. I have nothing on him. But I am glad to hear about *The Occult Underground* volumes you mention and will keep on the lookout for them.

Noticed something that might be of interest to you. The current (February) issue of *Human Behavior* (a kind of cheap production—black and white—version of *Psychology Today*) has an interview with Geller. In it reference is made to Harold E. Puthoff at the Stanford Research Institute which more-or-less endorses Geller. In this same issue, there is a short article on scientology by Eleanor Links Hoover in which she quotes Puthoff on scientology saying it is "a fine blend of Eastern and Western traditions," and that "Millions of carefully supervised research hours have gone into it and its success in rehabilitating people's abilities and emotional stability is, truly phenomenal." Christ, if he considers the work on scientology to be "carefully supervised research," what kind of controls did they have with Geller? Re Geller, you might have missed the following item in the interview with Geller in *Psychic* (July 1973), p. 8:

Psychic: Dr. H. C. Berendt of the Israel Parapsychology Society in Israel, writing in the *Parapsychology Review* (July/August 1972), said his group sent you eight personal invitations to demonstrate your abilities before them and that you postponed appointments and later refused to attend a meeting. Why?

Geller: Listen, I don't remember receiving any invitations from this group and don't even know who they are. They might have thought they got invitations to me, but I didn't receive any. It's too bad this thing happened, since I am not opposed to demonstrating for scientists. I've certainly done a lot of it in this country. So something is wrong somewhere.

I would think that Berendt would be a good person to contact re Geller. I presume you also know that the last three issues of *FATE* magazine have had a book-length series on Geller.

Two questions re Geller you might be able to answer. I knew that Geller was doing the photography trick (a la Serios), but you earlier mentioned in a letter that he was getting into healing (where the big money is). I have seen nothing about his healing that I can now recall. Where did you get that information? Secondly, I have now seen Geller numerous times on TV and the metal bending I have seen is always with his fingers on the objects. Yet the claim is constantly made that the stuff keeps on bending after he returns it or puts it down. I presume this is just cognitive distortion by the reporters. Are there any films or direct evidence of things bending out in the open, away from his person? The damned TV camera goes back to the keys and stuff later in the shows and the announcer claims further bending, but they don't keep the camera on the key or what-not during the alleged bending period. Any information on this? I saw Randi on TV allegedly duplicating Geller's effects, but I have to admit the effects were not exactly the same.

Christopher Evans will be down this way next month at which time we plan to get together here in Sarasota. He has relations in Ft. Meyer, which is about 50 miles from here. I am arranging a small speaking engagement for him at New College here. I have not yet been able to get his book, but I greatly look forward to and meeting him.

Glad to hear that you, Randi and Ray Hyman are meeting re the proposed investigative foundation. I had written to both Randi and Ray earlier about the proposal since I had something rather similar in mind. If you end up with room for me, I'd very much like to be involved.

I was pleased with your letter to Crowell. I wish more people had your integrity about such things. I gather that there are now various travel agencies promoting trips to the Philippines for psychic surgery! What sort book were you planning to with Crowell? No mention in your letter and I am extremely curious.

I'll see what I can find out about Webb from some friends of mine who might know.

Best wishes,
Marcello Truzzi

P.S. Could you give me Stefan Kanfer's address. I'd like to send him a copy of the Holmes piece I did since I suspect he'd like to have one judging from his *Time* pastiche.

[Euclid Avenue]   15 February 1974
Dear Marcello

Thanks for the *Human Behavior* reference, which I did not know about. I did see the *Psychic* piece on Geller, and the *Fate* series. Geller first tried (without success) psychic healing at an appearance in Stanford. He has since made several statements indicating that he expects to try again when the time is suitable.

Yes, one of Geller's ploys is to point at a bent key and say: "See, it continues to bend!" Of course it doesn't. However, if he gets a chance at the key, when you are distracted, he'll bend it a little more, then let you discover that fact later. Geller's methods are quite primitive, and rely heavily on the fact that he has plenty of time, and all sorts of things are going on at once. A magician has to do a trick, 1,2,3, but Geller may take thirty minutes before he has success with a borrowed key. It's hard to believe, but there are cases where a reporter has gone to the bathroom, then returned, and Geller picked up his key (which hadn't bent before) and then "bent" it for him! (He has a way of holding the key and sliding it on the table top so that it doesn't look bent; then he lets you hold one end while he gradually takes his fingers away from the other.) But I'll have to stop. I could write ten pages about Geller, nonstop. I'm currently reading the galleys of Puharich's book, *Uri*. They are hilarious, and will do Geller more damage than anything *Time* can do.

The Crowell contract was for an anthology of famous single poems (like "The Lost Chord," "Shooting of Dan McGrew," "Village Blacksmith", etc.) to be illustrated with nineteenth century engravings I've been collecting, and each with a short introductory essay. My working title is *Great Doggerel*.

Stefan Kanfer [address withheld].

Webb is a young Scotsman, living in London, and that's all I've learned about him. However, I've turned in my review (today) of the Evans, Webb books, so don't waste any time on him. He's less gullible than Colin Wilson, and less egotistical, but Wilson writes better (I'm referring to Wilson's book on occultism.)

Best,
Martin

[Euclid Avenue]   25 March 1974
Dear Marcello

Glad that you and Chris Evans met. I reviewed his book for *NY Review of Books*, but they haven't printed it yet.

The expose of Soal was by Hansel, and was many years ago. The details are given in Hansel's book. I have Hansel's original papers, but I think the summary in Hansel's book is adequate and more up to date on the controversy. Hansel caught a striking anomaly in Soal's worksheets (evidence of psi was confined entirely to certain lines on the sheet, which repeated regularly, suggesting doctoring.) His assistant, named Gretel something, said she saw him altering figures. Hansel and Gretel! When Hansel went to see Soal and asked to see the original worksheets, Soal told him he had lost them on a train!

The notion that Geller uses a chemical on keys or spoons is pure hogwash, and he must be laughing up his sleeve at such suggestions. There are *no* chemicals that do the trick, and Geller doesn't need them. He simply gets to the spoons in advance and pre-bends them almost to the breaking point. The key bit is more complicated. Geller has never yet bent a key on TV because he needs more time and misdirection. He keeps trying to bend it, and failing, then goes on to something else, comes back to the key, etc. All he does is wait for a moment when he's not being watched closely, and he quickly bends the key by pressing the point against something (side of chair, etc.) Most people don't know how easily car keys can be bent—especially if they have one deep cut. Once bent, he replaces the key so it is partly hidden, then he picks it up and holds it so it looks straight. Sometimes he pretends to slide it flat, back and forth) on a table top, keeping the end concealed. After he has bent, it, he may secretly bend it further at a later moment. It's all very crude magic. Geller is not skillful at manipulative magic at all—just at psychological misdirection. The key bit, however, is not something he can do except in informal circumstances when he has plenty of time. I have the *Der Spiegel* issue, in German, if you want to borrow it and return, but it really contains very little of interest.

I hadn't heard, that Price had become a Roman Catholic. Marvelous!

Ray Hyman has promised to type up notes about our discussion of a foundation, and will, he says, send you a copy. The basic idea is to get sufficiently well organized plans made so that Ray can apply for some sort of grant. We figure at least $50,000 is needed to get the thing going, pay for an office at some university, etc. But it is all in a very nebulous stage at the moment. The first step will be Ray's report, which he hopes to circulate among interested parties.

Puharich's book on Geller is now being serialized in England, on the front page of a newspaper, so Doubleday can't back out. This is the book in which Puharich claims that Geller gets all his power from computers in a flying saucer controlled by superbeings in another solar system. Geller meanwhile is trying to do his own book, with an English publisher, and Doubleday is suing to prevent it, on the grounds that Geller is under contract to promote the Puharich thing. So there is a falling out between Puharich and Geller that may lead to amusing consequences, or perhaps grim consequences in case Puharich is sick enough to believe everything in his book (as I suspect he is). The two of them appeared last week at the Goddard Space Center, in Baltimore, and hardly spoke to each other. Obvious hostility. Geller did his act for the NASA crowd and bombed out. They established a few elementary controls, so nothing worked except one test in which Geller guessed correctly a color someone thought of (yellow). The whole thing is getting wilder and more hilarious. Geller will either have to repudiate the Puharich book, or defend it and try to get a new religion going, with himself as the guru. Or maybe he'll compromise and try to dismiss Puharich without saying one way or the other whether his book is accurate. Whatever way he plays it, I think the Puharich book will do him much more harm than good. If he's smart, he'll imitate Marjoe and do a documentary exposing himself—but I don't think he's *that* smart, and his vast ego won't permit it.

John Wilhelm, of *Time* magazine, has a leave of absence to do still another book on Geller—for S&S. I'm reviewing the Puharich book for *NY Review of Books*, along with the Crowell book on Arigo, the Brazilian psychic healer, written by John Fuller, with a foreword by Puharich!

Best,
Martin

P.S. "Euclid Ave" is just a coincidence. We lived a few blocks away, on the same street, which changes it name to Bellair Drive when it crosses a village border. When we needed a bigger house (I occupy the entire third floor as my "office") the ideal spot proved to be down the street.
[Copy of Krippner article in *East West Journal*, 1974]

[Euclid Avenue]   12 April 1974
Dear Marcello

Just back from a week in Tulsa, to visit with some relatives there. I've put a copy of *Der Spiegel* in the mail, fourth class, and you can return at your leisure. No hurry, but eventually I'd like it back.

I'm now reading the two all-time nuttiest books of the past 20 years: *Uri*, by Puharich, and *Arigo*, by John Fuller, with afterword by Puharich. I'll be reviewing both for *NY Review of Books*. At the top of Puharich's mythology are The Nine. Under the Nine are the Controllers, who are in charge of millions of planetary life systems. Earth's controller is Hoova. Under Hoova is a spaceship called Spectra, filled with computers. Occupying the computers are intelligences from millions of "light years" in the future, who have gone back in time. Geller's power comes from the computers. Hoova has chosen Geller as the messiah for the next 50 years, and Puharich as his keeper and scribe. The book is filled with UFO sightings, Geller miracles, and tape-recorded messages from Hoova (Geller goes into a hypnotic trance and this voice comes out of his mouth, etc.) Unfortunately, the tapes later self-destruct (i.e., Geller steals 'em.) Arigo, the late Brazilian psychic healer, also gets his powers from Spectra. The two books interlock. The sad thing is, I'm convinced, that Puharich believes it all.

Geller is now at an amusing turning point. Will he repudiate the book and risk P's enmity? Or will he promote it, as Doubleday claims he is under contract to do, and try to start a new religion with himself as the central figure? It's hard to see how he can avoid doing one or the other.

I enclose a few other recent odds and ends I've acquired. Unfortunately, the material is cumulating so rapidly; and so many friends are asking for copies of this or that, that I'm wasting endless hours on this nonsense. When is the occult wave going to peak?

Best,
Martin

[Euclid Avenue]   21 April 1974
Dear Marcello

What a splendid article on juggling! Thank you for the copy. If you haven't already done so, I wish you'd send a copy to my old friend Roger Montandon [address withheld] to add his excellent collection of juggling books, papers, etc. You will recall him from the days when he put out *The Juggler's Bulletin*.

Another person who would like very much to have a copy is the eminent mathematician Elwyn Berlekamp, formerly of Bell Labs but now in the math department of Univ. of California, Berkeley, Calif 94720. His hobby is juggling (he does five balls with ease, he told me); but more than that, he has made a special study of the mathematical aspects of juggling. He would be the best of all people to ask about the theoretic possibility of 10 balls.

No, I don't know Syd Bergson. Nor do I know Kreskin personally, though I've seen him work many times. Your remarks about Barber explain to me for the first time why he makes such pretence of hypnosis as being nonexistent. I've seen him give a number of TV "hypnotic" demonstrations—or just the power of "suggestion" as he puts it—and they obviously used stooges.

I enclose some recent nonsense by Krippner. Since a magnetized match is such a great scientific breakthrough, you'd think Krippner would have asked to take it back with him, to be analyzed, but there's no indication he thought it worth the trouble! For decades magicians have pushed small needles into matches so that they could be used in various table tricks making use of concealed magnets. Krippner's gullibility continues to astound me—and his dream lab work continues to be the most respected of all work in modern parapsychology!

> Best,
> Martin

[Euclid Avenue]   12 May 1974
Dear Marcello

Enjoyed looking through the program of the Popular Culture Assn. No, I've never attended, but it certainly looks like I should sometime. Thanks for telling me about Foster Brook's Casey poem—new to me, and I will have to get it. The book is about to go out of print, and I'm planning to get reversion of rights and try to place it as a paperback. I have quite a few new poems to add, as well as some more references on poems by Thayer I hadn't been able to run down before.

I've been so busy writing to people about Geller, and answering phone calls about Geller, and so on, that I am hopelessly confused and do not recall if I did or did not send you my review in the current (May 16) issue of *NY Review of Books*. Anyway, I review the Puharich book and the book on Arigo, with a strong blast at Crowell for jumping on the occult bandwagon.

Geller, on Midday show two weeks ago, came out in support of Puharich. He said "Every word in the book is true," and that the *Popular Photography* piece this month is the "dying gasp" of his enemies. For those who didn't believe in him, he said, it "was their problem" not his. He said that the producer of the movie of Jesus Christ superstar has signed him up to play himself in a documentary of his life, to be filmed in Israel and Tibet. Last week he was the star of a big meeting on parapsychology at the Essalen Institute! So he is still making lots of money for his appearances, and it is too early to say whether Puharich has boosted his stock or damaged him. I have heard that Heinemann will not publish his autobiography, but that Putnam's will. It figures.

> Best,
> Martin

[Euclid Avenue]   11 July 1974
Dear Marcello

I certainly didn't expect a review of my novel, but thank you very much! Quite aside from this, I found the issue to be unusually good and informative. Geller is certainly kicking up a storm. I enclose the latest Geller victory, sent to me by Phil Morrison, the physicist at MIT. David Bohm, the expert on quantum mechanics (the number one defender of Einstein's approach, now very much in disfavor, which assumes that deeper understanding of the microworld will cause quantum mechanics to be replaced by a deterministic theory) was a surprise; I hadn't known of his leanings toward the occult. But Clarke's reaction was no surprise. He's been moving in this direction for many years. Koestler is getting in deeper and deeper. What he should do, of course, is give himself a crash course in modern magic, taking lessons from some professionals, so that he gets some insight into the psychology of deception; but I suppose there is little chance that he would consider this worthwhile.

Some random bits of information: John Fuller has a long rebuttal of my review in the current *NY Review of Books*, to which I appended a brief reply. I am sure there must be a wealth of literature in Brazilian newspapers and magazines reflecting the negative opinion of Arigo on the part of Brazilian doctors, but I can't read Portuguese and don't know anyone in Brazil who could dig it up. Fuller probably didn't even look for it, or, if he did, was careful not to mention any of it. Fuller, by the way, has contracted with Putnam's to ghost write Uri Geller's autobiography. Geller's former deal on this book, with a London house, fell through.

Bernard Geiss, one of the great charlatans of publishing (he introduced *Valley of the Dolls*, and many other best-selling pot-boilers) read my Dr. Matrix column on the pyramids, took it seriously, invited me to lunch, and offered me an advance of $15,000 to do a book called *Pyramid Power*! I have declined.

Geiss said he was just back from Canada, having signed up Xavier (the Happy Hooker) for her first hard-cover book: a hard-cover hard-on book, I suppose. He was complaining because she didn't offer him a free tumble in bed. I recommended to Geiss that he latch on to that up-and-coming black minister, Rev. Ike, who had been on Merv Griffin's show the day before. Ike is a marvelous stand-up comic, who ... has Red Skelton's technique of laughing at his jokes. There's a good chapter on him in Morris' *The Preachers* (St. Martin's Press) (Morris is doing a similar book on modern psychics), and *Esquire* had a good article him recently. Ike doesn't believe in black power. It's green power (I'm quoting Ike). He put on a fantastic performance for Merv. But back to Geiss. After I finished talking about

Ike, he told me he had Ike signed up for a book, but hadn't yet located a ghost writer for it.

But I've saved the big news for last. Dr. Martin Levy, Rhine's right-hand man (he was in charge of all the animal ESP work for Rhine) was caught flagrantly cheating a few weeks ago, and has been fired. It's an enormous scandal, but Rhine and his staff are close-mouthed about it so I can't give any details. Is there any way you can find out what happened? Martin Ebon told me this on the phone. He and Rhine are old friends. He said he had telephoned Rhine to talk about it, and that Rhine was "crushed" by the disclosures, and would be writing it up eventually for his journal. A friend of mine phoned Rhine's headquarters and was told that, yes, Levy had been fired for "unsatisfactory" work, but the lady clammed up when asked for details. *Scientific American* would like to print something in its Science for the Citizen department if we can find any trustworthy source. I can't even find Dr. Levy's current whereabouts. Levy, by the way, is famous for his experiments with chicks, live eggs, and a radioactive randomizer. The randomizer turns a heat lamp on and off. The room is cold, so the chicks are uncomfortable when it is off. When the heat lamp has nothing under it, it is on as much as off, as one would expect. But when the chicks are there, it stays on much longer that chance expectation. Live chicken eggs have the same PK ability to influence the randomizer, but when the eggs are hardboiled, no PK. Isn't that beautiful? There is a brief reference to all this in the recent book, *The Challenge of Chance*, of which Koestler is one of the three authors.

I have a curious book project underway. A selection of material from Hugo Gernsback's old *Science and Invention* magazine—probably all from the 1924 issues of just 50 years ago. Scribner's may do it. I have a full run for 1924, and will pay well for issues in other years to fill the gaps in my collection. If you know anyone who owns back issues of this magazine, please let me know!

Time has a big article on magic scheduled for this or next month, and *Not For Women Only* will devote a week to magic, with Randi, Christopher, and Doug Henning (star of the very successful magic musical) on the panel. Don't know yet if Barbara Walters is going to allow them to mention Geller by name; after all, this is embarrassing to Barbara, who tumbled completely for Uri when he bent her spoon.

All best to you on your new position,
Martin

[Press release on Geller sent by P. Morrison]

[Euclid Avenue]   5 August 1974
Dear Marcello

Here's the first news story on Rhine—indirectly from me because I told Rawlins and Rawlins called the *Baltimore Sun*. I gave the *Sun* Barry Poss's phone, but don't know if the reporter reached him or not. I've also tipped off *Time*, and the *NY Times*. Walter Sullivan phoned me at once, and said his assistant would follow it up, so more will probably be appearing soon. *Scientific American* is covering it in the next Science and Citizen column. I've not talked to Barry Poss myself, but have been giving his name as a source of information. The *Scientific American* editor got hold of Rhine, but all he could get out of him were remarks about my evil influence on the magazine.

Ted Serios was back in Denver, sponging off poor old Eisenbud, a few months ago. I heard Eisenbud lecture on Ted at an ESP meeting at N.Y.U. Did I write you about it? If so, forgive repetition. Anyway, Eisenbud said he was working with Ted again, trying to get him to produce pictures, but so far nothing but blackies and whities. I think Eisenbud is still convinced Ted is genuine, although he now knows Ted is capable of cheating. And of course Ted gets free room and board and probably pay for his visits. But Ted hasn't thought up any new gimmick, and Eisenbud now exams his gismos, so results are negative. Eisenbud is now lying to save his reputation. He flashed on the screen a photograph showing the results he got when he, Eisenbud, tried to duplicate the *Popular Photography* work with the optical gimmick. He showed a picture of a room in perfect focus, with a small gray circle in the center. Of course the doctor knew full well that Ted always had the lens set at infinity, to prevent a picture of the room and to allow the optical device to cover the entire film. So this little bit was deliberately designed for his audience, most of which had never heard of the *Popular Photography* piece. It was a sad and pitiful performance by an aging man who is desperately trying to salvage a well shattered reputation. The height of absurdity was reached when he showed a picture Ted had obtained once of a car. The "target" word Ted had been trying to hit was "flat iron." Was this a miss? No, it turned out the car was a Fairlane. Eisenbud then explained how Ted's unconscious had rearranged the letters to make rude anagram of flatiron, producing a picture of a Fairlane! I could hardly believe what I was hearing. Nobody in the audience laughed.

Don't know anything about Hans Bender. There are so many of these fellows now in the act that new names pop up almost daily. I heard Judy Skutch on the radio recently saying that Puthoff and Targ had found a marvelous new sensitive, getting sensational results, etc., but she wouldn't identify him except for saying that he was a former mayor of a city or town in Calif.

Irving Geiss, the publisher, didn't know my column on the pyramid was a joke. He offered me an advance of $15,000 for a book on Pyramid Power. I have declined. (It was Geiss who published *Valley of the Dolls*, etc.) A reader in Honolulu wanted to pay my expenses there to lecture to his group on the sensational new discoveries I reported about the pyramids, which he says are on sale in all the natural food stores. Another reader drove all the way to Pyramid Lake to see Dr. Matrix's pyramidal factory, and wrote me a long angry letter about it when he discovered that nobody there ever heard of a Dr. Matrix.

It suddenly occurred to me that animal ESP has been Rhine's nemesis. His first major paper, which has done him great damage, was his paper on Lady Wonder, the mindreading horse. And now, what may be his last major paper, is a defense (probably) of the earlier work on animals by Dr. Levy.

I wonder if Fuller of *Fate* is related to Fuller of flying saucer and Arigo fame.

I'm thinking of doing a special issue of the *Pallbearer Review* (magic periodical) on "The Secret Notebooks of Uriah Fuller," an Arabian psychic, in which I will go into great detail about all of Uri's techniques. The magic crowd can't raise a cry of "exposure," but the issue would fall into the hands of lots of scientists who might otherwise be taken in.

Best,
Martin

P.S. Off on a vacation in a few days, and won't be back until about August 25.

[Euclid Avenue]   14 June 1974
Dear Marcello

Yes, *Der Spiegel* is back in my Geller files. Geller's act in Israel was pretty much the same as what he has been doing in his stage appearances here. To professional magicians, it is all rather stale and dreary stuff. My impression is that Geller more or less improvised his own methods, without much knowledge of modern mentalism. He can't afford to use such things as clipboards or nail writers, because he might get caught with the incriminating evidence, so he relies mainly on guts and acting the role of the innocent sensitive who is somewhat awed by his own powers. In his stage appearances here he has performed as follows:

1. He will turn his head away from a blackboard. Someone from the audience thinks of a color (or city, number, etc.) and writes it on the blackboard. Without looking around, Geller manages to name the color. Wild applause, while magicians in the audience stare at each other in disbelief that he could get away with it. All he needs is a signal from his Israeli pal, Shipi Strang, who is seated on the front row. I'm not sure, though, that he even bothers to use an audience stooge. There is so much confusion on stage, and it is all so loosely controlled, that he may just get a quick peek. He is always covering his eyes with his hands, for example, but of course can easily see between fingers.

2. Rings are collected from the audience. Geller picks out a ring and makes it bend or break.

3. Watches that won't run are collected from audience. He handles various watches and occasionally one of them starts running again. Does Geller bother with stooges who supply watches? He might, but I doubt it. I think he just relies on the fact that lots of watches that "don't run" will run for a minute or two if you give them a shake. Geller shakes them when he puts one to his ear to listen for ticking. Maybe warmth of hands helps.

4. He picks which little can of ten contains water, or nails, or whatever. He makes a big show of never having seen the cans before, but actually either he or a friend has inspected them beforehand, learned which can is filled, then Geller simply memorizes some slight imperfection that enables him to recognize the can. This is an old magic principle. Card men have been identifying cards by little marks of dirt on the back for hundreds of years. Each can is usually taped shut; you have only to notice something about the taping, or a dent on the side of can, or a tiny discoloration, etc. Obviously no two cans are going to appear identical to a sharp eye. Then when time comes to perform, Geller turns his back and asks someone from audience to move the cans around, etc.

5. He duplicates a drawing made before the show, and sealed inside envelopes. Geller has already learned what the drawing is. I don't think he relies on any subtle method. Sometimes the drawing is made in his presence and he watches the hand movements. Sometimes the envelopes are brought to him before the show and he has time to open them and reseal in new envelopes. Sometimes he can hold them up to a strong light and see the drawing. I would be very much surprised to learn that he ever used a clipboard, or any other clever technique (of which there are dozens)—just crude methods he has hit on himself over the years.

6. The hands of a watch are set to a certain hour. Geller causes the hands to move to a different setting. This is probably Geller's cleverest trick. I've not seen him do this, but from what I've heard, he simply pretends to wind the watch, but actually pulls the winder to the hand-

setting position. He shows the time on the face. Then as the watch is turned face down by revolving it, the fingers of the other hand retain their grip on the winder. In other words, the act of turning the watch face down automatically advances the hands a few hours. Geller then has a nice throw off. He stands across the room for a while, then comes back, picks up the watch, glances at its face, looks crestfallen, and says that nothing has yet happened! It's already occurred, but he doesn't let anyone else see the face of watch, and of course Geller's admirers believe everything he says. Five minutes later he tries again. This time, someone else picks up the watch. Lo! the time has altered! They swear afterward that Geller never touched the watch, was ten feet away when the hands moved, etc, etc. Of course they actually believe this, because after all nothing happened when Geller last touched the watch, so it must have happened later when Geller didn't touch it. It is all so obvious!

There is nothing really interesting about Geller's methods. What is interesting is the extent to which intelligent people will fail to notice crucial things, and to give inaccurate descriptions later of what they saw. Over and over again people will swear that they saw a key bend in their hands while Geller was gently stroking it. They actually believe this because they assume that when Geller picked up the key—perhaps for the sixth time—it was still straight. Of course Geller is telling them that it is, and saying "Let's try it again. Maybe it will work this time. Do you feel it bending?" and so on. So, as Geller slowly lets his fingers lower while he is stroking the key, they see the slight bend for the first time, and naturally assume that the key has bent at *that moment.*

To sum up: Geller is doing very low level magic. He has no special manipulative skills. His knowledge of modern mentalism is probably very low. He relies mainly on his ability to act a role, to lie convincingly, and to use verbal misdirection. A typical example of verbal misdirection: He picks up a bunch of nails that are taped together. As he undoes the tape to get a nail, he will say: "Who put on this tape? I've never seen tape like this before." In other words, he makes some remarks to leave the impression that he is seeing the nails for the first time, when actually he had access to them before the show and has loaded in a bent nail. After bending the nail, he sometimes points to it, as it lies on the table, and says: "Look! It's still bending!" This makes a strong impression on his audience. They "Oh" and "Ah" at the great miracle. Of course nothing is happening to the nail, but it is inconceivable to the true believers that Geller would lie so blatantly. When he bends a key he always tells the owner that it will continue to bend for several days. In many cases the owner looks at the key next day and swears that it has bent a little bit more! The power of suggestion is strong, and people so naive, that when he bent silverware on

British television, hundreds of people reported that they found their own spoons, etc., had bent.

As you can see, I could run on for hours about Uri, but I must stop somewhere. You ask about my U of Chicago days. I majored in philosophy, and hung around for a year or two of grad work without getting a master's. Took some classes from Carnap. If you don't know Carnap's one popular book, which I edited, I think you'd enjoy reading it—just out in a new paperback with the new title: *Introduction to Philosophy of Science* (formerly *Philosophical Foundations of Physics*, which frightened away most purchasers of the hardcover).

Best,
Martin

[Photocopy of article on Levy from *Balt Sun* 7/24/74]
[Photocopy of article on Geller from *Science News* vol 106]

[Euclid Avenue]   24 August 1974
Dear Marcello

Just back from a trip to Ohio. A son got married in Bowling Green. I almost telephoned you, with the thought of driving to Ann Arbor, then I recalled all the weeks of confusion every time I've changed an address, so we (wife was along) decided against it. However, I'll probably be in the area about once a year, so maybe next time.

The story about Levy is in *Time* (Aug 26), *NY Times* (Aug 20,) and *Science News*, (Aug 17). *Scientific American* will have it in September issue. I may "review" Rhine's paper on it, for *NY Review of Books*. Rhine, of course, is trying to capitalize on how quickly he fired Levy, etc., but of course it wasn't Rhine who thought of setting a trap for Levy; it was two assistants. To me, the main point of the story is that unless you set a careful trap, you never uncover this kind of fudging.

I think Dover would indeed be interested in your project. They are reprinting public-domain magic books like crazy, to meet the popular demand. The person to write is the president of Dover, Hayward Cirker [address withheld]. He's an old friend, so you can say I suggested it, if you like.

I don't know of any pieces by psychologists other than those you mentioned. Ray Hyman might know of some. I have a note in the files saying that there is something in *Genii*, October, 1940, about Houdini's criticism of the Binet article.

One aspect of the psychology of deception that would be good to emphasize in such a book—if not covered by any of the papers, Hyman might cover it—is the inability of spectators to recall exactly what the magician did. I have some amusing examples of this on file. Of course magicians often consciously do things and say things that are explicitly designed to create false memories of what really took place; Geller does this constantly, and with great effectiveness. I once saw Dunninger collect envelopes in which spectators had put billets with questions. Dunninger poked a thumb through the back of one envelope, stole a handful of billets. Since he had to keep his fist closed, I wondered how he was planning to get them into his pocket. What he did was this. He raised his hand in the air, shaking his fist in simulated anger, and shouted that he would give ten thousand dollars to anyone who could prove that at any time he touched any of the billets! Then he stuck his hand in his pocket.

Horowitz (a NYC magician) once told me that he did the old illusion of bending a half-dollar for a lady. Later he heard her describe it to someone else by saying that he borrowed her half dollar then pulled it out like taffy!

Best,
Martin

[Euclid Avenue]   21 September 1974
Dear Marcello

The president of Dover is: Hayward Cirker [address withheld].

He's a good friend, so you can use my name.

I've finished a 40-page mss titled *The Secret Notebook of Uriah Fuller*, in which Geller's chief rival explains all his secrets in vast detail. Geller's cousin Shipi (his number one confederate) is in it as Schleppi, and his Japanese secretary (No. 2 stooge) Miss Toyofuko, is there as Miss Gofukyu. At least *I* think it's funny, and if it gets printed, it should circulate widely among parapsychologists and put a big dent in Geller's career. After all, he has a limited repertoire!

I'm waiting to hear from Karl Fulves, of *Pallbearer Review*, who expressed interest in it several months ago when I told him I planned to write it. If Fulves gets cold feet, is this anything you might handle in some way as a Xeroxed mss to sell to your mailing list? It's all inside stuff, written primarily for magicians, and not for the general public. Besides, if it got to the public, the magicians would be down on me for "exposure."

Don't know if your local TV carries the Barbara Walter's "Not for Women Only", but if so, next week's topic is magic, with Randi and others as panelists, and Uri coming in for heavy fire.

All best,
Martin

[Euclid Avenue]   8 January 1975
Dear Marcello

When Bernard Geiss found out that I wasn't serious about Dr. Matrix and his pyramids, he immediately thought of a hoax book, and cited *Naked Came* ... as an example of such a book that had a new spurt of sales *after* being disclosed as a hoax. He was willing for me to go ahead with *Pyramid Power* on that basis, provided I gave him a full year to promote it before revealing it as a hoax.

I declined for several reasons. I didn't want to spend all that time on what would have been essentially a worthless book; I was afraid that if it were successful and I made a lot of money on it, people would feel I had ripped them off; and finally, I was (and am) very dubious about such a book being successful. *Naked Came* ... was hard-core pornography, and readers didn't care who wrote it. But pseudo-science is hard to fake, and it would be difficult to keep news from leaking out in advance that it was all a put-on. The occultists all know each other, and the network might kill the book before it got off the presses. Also, it seems to me that occultism has peaked, and is starting down. I could be wrong. Anyway, I have been told that recent books, for which publishers had high hopes, have been commercial flops: the Puharich book on Uri, Arigo, Ed Mitchell's big anthology, and so on. It's hard to see how a put-on could be more outlandish than *Uri*, or *Secret Life of Plants*. My feeling is that our timing is off, and that the time for such a book would have been a year ago.

Having said all this, I would be happy to collaborate in any way feasible if you found a publisher. You might try Bernard Geiss himself [address withheld]. The idea also might appeal to Lyle Stuart, who did *Naked Came* ... but I don't have his address handy.

My Uri booklet is in the works—to be published anonymously by Karl Fulves. I will, of course, rush you a copy as soon as available. He's worried about lawsuits. I've already agreed not to refer to Shamford Institute of Technology (acronym: SHIT) and a few other changes have been made, but I imagine it will be printed soon.

Charles Fair's *The New Nonsense* (good), and a prime example of non-sense, Panati's *Supersenses* (published by *NY Times!*), are two books I reviewed jointly for *Washington Post*, but I don't think they've published it yet. Panati is a science writer for *Newsweek*, and *Washington Post* and *Newsweek* are jointly owned (are they not?); so it will be interesting to see if the *Post* publishes my attack on Panati. He's a very stupid fellow, trained in physics, and as naive as they come with respect to parapsychology. News of the Levy scandal broke just in time for him to remove Levy from the text, although all the references in back are to his papers!

Thelma Moss has a big book out (on sale in stores, but pub. date is April) on *Kirlian Photography*. She's still insisting Ted Serios is genuine!

I'll be attending the Baker Street Dinner this Friday with Steve Kanfer, of *Time* (who did the *Time* cover story on Psychics, and on Magic). Do the two of you correspond? He has a contract for a book on the history of US child rearing, and would welcome any leads on early books on the subject.

I'm doing a hoax column for April (Fool) issue of *Scientific American,* a report on scientific and mathematical discoveries of 1974 not reported in the media. I have a marvelous lost page from Da Vinci's notebook proving that he invented the flush toilet, and few other similar bippies.

Best,
Martin

[Eastern Michigan University]   22 January 1975
Dear Mr. Geiss

I am writing you at the suggestion of Martin Gardner. I am a sociologist specializing in the study of the sociology of occult movements. I proposed a book idea to Martin and to an archaeologist-historian friend, Dr. L. L. Orlin, which we have all agreed to collaborate upon and in which Martin thought you might be interested.

As you know, a few years ago a book hoax called *Naked Came a Stranger* was put together patterned after the Jacqueline Suzanne novels. I wish to do the same sort of thing patterned after the work of people like von Däniken, the pyramid power exponents, and others. In short we want to write the "crackpot science" book to end all crackpot science books. The idea would be to put together a book containing all current elements plus some of our own. This would include UFOs, pyramid power, ancient astronauts, plant communication, auras, etc., but the whole thing would be done with a very serious presentation full of appropriate scholarly citations etc. The booklet would also contain elements of a conspiratorial view of history tying in UFOs with recent political assassinations, the energy crisis, and other current issues including the international monetary crisis.

It would be our hope that the book would be accepted as sincere by the general occult press. We would remain in anonymity presenting the book is being by a group of academicians-scientists (possibly committee X) who feel compelled to reveal the truth at last but do so in fear of our own jobs and the sanctions of our universities. After the book is been out for a while (circa 9 months), we would then expose it and hopefully gain a new market of readers who want to enjoy it as a hoax. From our standpoint we feel this could be valuable scientifically in so far as it would damage the initial credibility of future crackpot books. Ultimately, laughter may be a more potent force than argument when dealing with a large volume of crackpot books currently on the market.

When I mentioned my idea to Martin, he indicated that you had coincidently mentioned something very similar to him in relation to his doing a pyramid power book, So, he thought you might be interested in this venture. Our book should be far more marketable and timely than the pyramid power volume since we can make a point of making it as up-to-date with the current fads as possible.

I look forward to your reaction, and, if it is one of interest, will be glad to give you further details (to a large degree we can adapt to whatever direction you think would be best). I possibly should add that I plan to be in New York City in April at which time you also might wish to meet and discuss the project (again, presuming you are interested).

Sincerely,
Marcello Truzzi

[Euclid Avenue]					28 March 1975
Dear Marcello

I'll be home April 18-20, so I see no reason why we can't get together some way then, and I am looking forward to it. My phone is [phone number withheld]. If you'll let me know the hotel where you'll be staying, it would be good information to have in case, for some reason, nobody is home when you try to reach me. I don't know how interested you are in checking material in my files; but if you are, it would mean a 40-minute ride, up the Hudson, from Grand Central. I live only a few minutes by car from the Hastings station. Trains leave at hour intervals, and more frequently during rush hours. I have a photocopy machine, so I can copy anything of special interest.

Reader's Digest, after carefully avoiding the occult revolution for years, jumped into it suddenly in their March issue by condensing Fuller's awful book on Arigo. Since *RD* has 18 million US readers, most of whom believe everything in the magazine, I find this particularly reprehensible. Dr. Nolen, who did the recent (sceptical) book, *Healing*, tells me that they kept his book six weeks, then returned it to his agent saying that couldn't condense it because of a "prior commitment to a similar book."

My current April Fool column is kicking up a storm. As I expected, lots of readers are taking it seriously in spite of the ridiculous names like Ms. Birdbrain and Prof. Macaroni. Fulves hopes to have my Geller mss of the press next month.

Best,
Martin

[Letter from Truzzi to Bernard Geiss proposing a hoax book on "crackpot science" 1/22/75]

[Euclid Avenue]   27 April 1975
Dear Marcello

Needless to say, it was a great delight to meet you at last, and I hope something comes of our plans for a society.

I have been enjoying the stories in your chess anthology that are new to me, and your comments on them. I enclose a letter just received from my artist friend in Woodstock, with an interesting enclosure about a chess painting. Please keep both. Which reminds of something I forgot to mention.

Unamuno, the Spanish philosopher, wrote a novella about chess, called *Don Sodalio*, that is virtually unknown outside of Spanish speaking countries. I tried to interest a publisher in a translation (it's not been translated), to which I would contribute an introduction, but the deal never went through. I did interest John McClellan in doing a rough translation (to submit to the publisher), and the plan was for John also to do illustrations. (Some of his lithographs have chess themes.) To give you a rough idea of the story: it's about a man who goes to a chess club every day, for years, to play chess with a stranger, whose name he later learns is Don Sodalio. The two never speak. The narrator knows nothing about his life, and doesn't want to know. The man is arrested for murder, and is executed. Later, the man's son visits the narrator, thinking he would like to know something about his father (who often spoke of the narrator when he was home). But the narrator sends the son away. He doesn't want to know *any* details about Don's private life. He wants to remember him only as his chess opponent. It's a novel about loneliness, and the contrast between the ideal world of chess and fantasy, and the irrelevant world of reality, etc. If you're interested in promoting this project, it's all yours. I think it's scandalous that it's never been translated.

I've turned in my review of Thelma Moss's new book, and the one by Ed Mitchell (to *Book World*), and will try to remember to send you a copy when it appears.

Best,
Martin

[Letter to Martin from McClellan, with xeroxed chess related item.]

[Euclid Avenue]   22 May 1975
Dear Marcello

Enclosed is a review. My booklet on Uri is in press, and copy should be available in about a week. I gave Fulves (publisher) a list of names, and he's promised to send copies to all on the list, so you should be getting it one in a couple of weeks.

I wish I had some good suggestions about the proposed association, news letter, etc., but that I have no experience with such things. Maybe Klass is the key to all of this. He seems to know quite a bit about such things. I, too, would be willing to contribute some money to get things going. I suspect that what is most needed is one person who is the "prime mover" or "chief coordinator" or some such. For a while it was Ray, and we were dreaming about a grant from a foundation—but now that seems to have turned out to be impractical, maybe the mantle should fall on your shoulders? Or you and Klass? I can help financially, and editorially, and so on—but as an organizer, I have neither the time or the ability.

Thanks for the limericks and the addresses of Crist and Boardman. I'll mention them to Asimov, but my guess is he's not interested so much in limericks by others as in writing his own! Apparently all of the limericks in his forthcoming book will be by Asimov!

Best,
Martin

[Euclid Avenue]   17 July 1975
Dear Marcello

Thanks for your remarks about the Fuller booklet. I've learned so much sense about Uri's methods that I hope I get a chance soon to revise and expand it. You mentioned Playboy. Funny thing, but it was submitted there and Playboy *did* want to publish it but they demanded of Fulves a statement absolving them of all legal responsibility and he refused to sign it. The book is on the verge of being libelous. I myself hope that SRI sues, but so far, nobody has.

Thanks for sending me the stuff on that Yogi teacher who died of cocaine. A great story, and I'm glad to have it in the files.

Taylor's book *Superminds*, is now out in a US edition. I'll enclose a review by Chris Evans that you may not have seen. Randi heard Taylor lecture a few weeks ago in London. During the question period, several gentlemen stood up to accuse him of being "close minded." Why? He expressed some skepticism about Kirlian photography and precognition!

Just found out that Uri, for $250, gives a private demonstration. Ray Hyman reports that a friend of his, who knows magic, had just such a session (paid for by some backers). Uri specified that he draw a picture, seal it inside an envelope made by so-and-so, size such-and-such, obtainable at such-and-such a stationery store and bring it along. Uri took the envelope, put it in his inside jacket pocket, then immediately withdrew it, saying "Sorry—I forgot. This must be in view at all times." He then put the switched envelope on the coffee table. A few minutes later, Uri had to go to the can. No self-respecting magician would stoop to this kind of switch, but Uri gets away with it!

All best,
Martin

[Euclid Avenue]   17 July 1975
Dear Marcello

Thanks for passing along the nice touch on that number trick—I've carefully filed for future use.

Yes I wrote the entire Uriah Fuller booklet except for Karl's introduction. I wanted to remain anonymous although it's no great secret that I'm the author. The plan is just to confuse SRI as much as possible.

Your piece on astrology is great and enormously valuable. It's the one aspect of the occult revival that I've never made an attempt to keep material on except in a very unsystematic way, and now you have pulled it all together in a marvel of compression. I'll be giving away copies for years.

I, too, noticed the item in *Fate*, and so far have no confirmation. Chris Evans would probably know.

Randi pulled off the great coup last week. He's probably written you about it but if not here's what happened. To get material for his Ballantine book on Uri (it goes to press this week! They're rushing as fast as they can), Randi made a trip to London where he had sessions with Hasted and Taylor and all those other British physicists who have been fooled by Uri. Joe Hanlon, of *New Scientist*, and Randi and Evans cooked up a joke to play on *Psychic News*, the leading spiritualist periodical in England. Using his real name of Zwinge, Randi had sent by Hanlon over to the *Psychic News* offices and the editor told that he was a new psychic from Canada who would baffled everybody at *New Scientist*, etc. Randi was in top form. Last week the issue came out with this picture on the front page, and a big story about all the things he may bend in the office without touching any of them. Indeed things bent in rooms which Randi didn't even enter (according to the story). I can't imagine how the editor will back out of it next week, now that news is out about the hoax. What can he do except state his belief that Randi has genuine powers? His description of what happened leaves no other loophole. I phoned Walter Sullivan of NYT about it but don't know if they'll follow it up or not.

SRI will receive at least strong body blow in October. I got permission to devote an entire column (*Scientific American*) to report on the SRI test (with $80,000 from NASA) of Targ's ESP teaching machine. This is not to be noised around—if SRI finds out it's coming up they'll try to forestall it, want to check copy, and so on. The report is supposed to be available to the general public, but the big brass at SRI are making it as hard to get as possible. My four letters asking for copy brought no response. I finally was able to obtain the loan of the copy belonging to Earl Jones, head of the Electronics Laboratory where the Geller tests were done, and for whom Targ and Puthoff work. He assured me on the phone that there was no "suppression," so he was trapped into sending me his copy. I have promised return it as soon as I get one of my own.

This 61 page report is so hilarious in its final negative results, and Targ's explanation of why (keeping a computer tape printout made it so complex that disturbed psi) that all I have to do is just describe experiment accurately, and it will do SRI almost as much damage as Uri has done. I'm urging readers to write SRI for the copy of the public report, and I imagine SRI will get at least 2000 requests for it.

All best,
Martin

P.S. Will enclose a copy of title page of the report. It's one of the funniest (unintended) documents in recent psi history.

[Written on the attached title page of *Development of Techniques to Enhance Man/Machine Communication*: Dig this title! Nothing to suggest that it was an attempt to *teach* clairvoyance by instant feedback by electronic machine! (For use of the machine to teach *precognition*, see paper by P and T in Mitchell's big anthology.)]

[Euclid Avenue]   22 August 1975
Dear Ray

I don't know how much Randi is sending you, but if you don't have a copy of the *Psychic News* story about how, as James Zwinge, he fooled the *News* or the *New Scientist* report about it, or the *Psychic News* follow-up story, let me know what you lack and I'll photocopy. I'm under the impression that Randi has mailed you copies of all this, but maybe not.

Uri was on the Tom Snyder "Tomorrow Show" at 1 A.M., morning of August 14. Randi is scheduled to be on the same show, morning of September 2. Uri opened by sending a picture to Snyder. Damn if it wasn't a circle inside triangle. Very effective, because Uri clearly told him to just "draw anything you like." However, after they hit, Snyder seemed unimpressed and mumbled something about "I drew what you told me to draw," but Uri quickly shifted the conversation. My guess is that before the show he told Snyder he would ask him to draw a geometric figure inside another, and to think of it now, but to draw it later. Then on the show he didn't mention the geometric figures! He fixed watches, bent a key (the bending occurred during a commercial. Nothing happened, cut for commercial, then Uri is stroking the bent key). He spoke bitterly of the magicians' campaign against him and a small group of enemies including a man at *Time*, a psychologist and someone at *Scientific America*. He made a big thing about his certification by committee of magicians in Georgia and said the SAM was planning to "kick them out" for it. He made a compass needle move. From his gesture of darting his head forward at just the right moment I guess the magnet was either in his mouth or under a shirt collar.

The *Parapsychology Review* mentioned the projected society. I received two letters from persons who are enthusiastic about it and want to join. One is a Dr. Feola, PhD, of Minneapolis. He is convinced that parapsychology desperately needs an organization to expose the fraud such as Dr. Rhine, whom he calls "self-deceiving." He (Feola), however, has been doing *legitimate* research which positively establishes the existence of telepathy, clairvoyance and PK, and he sent me one of his articles, but since it's in Spanish, I can't read it.

The other letter is from Helen Solem [address withheld]. She too wants to join and thinks organizations a great idea. She knows this because two of her spirit guides approached her recently and told her so, and asked her to write to me.

I've come to the conclusion of the proposed organization is a fruitless endeavor. There seems to be no way to get it funded, no way to exclude the nuts, and nobody was the time and money to invest in handling correspondence etc. I don't even have time to answer the two letters I received, and believe it would be would be fatal to do so. I'd just have them telephoning me or popping by on a visit.

As I see it now, the best plan is for those of us were sceptics to keep making our voices heard now and then, and forget about any formal organization. Marcello's newsletter can, of course, continue to serve as a way of keeping readers informed of what's going on. Possibly if the number of subscribers which increased dramatically, and it could be put on a paying basis, it might become a regular periodical that could be the focal point for society, but even this now seems dubious. I think we best just enjoy the spectacle while it lasts, and wait for the pendulum swing .

Best,
Martin

[Euclid Avenue]   8 September 1975
Dear Marcello

Just back from 10 day vacation in Canada.

Current issue of *Reader's Digest* has a pro-Uri article by his ghost writer, Fuller. Taylor's crazy book, *Superminds*, is now out in a US edition, filled with photographs, and incredible nonsense that makes Zollner's *Transcendental Physics* seem like sane reporting!

My blast at SRI's testing of Targ's ESP teaching machine will be in the next (Oct) issue of *Scientific American*. The legal department toned it down considerably, so I don't think SRI will sue. But they're worried. I've had several phone calls from the PR man, who wanted to check copy in advance but I wouldn't let him.

One more thing. Gershon Legman's sequel to his *Rationale of the Dirty Joke* has just been published. It's a monumental work, analyzing the "dirty dirty jokes" (in contrast to the "clean dirty" jokes in Vol. 1). The tome (992 pages!) is more or less privately published by an old friend of mine (so is Legman), Osmond Beckwith, who calls his firm Breaking Point, Inc. [address withheld]. He did a very professional job, and the book sells for $18. If you write him a note, maybe you can get a review copy from him. I know he'd appreciate any boost you can give it. The text is a great "read", as the reviewers like to say these days. It is marred, in my opinion, by Legman's old-fashioned Freudianism, but in spite of that, filled with remarkable insights and erudition, and of course—thousands of choice jokes.

Best,
Martin

[Euclid Avenue]   25 September 1975
Dear Marcello

Very glad to get the article on Velikovsky, and copy of your letter to Randi about role of magicians in observing psychics. Did you see the letter in *Nature* about the tests at Bath University? Six young spoon benders were put in a room with an observer instructed to allow loose controls. Spoons bent, etc. But through one-way glass they saw the kids putting in every bend!

NY Review of Books has given me 2500 words to do a double review of Taylor's *Superminds*, and Randi's new book, and I've been writing it with great gusto and cackling laughter. It's hard to say which book is the funnier, though of course Taylor has no sense of humor at all, and Randi's humor is all deliberate. They make a marvelous contrast.

Randi is in London this week to appear with Taylor on a big BBC TV show about Uri. I can't wait to get his report. Long John's guest last night was Mrs. Edgar Mitchell. I stayed awake only for the first two hours. She said she had a "terminal kidney disease," though not cancer, until she went to a psychic healer in Houston. She refused to give his name. After some grape juice, and laying on of hands, the pain "instantly went away." She is not a believer, she told John, but a "knower."

She defended Uri as a great psychic talent, but of course when he is "entertaining," he uses sleight of hand. Describing the dice box test at SRI, she said the box was shaken only 10 times (2 passes), no mention of hundreds of others, and that "Uri never touched the box at any time." (Targ told Wilhelm, of *Time*, that they let Uri shake it after one of them shook it first.) She is bright, quick-witted, articulate, and, as Candy put it on the show, "high strung." I'm wondering if she hasn't had more effect on Ed than vice versa. Pat Price, the star subject at SRI after Uri, she said died this year of a heart attack. I hadn't known that.

I'm having an amusing correspondence with Cox over that watch of his, and will report later when we've finished. I'm trying to get him to answer a few simple questions about the circumstances; so far, without success.

Best,
Martin

P.S. Ron Graham, the mathematician juggler at Bell Labs, would like to meet if you're up this way again soon. Had lunch with him last week.

[Euclid Avenue] 17 November 1975
Dear Marcello

That paper with Harary is indeed a weird one. I found it both interesting and informative—the bibliography alone is valuable—but it's hard to think of what journal would be most interested. Among the math journals, the best bet (I think) would be *Mathematics Magazine*. The editor to send it to is Lynn A. Steen, Math Dept, St. Olaf College [address withheld]. He's looking for off-beat articles, and this one might amuse him.

I suppose you saw the three-page review of the Legman book in *Time*, Nov. 10—largely the result of a phone call I made to the book editor (of the moment), Steve Kanfer, a good friend. I can't recall when *Time* ever reviewed a privately published book. Poor Beckwith (a friend of Legman's who pub. it) is now being swamped with orders, passed along by *Time*, from people who couldn't find the book in a local store. Beckwith has no distribution facilities. I phoned a vice president at Crown, and maybe they'll make a deal for taking over distribution.

Do you know anything about Dr. Aman, whose card I enclose? He is planning a periodical called *Maledicta*, devoted to "research on verbal aggression, pejoration, value judgment, and related subjects in all languages, dialects, cultures, religions and ethnic groups." Write him for descriptive literature. I've joined his tax-exempt society, to get the journal—and hope I haven't fallen for a confidence swindle. Is the guy legit?

I've reviewed Adam Smith's new book for *NY Review*—probably in next issue. Random House has just published a giant paperback tome called *Roots of Consciousness*, by Mishlove, that you'll have to get, if only for the photographs of characters on the psi scene.

Best,
Martin

P.S. My Uriah Fuller is sold out—Fulves is printing another 500. When they get low, I'm doing a Volume 2, *Further Confession* ...

[Euclid Avenue]   8 December 1975
Dear Marcello

Many thanks for the remarks about Uriah Fuller. Someone at the *NY Times* (Boyce Rensburger) got hold of a copy, along with Randi's book, and has been working on a feature on how the magicians are beginning to fight back. It's supposed to run sometime this week. Randi visited the *NY Times* offices and wowed 'em with his demonstrations, so hopefully he will get his picture in the story. Rensburger talked to Puthoff on the phone, P. invited him to SRI, he flew out there, and was turned away at the door (by SRI officials, who wouldn't let him see P and T on the grounds that SRI didn't want any more publicity about Geller). How about that for a piece of PR, to help influence a *NY Times* story!

I don't think it's worth copying my voluminous exchanges with Cox, but I can do so if you like. He writes rambling, semi-coherent letters. I asked lots of detailed questions about the watch, and have reached the conclusion that while Geller was handling the watch (something Cox readily admits) he obtained enough misdirection to pry open the case with a thumbnail, and move the balance wheel with a thumb. Cox denies this of course. It is possible that he is consciously exaggerating to build up his case, and also possible that he was misdirected so cleverly that he does not even now recall what it was. In any case, judging solely from his replies, this seems to me the simplest explanation of how Uri started his watch.

One amusing thing. Although I readily admitted that Uri might have started the watch by PK (low probability of course), Cox flatly refused to admit the possibility that Uri could have deceived him and done it some other way. I then asked him how Rhine felt about this: i.e., did Rhine also rule out the possibility, or did Rhine agree with me that there is a "possibility" he was fooled. To this he has given hilariously evasive replies, which say nothing. Actually, corresponding with him is a big waste of time. The man is so gullible and incompetent that it is a mystery to me why Rhine keeps him on the staff. Maybe it's getting harder for Rhine to find acolytes. I hope something works out with the *Humanist*. With their backing, a substantial journal might result, and grow rapidly in influence. There's no reason for me to be on a free mailing list, so I enclose a check for $3.00. If money is a problem, I could even make a donation later.

I enclose a copy of my recent blast at Adam Smith's book. The Scientology church sent me a copy of the letter of expulsion for Erhard, so I'm safe on that remark. The organization he has funded is Sarfatti's

Physics/Consciousness group. Sarfatti has been "processed" by est, and has great admiration for Erhard. Randi had lunch with Sarfatti and convinced him that Geller is a fraud—he has even sent out a release retracting his earlier one from Birbeck College. And I'm working on an article called "Magic and Paraphysics" for a journal—but more about that later. On est there are two basic articles so far—the one in *Harper's* ("The New Narcissism") in October, and one in *Psychology Today* in August. The cult is growing like wildfire, and even attracting intelligent people like Robert Fuller (another Fuller!), formerly president of Oberlin.

Best,

Martin

P.S. Thanks for astrology paper and anything else I've forgotten to acknowledge. I'll send Legman a copy of your good review.

[Euclid Avenue]   6 February 1976
Dear Mr. Thompson

I saw in today's *NY Times* that you are the new *RD* [*Readers' Digest*] editor. My congratulations.

As a former contributor, friend of Dan O'Keefe, and the author of many books in the science field, I have been greatly disturbed by the recent turn of *RD* toward occultism and irrationalism, starting with the condensation of Fuller's shabby book on Arigo, then his defense of Geller, followed by an article praising Velikovsky. And now an offer of $3,000 for "true" psychic experiences.....

Best,

Martin

[Letter in response from Ed Thompson.]

[Euclid Avenue]   25 February 1976
Dear Marcello

Thanks for the updated directory.

Current *Esquire* has one of the silliest articles yet on Uri. Objects kept falling from the ceiling, etc., when the writer visited Uri in his NYC apartment. Apparently the idea never crossed his mind that there could be *someone else* in the apartment who occasionally tossed something when his back was turned. And there is a boxed comment (irrelevant of course) by some medical doctor pointing out that quite possibly the writer had been *hypnotized* without knowing it!

I enclose a recent exchange of letters with the new editor of *Reader's Digest*. I can't believe that his letter is an honest one. The turn at *RD* has been quite marked, with an emphasis qualitatively different from previous policy. From the new editor's evasive letter, I am beginning to suspect that maybe the Wallaces (are they still alive?) are into occultism and have proposed the shift. Or maybe it's just that the magazine is moving with the trends. For $3,000 I'm tempted to fake a "true psychic experience" myself!

Had hoped to hear from you on 23 or 24. Did you get to NYC?

The "Endless Octave" record is available from Doug Clark, of Clark Systems [address withheld]. I don't know his current prices, but he has the illusion on a small lp record, and also on tape, and a third version on paper records that he can supply in quantity. This is the *discrete* tone illusion. The Bell Lab engineers have two other versions, but neither is commercially available. One is a continuous rising or falling tone (no jumps). The other is a rhythmic illusion in which drum beats seem to be increasing in tempo (or decreasing if you play it backward). I've not heard it but am told it is very effective. You try to keep pace with the beats by clapping your hand or tapping a foot, and it gets faster and faster until you can't keep up with it, then you realize that the sound is the same as before. (I'm told that the illusion is known in India, where it is used in certain ragas.) It's a technique of superimposing several different rhythms, and damping out certain ones as they get faster, and strengthening the slower beats. I'm currently trying to run it down. Clark knows about it, and may be planning to issue versions of it soon.

Best,
Martin

[Euclid Avenue]   10 March 1976
Dear Marcello

Glad to hear that the new magazine is coming along well.

P and T did not send me a copy of their reply to Randi's book, but I do have a copy that someone sent me. I should think they would be delighted to send you one; if you don't have one by now shall I photocopy for you?

On specially printed books and magazines, I enclose a bibliography that will give you the date of the *Ellery Queen Mystery Magazines*, plus more about other similar items than you probably care to know! I've not met Kuethe—I think he's a bank official of some sort. He collects magic books and periodicals, and sells his duplicates. To me one of the most unusual of the special books is *Shadows in the Moonlight*. It contains original poems by T. Page Wright, with the words "love" and "rose" at the same positions in every poem; and the remarkable thing is that the poetry is by no means doggerel. Some of it, I think, quite good. I used to own a copy, but somehow it slipped out of my hands. Please keep all the enclosed. Kuethe's phone is [phone number withheld]. I don't know how any of the magazines are gimmicked.

I like your newspaper number force, which someone told me about (maybe you?) last year. The idea of putting a number from thumb to the chosen page is certainly new to me, and worth experimenting with. I should think one problem would be making sure the number was correctly aligned. It's an amusing idea and just might be made to work.

Re: your projected mentalism book. I assume you have the standard references: Annemann, Corinda, etc. Dr. Jaks kept a mammoth set of notebooks in which he pasted down instructions for thousands of mental tricks he clipped from magic magazines, instructions sold with equipment, ideas of his own, and so on. They have passed into the hands of someone here in NY (I've forgotten who, but can find out if you like.) Whether it would be worth your while to arrange to spend a few days poring over these books, I don't know.

Also, the west coast magazine *Magick*, going for several years now, publishes only mentalism material. You probably know about it, and may even subscribe. Bascom Jones, the editor, is a strong defender of the right of a mentalist *not* to make a "disclaimer" speech before his act, and he even takes frequent pokes at Randi and me for attacking Uri Geller.

NY Times last Sunday (Mar 7) had a long article in theatrical section on juggling as a growing campus hobby. I'd copy and send except I mailed it off yesterday to friend Roger Montandon, in Tulsa, who started *The Juggler's Bulletin* and sells books on juggling. The article had mentioned Roger and his address [address withheld]. (I grew up in Tulsa with Roger.)

Best,
Martin

[Eastern Michigan University]   18 April 1976
Dear Martin Gardner, Ray Hyman and Randi

Thought you might like to see the continuing exchange between Cox and me now entered into by Randi and Theodore A. Dunst, whoever the hell he may be. Any of you know who or what a Theodore A. Dunst is?

I have no further interest in answering Cox, but I might answer Dunst since he asks numerous questions that may not all be rhetorical and it would be a way of further advertising RSEP and the new journal, etc.

Since Paul Kurtz has not yet called me back (he was supposed to do so after leaving the hospital re the parasite he picked up in India), I presume he is still ill. In any case, the plan to get the new *Zetetic* out in time for the Conference of the AHA [American Humanist Association] at the end of this month is obviously out. I have most of the material for this first issue—it looks very good to me—and hope to get the thing into production before very long (hopefully right after the meeting with Kurtz at the Conference).

I also enclose a copy of Dingwall's review of Randi's book, which is in the same issue of *Parapsychology Review* as my exchange with Cox. et al.

The more the Geller thing goes on the more I think SAM should simply make him Magician of the Year, comment on the fact that many magicians (Kreskin, etc.) pretend to real powers to enhance their bookings and attraction for the public but, of course, no one who knows about such things would take his claims seriously given his background, etc. This way we, as magicians, would talk of his deception as "showmanship" and we would speak of him admiringly. Of course, he would deny being a magician, but we would not be his "enemies" and thereby (a) hurt our credibility and (b) indirectly help his career. It would bring all the pro-Geller magicians on our side and it would simply be "well-known and accepted among magicians that Geller was a magician." This would get to anyone that really mattered (i.e., government agencies, and funding sources for research) and would block most of his effectiveness among scientists. Geller would be put in the odd position of dissociating himself from those who admired him and were his "friends" which is much harder to do than dissociate yourself from "jealous enemies." What do you think?

Sincerely,
Marcello Truzzi

P.S. R. A. McConnell tells me that there is a new experimental report out by Targ and Puthoff in the March 1976 issue of the *Proceedings of the Institute of Electrical and Electronic Engineers*. Any of you seen it yet?

Also, I presume you have all seen the recent attack on Martin and Randi by *Fate* magazine editor Curtis Fuller.

P.P.S. Enclose a copy of the letter re Dunst I just wrote since it might interest you.

[Euclid Avenue]   25 April 1976
Dear Marcello

Much thanks for the further exchanges with Cox. Never heard of Dunst before. Thanks, too, for Dingwall's review of Randi, which I had not seen.

I agree with you that if the SAM made Uri magician of the year it would be marvelous. I'm no longer a member, so I have no clout with the powers that be, but I'll certainly urge it on members when I see them.

A copy of the new Put and Take paper, on remote viewing, came to me by way of *Scientific American*, which was favored with a copy. If you don't have one by now, I can copy it for you—but my machine is badly in need of cleaning at the moment, and is giving fuzzy copies. The report is 26 pages, and P and T must have sent out an enormous number. *Time* is thinking of doing something on it, since the work was financed partly by NASA. The authors list three supporting organizations: Parapsychology Foundation (NY), Mitchell's Institute of Noetic Sciences, and NASA, (under Contract NAS 7-100). A wire service must have sent a release on it because I saw a story about it in the Tulsa paper when I was there on a visit last month. I have two copies of the National Enquirer spread, and will enclose one for you to keep.

Here are a few random comments that occur to me after a hasty reading of the original paper. The star subjects, Pat Price and Ingo Swann, both Scientologists and friends of Puthoff, are not, in my opinion, to be trusted. As usual we know nothing about the conditions under which the tests were made, so the opportunities for both men to cheat (leaving aside the possibility of collusion on the part of Puthoff, now desperately anxious to get away from Geller) may have been considerable.

In the case of Phyllis Cole (coauthor of the report on the ESP teaching machine), her one big "hit," which they reprint in full, is preceded by a statement that Targ was ignorant of the target. I wonder. We know now how P and T word things in such a way that they can later wiggle out of ambiguities. If Targ knew the target, Phyllis would be getting cues from him. Note that the paper gives away the nearness of the target (20 minutes). Phyllis may simply have made a shrewd guess. If she made shrewd guesses on all the targets, and missed on all but this, then we have been given details on the one hit. Of course we lack enough information even to set up a good probability estimate. How many suitable targets are 20 minutes away? How many times did Phyllis guess and miss? If we knew all the facts, it might be that she simply was lucky on one guess, in a situation where the probabilities against success were not as high as they seem to be.

We are given only one transcript of what a subject said. What were the others like? Judging from this one, the subject rambled on and described many things. One longs to have all the transcripts to look over. If a typical transcript contains, say, 50 different images, and the experimenter is viewing a complicated scene, the probability of finding a match is probably quite high. Remember the two decks of playing cards, each shuffled, then dealt simultaneously. The probability of two identical cards falling at the same deal is slightly better than 1/2.

Note that the photos, made afterward, are carefully made to emphasize the accuracy of the hit. By taking the photograph from the right spot, at the right angle, camera aimed just so, etc., the result can be made to look very much like the drawing. The drawing, alongside these carefully contrived photos, gives a strongly biased impression of similarity.

Someone should make a series of tests, similar to the P and T tests, but using descriptions selected at random from famous novels. Select 50 target spots at random. Then take 50 descriptive scenes from novels, and randomly match them with the targets. Then look for coincidences. Have someone draw a sketch based on each scene, then go to the matched target and take a photo from the best angle. Etc. I suspect one could come up with an excellent parody of the P and T work.

Best,
Martin

CC: Hyman, Randy, Jaroff

[Euclid Avenue]   5 May 1976
Dear Marcello

Ray Hyman is coming out tomorrow, so I'll get from him a report on the conference. Here's a copy of my contribution to the just published book of essays on Holmes. Is this something that would be suitable for the new periodical, on the assumption that too few readers would wish to spend ten dollars for the book? You might also consider reprinting Sanderson's lecture (see my footnote 8), which is excellent, and certainly not accessible to US readers. (I have a copy.) The parallels between Doyle and Eisenbud are very marked: in both cases we have intelligent men taken in by a rather crude kind of deception, and both absolutely unable to alter their opinion in the face of mountainous evidence. Eisenbud, of course, believes that the spirit photographs of the time (perhaps even the fairy photos) are simply a manifestation of the Ted Serios phenomenon—a projection from the subconscious of persons near the camera.

Bobbs Merrill reprinted only 2 of the 5 photos, but they are readily available for reproduction from E. L. Gardner's book, which I must now try to retrieve from Bobbs Merrill.

Best,
Martin

[Euclid Avenue]   10 June 1976
Dear Marcello

I don't know whether you are planning a review of Panati's *The Geller Papers* or not, but if so, here is some amusing information that should be passed on to the reviewer.

Panati has been saying on TV and radio talk shows that the "strongest" chapter in his book, and the most irrefutable, is the one by Eldon Byrd. He has been saying that this is the first time that such work has been done in a laboratory of the US government, in this case the Isis Center of the Naval Surface Weapons Center, Silver Spring, Md. (This is stated in Panati's introduction to his book, and on page 67 in his introduction to Byrd's paper). I have been in correspondence with Byrd, who does indeed work for the Naval Surface Weapons Center. But he tells me that Panati mistakenly assumed that the Isis Center was part of the Naval center, and has since apologized to him for this error. The Isis Center no longer exists. Its full title was "The Isis Center for Research and Study in the Esoteric Arts and Sciences." They had their own laboratory, and it was *there* that the great test was made. The Isis Center was run by occultists. It was this Center that had booked Uri to speak in the area. Byrd had obtained from them permission to test Uri, with nitinol, in *their* laboratory. No connection with the Navy or the US.

Byrd has a master's in medical engineering. See page 40f of *The Secret Life of Plants* for Byrd's experiments with plants, confirming Backster's results. Byrd proved that plants responded with fear to his "intent" to burn them. They were similarly frightened when he shook a matchbox containing a spider. And so on.

Byrd writes that Uri had no way of knowing about nitinol before he and Byrd got together in October, 1973. But Charles Kalish (amateur magician), who heads the US Photographic Equipment Corp [address withheld], experimented with nitinol in 1972 and developed a magic trick which he showed to numerous magicians in the US, Canada, and England. You had someone select a digit. Then you put a paperclip (made of nitinol) in an envelope, burned the envelope, searched the ashes, and found the wire in the shape of the digit.

The trick was sold by Davenport Magic Company in England, although I have not yet obtained the date of their advertising in English magic periodicals (probably 1975).

In October, 1973, at the Isis Center, Uri produced a kink in one wire of nitinol. Later, he put kinks in nitinol wires at the home of John Fuller, a mutual friend of Byrd and Geller.

The best theory at the moment is that Uri knew in advance, through mutual friends, of Byrd's interest in nitinol. So he came prepared in October, 1973, with pieces of wire he had previously kinked by heat. Byrd presented him with two sample wires. Uri had no success with one of them. The other matched one of the wire's Uri had brought with him. He simply switched wires, then followed with the stroking routine, giving Byrd the impression that the wire bent while he was holding it.

A second theory is that Uri bent the wire by mechanical pressure. Byrd tells me in a letter that strong mechanical pressure *can* alter the "memory" of nitinol, but of course he is convinced that Uri could not possibly have applied such pressure because he was watching him closely. He also claims that if Uri had used such pressure, there would have been marks on the wire.

I mentioned in a letter that Uri often lied. Byrd responded by saying that he had no evidence that Uri had ever lied. I assume, therefore, that he believes Uri's claim of teleportation from Manhattan to Ossining.

I send you all this because Panati has been stressing that Byrd's paper is the strongest in his book.

Be sure to check the July issue of *Technology Review*, due out next Tuesday. It contains my article on "Magic and Paraphysics," with photos of Randi.

Our next president is likely to be a Baptist fundamentalist. Isn't that great? Mencken must be howling with laughter in his grave!

> Best,
> Martin

[Euclid Avenue] 18 June 1976
Dear Marcello

Glad to learn that Ray is reviewing *The Geller Papers*. I am almost sure I sent him a copy of my letter to you about Byrd and the Isis Center, but I forgot to note CC at the bottom. I think I sent another copy to Randi.

Technology Review is reportedly out, with my article (pix by Randi!), but I haven't yet seen a copy. Randi actually gives away an excellent key bending technique that I've seen him use; this, no doubt, will raise a howl from Bascom Jones!

When I first heard that Uri is doing the old finger lift bit, it broke me up. This, more than anything, convinces me he is going downhill. I had lunch with Chris Evans yesterday, and he tells me that things are not going well on the big movie starring Geller. The script was originally written by Colin Wilson (it figures), but Colin, wised up a bit (not much!) by Randi and Evans, said some critical things about Uri in a new book, and Uri is so angry that the script may be returned.

Back to the finger lift. Geller probably learned about it from Thelma Moss, who has a section on it in her big book (forget the title – her last one), p. 133. Uri used the routine she gives for it. Coincidentally, the identical trick is given in Bascom Jones' sheet, *Magick*, No. 152, 1976. I asked Bascom in a letter if Uri or a friend of Uri was a subscriber, but in his reply, he ignored the question. Could be Uri saw it there.

The trick can be presented with all sorts of preliminary rituals. They have no function except to get the four lifters coordinated so that they will all lift at precisely the same time. The first time it's tried, the subject (if the subject is presenting it as a psychic levitation) sits loosely. Even though there may be a count of 1,2,3, the lifters rarely get synched, and the lift fails. Each person has the feeling that it can't be done because the weight on his finger is about equal to the weight of the subject. Then the ritual is observed with instructions to lift precisely on the count of 3. This time the subject makes himself rigid on the chair. When we did it in Tulsa as grade school children, the 4 lifters and the subject were instructed to take three deep breaths and hold the last one, while a fifth person on the side counted 1,2,3, LIFT! When the coordination is accurate, the weight is distributed four ways (from the outset) and the subject rises easily. The contrast between the two lifts is very spooky, and as mystifying to the subject as to the lifters. (The limp vs. rigid bit isn't necessary, but it helps. I didn't see Geller do it, so it may be that by sitting limp he can even stress a coordinated count on the first lift.)

The trick has been written up many times, but I can't remember where and I don't have a reference on it.

Do you think a review by me would be appropriate of the book on faith healing by Jimmy Carter's sister? I haven't read it yet, but have it on order. The sister speaks in tongues, and is a full-time faith healer. I predict that after Jimmy is president, she'll replace Kuhlman as the best known lady faith healer in the country, perhaps the world. I heard her say in a TV interview that a boy had come to her who had been born with no inner ear organs of hearing. After praying over him, Jesus healed him and he began to hear immediately. Then she was off on something else, as though that was something typical and commonplace. If I reviewed her book I'd probably want to work in a few speculations as to how evangelical Jimmy's evangelical faith is. I *have* read his autobiography (which took about 20 minutes).

Had dinner two nights ago with Ron Graham, the Bell Labs juggler. He is really very good. He also is a trampoline expert, and can do one arm handstands—in fact he did one for me—ride a unicycle, etc., etc. The other good juggler in math is Elwyn Berlekamp, a top west coast mathematician.

Best,
Martin

[Euclid Avenue]   23 July 1976
Dear Marcello

Thanks for the copy of the P and T letter to MIT. They did not send me a copy. I enclose a copy of my reply. Sarfatti also wrote to MIT, a long letter of about 7 pages, in which he agrees fully with everything I say about Geller. He corrects one or two minor errors about himself—e.g., he did not change his name to Sarfatti, he restored a family name he had lost. I have him getting a degree from the wrong college. He also says he is now getting funds from other sources than est. He ends his letter by saying he and Erhard are "warm personal friends."

A day or so ago the director of est, a physicist named Robert Fuller (another Fuller!), and former president of Oberlin, telephoned to tell me how much he liked the MIT piece, and to let me know that est is not giving Sarfatti "another penny." When I asked if Sarfatti and Erhard were indeed "warm personal friends" he laughed and said that Sarfatti is a name dropper who says that about everybody he has met. He added that Erhard "has no warm personal friends." I said I had been fearful that est was moving into parapsychology when I heard est had funded Sarfatti. Fuller's response was "over my dead body."

Item 3 in your letter is typical Puthoffian. I never received a single letter from either Puthoff or Targ, although I did indeed write them. I had a couple of letters from Jones, and a couple of phone calls from SRI's PR man. *Scientific American* published their letter, and my reply in which I spoke of writing to Puthoff (copies to Jones and the *Scientific American* office) proposing that he and *Scientific American* agree on an outside statistician to look over the records of their ESP teaching machine test to determine the accuracy of my description of the tape as in bits and pieces. Nobody replied, nor have I heard from anyone at SRI to date. It was obvious that they would not agree to inspection, and it is also embarrassing to go on record as not agreeing—hence silence.

The contents of Vol. 1 of *The Zetetic* look great.

> All best,
> Martin

[The letters mentioned above follow. The first was by P and T.]

[STANFORD RESEARCH INSTITUTE]   July 15, 1976
Dear Mr. John I. Mattill, Editor Technology Review

In Gardner's article on "Magic and Paraphysics" (*Technology Review*, vol, 78, No.7, June, 1976) some references were made to the work with Geller at Stanford Research Institute. Unfortunately, Gardner's statements concerning, what happened at SRI and what we published are grossly in error. We therefore wish to inform your readers of the facts involved, all of which can be independently verified on the basis of information available in the public domain.

To begin, Gardner states that "Although Puthoff and Targ are personally convinced of Geller's ability to bend metal by PK (psychokinesis) and to perform even more remarkable miracles, their *Nature* report was limited to Geller's ESP (extrasensory perception)."

Gardner is wrong on both counts. We are in fact *not* convinced of Geller's ability to bend metal, and our negative findings *were* reported in the very *Nature* article to which Gardner refers (R. Targ and H. Puthoff, vol. 252, No. 5476, p. 604); "It has been wildly reported that Geller has demonstrated the ability to bend metal by paranormal means. Although metal bending by Geller has been observed in our laboratory we have not been able to combine such observations with adequately controlled experiments to obtain data sufficient to support the paranormal hypothesis." A more detailed statement is found in the SRI film, "Experiments with Uri Geller," the text of which was released accompanying a March 6, 1973, presentation at a Columbia University physics colloquium. With regard to metal bending the text states: "One of Geller's main attributes that had been reported was that he was able to bend metal. . . In the laboratory we did not find him able to do so. . . becomes clear in watching this film that simple photo interpretation is insufficient to determine whether the metal is bent by normal or paranormal means. . . It is not clear whether the spoon is being bent because he has extraordinarily strong fingers and good control of micro-manipulatory movements, or whether in fact the spoon "turns to plastic" in his hands, as he claims."

In discussing our dice box experiment, Gardner goes on to claim that the reported successful run (of correct guesses as to which die face was uppermost) was selected out of a longer run, which included "many prior trial run's." That is completely false. The facts are exactly as reported in the *Nature* paper and in the SRI film. The experiment was performed ten times, with Uri passing twice and giving a correct response eight times. These ten trials were the *only* ten; they were *not* selected out of a longer run; there were *no* prior trials nor follow-up trials, as Gardner claims.

Gardner's errors appear to be due to his taking at face value the erroneous speculations of Geller's self-appointed debunker, the Amazing Randi.

Yours truly,

H. E. Puthoff
R. Targ
Electronics and Bioengineering Laboratory

[Gardner's response was in the files but was undated, without letterhead and unsigned.]

I have enormous admiration for the expertness of Puthoff and Targ in one field—verbal obfuscation. Let me take each of the two counts in turn:

Count 1. When I say that P and T are "personally convinced" of Uri's ability to bend metal, I use the phrase in the ordinary language sense, as when an astronomer says he is personally convinced that quasars are not within our galaxy. It is a probability estimate, as are all scientific beliefs. When P and T deny they are convinced Uri can bend metal, they mean they are not "convinced" because they have not proved such ability in their laboratory. Privately, in letters and conversation, they have expressed their personal beliefs that Uri has such ability. If P and T wish to make precise their present beliefs about Uri's PK powers, let them give it as a probability estimate. How do they rate the probability that Geller has PK ability? If they rate it low, it means they have changed their minds.

In my article I said that Jack Sarfatti, the paraphysicist who first staked his reputation on the PK powers of Uri, but who recently branded Uri a fraud, "did not doubt" that others have PK ability. Sarfatti telephoned to say that this is not true. He does doubt the existence of PK. He then added that he doubts "everything" except the existence of himself. This surprised me because I doubt even my own existence—for all I know may be just a figment in the Red King's dream. When I said Sarfatti does not doubt PK, I meant it only in the ordinary language sense, as when a physicist says he does not doubt electromagnetism.

Count 2. I did *not* say in my article that in the dice box test P and T selected 10 guesses out of a longer run. I said that many "prior trial runs" had been made. A trial run, in ordinary language, is a practice run. Every time someone points out that practice runs were made with the dice box, P and T obfuscate by denying that the 10 guesses were "selected" from previous trials. They were not so selected.

But that is not the point. The point is that a very large number of practice runs were made during which Uri was allowed to handle the box. This gave him all the time he needed to devise a method of cheating when the final test of 10 trials was made. That was all I said and all I meant.

Anyone reading the letter from P and T would assume from their last paragraph that there were no practice runs. Without drawing upon private information, I content myself with the following published data:

In his autobiography, *My Story*, Geller himself speaks of the practice runs. "Time after time," he writes, "without fail I was able to tell what number was on the face..."

In the July, 1973, issue of *Psychic* there are two photographs of Uri performing the dice box test. In the first picture we see Uri recording the die's face, the closed box about four inches from his hand, while Targ watches. In the second, we see Uri opening the box to check on his guess. We know it is the box used in the famous test because we see SRI printed on top. In the obfuscatory Puthoffian-Targian dialect, this is not a "follow-up trial" because it is not part of the test they reported.

In John Wilhelm's carefully researched book, *The Search for Superman*, just published by Pocket Books, he reports on dice box tests conducted in Geller's motel room. Geller was allowed to do all the shaking, although Puthoff insists it was vigorous enough to "ensure an honest shake."

Commented Targ: "He's like a kid in that he had something that made a lot of noise and he just shook and shook it." Targ also told Wilhelm that in the famous run of 10 Geller was allowed to place his hands on the box in a dowsing fashion.

Targ told Wilhelm that SRI has a good-quality videotape of another dice test in which Uri, five time's in a row, correctly wrote down the die's number *before* the box was shaken. Targ first shook the box then Geller took it and dumped out the die. Since magicians familiar with dice cheating know a variety of ways to control a fair die when it is dumped out of a box, how about letting magicians see this valuable videotape? Why keep it top secret? Targ was so impressed by *this* test that he told Wilhelm he suspects that, even in the run of 10 that they reported Geller probably used precognition, not clairvoyance, to guess the number he later "would see when he opened the box." Note: Targ said it was Uri who opened the box!

In the sound track of the same SRI film; from which P and T quote in their letter, the following occurs:

"Here is another double-blind experiment in which a die is placed in a metal file box.... The box is shaken up with neither the experimenter nor Geller knowing where the die is or which face is up. This is a live experiment that you see—in this case, Geller guessed that a four was showing but first he passed because he was not confident. You will note he was correct and he was quite pleased to have guessed correctly, but this particular test does not enter into our statistics."

Now P and T have not, so far as I know, revealed whether this single trial, which was recorded on videotape, was part of *the* test of 10 trials. If it was not, then surely it was in a practice run. If it was, then presumably the entire dice box test was recorded on videotape. In the interest of scientific truth, P and T should make the entire videotape available to inspection by magicians. It then could be determined unequivocally whether the theory suggested by James Randi, as to how Uri might have cheated, is a viable one. Come, gentlemen, let us see the entire tape! If we are wrong, we will humbly apologize.

[Euclid Avenue]   10 September 1976
Dear Marcello

Sorry I had to run the other day, when you called. I'm still in the middle of one of those periods in which suddenly half a dozen friends from out of town are in the NY area, and I find myself trapped in endless socializing, combined with all sorts of writing commitments.

Do I recall correctly that you'd like for me to review the Wilhelm book for the *Zetetic*? If so, let me know the approximate length you want, and when you want it. It is far and away the most revealing book yet on the incompetence of P and T, but then after demonstrating their inability to conduct a controlled experiment in ESP, Wilhelm winds up by saying that he thinks their remote viewing stuff is probably sound!

Klass has been favoring me with his correspondence with you. I think he is unduly alarmed, although I agree that the probability of anyone changing his mind about anything is extremely remote. As soon as the true believers who want to affiliate find that they're not getting anywhere in changing the minds of the sceptics, they'll raise a huge outcry about the committee's prejudice, and make a noisy exit. Sarfatti has a new source of funds (Erhard wrote him off as a bad bet), and is hooked up with Timothy Leary!

As I see the next few years, the occult revolution will be going down, but not because of any rise of rationality, but because of the rapid rise of Protestant fundamentalism. The extent to which this is happening is beyond the belief of those who are out of contact with the masses. Nothing can prevent JC from becoming president, and we're in for a flood of books and magazine articles and TV talk shows about the power of Jesus to transform our lives. Moon just bought Manhattan Center for a few million dollars—and he's way out on the fringes of the Great Awakening. I see that Billy Graham's topic this Sunday is "The Antichrist." Did you see Tom Wolfe's article (*NYT Magazine*, August 22) on the "me, me, me" generation? It's all beginning to come together. The Seventh Day Adventists have been trying to say all along that one of the great signs of the Second Coming will be the rise of occultism and Spiritualism. The occult is about to go down the drain not because the American people are ceasing to believe in psychic phenomena, demon possession, etc., but because they are beginning to view it all as Satanic!

Best,
Martin

[Euclid Avenue]   25 September 1976
Dear Marcello

1. Enclosed is a copy of the latest "letter" from Eldon Byrd. He always answers with brief marginal notes on my original letter, which he returns. As you can see, he misread a previous question. His paper reports on 3 different tests.

1. At Isis Center. Present were Jean Byrd (no relation) who is the head of the center (she thinks she is a reincarnation of Isis), Byrd, Uri,

Shipi, and Puharich, and 2 secretaries. Uri bent one wire. This on October 29.

2. Plates 6 and 7 (in Byrd's paper) show results of second test. No date, or place—but sometime in November. Uri had taken the wire home with him, later brought it to Byrd!

3. At home of Fuller. Byrd considers 1 and 2 uncontrolled, but 3 is "very controlled."

I hope that the error caused by Byrd's answering my earlier question wrong (under assumption I had asked about 1) isn't reflected in Hyman's review. I've just sent him a copy of the Byrd letter here enclosed.

2. If Hyman has not gone into much detail about the Byrd paper, would you be interested in a short article about it? I now have a mass of data, based partly on my correspondence with Byrd, and partly on correspondence with a scientist at the Naval Weapons Center who is an expert on nitinol.

3. I've been debating with myself about the advisability of allowing Panati, Hynek, Puthoff, etc., becoming members. Here are my views at the moment. I don't hold them with any high degree of conviction, so I merely pass along my arguments to further the discussion.

I think we should abolish the category of "member" as separate from subscriber and, like most scientific societies, distinguish only between (1) member-subscriber, (2) the controlling committee (fellows). Let the price of subscription-membership be $15. This includes getting the journal, announcements of meetings, privilege of attending gatherings, etc., but nothing else.

I am deeply pessimistic about any good whatever coming from having special members (distinct from subscribers). Clearly those who pay an extra five bucks will expect to get something more than the subscribers. Presumably, to establish a dialog with them, they will become privy to plans that are not available to mere subscribers.

If they are not privy to such plans, they will be indignant and accuse the fellows of concealing information from them. If they do get such plans, I foresee nothing but trouble.

The chance that a dialog with parapsychologists such as Puthoff and Panati, and characters like Hynek, will alter their beliefs I regard as too minimal to be considered. What is much more likely is that as soon as they discover that their dialog is not affecting the fellows, they will make an enormous amount of noise about it in their publications, accusing the committee of pretending to be open-minded, but actually a group of sceptics who are not permitting the group to function democratically.

I think it can be assumed that the parapsychologists will instantly exchange information of any value to them. For example, if there are plans to test Geller, or any other psychic, and they are given information not available to subscribers, the information will instantly find its way to the psychic.

I cannot see that having Panati and Puthoff as members will make it an iota "harder" for Geller to refuse a test. He as already refused to be tested by Rhine, and by Honorton (and many other parapsychologists in the US and abroad), so there is no possibility whatever he will be tested by us.

Although I understand your point about calling yourself a "parapsychologist" and Klass a "ufologist" isn't this a Humpty Dumpty misuse of common speech that can only cause confusion? A "creationist" attacking evolution shouldn't be called an "evolutionist" because he is studying evolution, any more than an atheist writing about, say, proofs of God, should call himself a "theist" because he is studying theological claims. I see no advantage in going against common usage on such matters. I doubt if Ray Hyman would like to call himself a "parapsychologist." Nor would I, because I wrote a book on pseudoscience, wish to call myself a "pseudoscientist." Scepticism *is* relevant to definitions. A doctor sceptical about acupuncture wouldn't call himself an acupuncturist. But I'm belaboring an obvious point.

I think we should drop our pretense of "objectivity" except in the sense that we believe, as do all scientists, that all scientific knowledge is probable knowledge, not certainty. Of course we are open to the *possibility* of every variety of psi phenomena, but in the absence (in our opinion) of compelling evidence, I think we should openly announce ourselves as sceptics, as opponents of the true believers, and not create an in-between status for them. Let them subscribe to the journal if they wish. Nothing more. If Puthoff can become a member, then so can a hundred others (Eisenbud, Thelma Moss, Krippner, etc.) In my view, the inquiries received are from those who are testing the committee. If word gets around that we are accepting them, I foresee the danger of a flood of applicants—all convinced believers—until the sceptical "members" are outnumbered. As soon as they discover they are out-maneuvered on anything, they will raise a great hue and cry. I think it best to let them raise a lesser hue and cry now by letting it be known that the committee is a committee of sceptics, in complete control, though of course anyone is welcome to a membership-subscription status.

A final point. I am dubious of any sort of better communication coming from having the believers in a special in-between status. Indeed, I think we can communicate better with them as outsiders than with them as partial insiders. As partial insiders, they will be able to complain that we do not listen to them with much greater force than they can complain of this as outsiders. If I do not listen to a man who does not come to dinner at my house, he has little to complain about except that I didn't invite him to dinner. But if I invite him to dinner and then don't listen to him, he has a legitimate complaint. By "listen," of course I mean more than just *hearing* what he says. I mean listen and modify my views as a result of what he says. We already can listen all we like to what the believers say, since they flood the media with their views. As partial insiders they would expect us to "listen" in a different sense. When they discover we don't, they will become very noisy about it.

To sum up. I think we can function best if we call ourselves sceptics, admit that we are opponents of the psychic revolution, and let it be known that we don't want any Panati's and Puthoff's in our hair.

Best,
Martin

CC: Randi, Hyman, Klass

[Attached was a xerox of letter to Byrd with Byrd's writing on it]

[Euclid Avenue]   21 October 1976
Dear Marcello

Many thanks for an advance look at your review of the Houdini book. A splendid review. I haven't yet read the book, but I'm sure I would have precisely the same reactions. And I'd not heard before that answer to the riddle about the difference between psychologists and magicians!

Thanks too for your reactions to my letter about the membership levels of the Committee. We'll have a chance to talk about it soon.

The *NY Review of Books* has asked me to review the new book by Put and Take. I have agreed, and asked if I could combine it with a review of Wilhelm's book. The two go together beautifully, and I have about 2500 words to move around in. Library of Science wrote to ask me to evaluate for them the latest book of paraphysics nonsense: *The Dark Side of Knowledge*, by two PhD physicists, Shadowitz and Walsh. Addison-Wesley is the publisher. It covers the whole psi scene, but in such a quick superficial way that it isn't of much value to anybody. They pretend to be objective, but end up by suggesting that there is "something there" in everything, including Geller and astrology.

To me, the most hilarious aspect of the continuing decade of gullibility—the 70's—is the fantastic rise of Protestant fundamentalism. Today's *NY Times* has a story of the most important trend in US elementary education. Guess what it is. It's the rapid increase in private Bible-centered schools! After Jimmy gets elected, my precognition tells me that the occult revolution is going to be swamped by the Jesus revolution.

Back to our argument about words. Dictionary definitions of "-ologies" never mention belief. An astronomer is a person who studies celestial bodies. It is tacitly assumed that he believes the bodies exist. Today, in the common usage of the word "'parapsychologist," it is assumed by just about everyone who uses the word that any man who calls himself a parapsychologist is studying psi phenomena, which he believes to be genuine. There is nothing "wrong" about a sceptical psychologist calling himself a parapsychologist (if he is investigating parapsychological claims) but at the moment this is not the common usage of the word. The matter is easily settled by asking 50 top psychologists, who have sponsored tests of psi phenomena with negative results, if they'd like to be called "parapsychologists." I could be wrong, but I would guess that the majority would not want to be so called.

Of course it's all one of those trivial arguments about words. The question is one of tactics. I don't think anything is gained by sceptics adopting such words as "parapsychologist" and "ufologist." But we'll argue about it next month.

Best,
Martin

[Euclid Avenue]   5 December 1976
Dear Marcello

That was a good letter to Leon Jaroff. It's just possible he may be able to get something into *Time*.

Mind-Reach, by P and T, is to be published by Delacorte; Delacorte books are distributed by Dial Press [address withheld]. The publisher is given on the book as Delacorte/Friede, which means that the editor, Eleanor Friede must have gotten the book together and then sold it to Delacorte. Friede is the former Macmillan editor who pushed Jonathan Livingston Seagull to its great heights. She and the author, Bach, became lovers, and he persuaded her to leave Macmillan and go into business for herself. Bach provides a foreword for the book. According to Wilhelm, he gave $47,000 to P and T, and in the book he is cited as one of their most gifted psychics.

I enclose summaries of papers that P and T will give at a conference of the American Physical Society on December 20-22 at (where else?) Stanford University.

For an issue or two Panati was (I was told) off the masthead of *Newsweek*, but now he's back on again. Don't know what the story there is.

Straws in the Pentecostal wind. Suzy Smith is now deep into the charismatic movement and writing a book about it. Judy Skutch has abandoned parapsychology and has also got religion. She's changed the name of her foundation to Foundation for Inner Peace, and is financing a big book called *Course in Miracles*. It is rumored to have been dictated to the author by Jesus himself. All this from Martin Ebon, whose latest book, on Pyramid Power, has just been published in soft covers. Note the final paragraph in his introduction. I'm the unnamed "friend" who advised him not to fence sit on the pyramids. He quotes from my letter, then immediately fence sits.

From Eldon Byrd comes news of hundreds of psychic children in Canadian mental institutions who have been found to bend spoons better than Geller, and to "babble in tensor equations." A committee has been formed to investigate, with Byrd as a member. It is being pushed by, of all people, the recently resigned president of Kent, a somebody Olds, who is deep into the paranormal. The committee is seeking funding. Byrd is excited about it, and thinks maybe we have here a new mutation, a new breed of superkids with PK powers.

But the big trend is still, I think, toward fundamentalism. Ebon says his next anthology will be on the Apocalypse (he's started work on it, and has a contract).

Best,
Martin

[Euclid Avenue]   17 January 1977
Dear Marcello

Thanks for copies of those letters about Velikovsky. In my experience, the V believers are instantly angry every time anyone attacks the master, or says anything negative about him. Indeed, I find that true (don't you) of most believers in the far-out fringes of pseudoscience.

I liked the first issue, and where I have reservations it's entirely a matter of tone and emphasis. This brings us back, of course, to the old argument of how a rational person today, who understands modern science, should talk about pseudoscience. The line between science and pseudoscience is obviously fuzzy, and meteorites do sometimes fall from the sky, but in spite of twilight there is a difference between day and night, and when pseudoscience gets as far out as V and astrology, I don't think believers should be treated the same way one would treat a scientist who holds an opinion one believes to be wrong.

I disagree with all anthropologists who are cultural relativists, but I treat them with respect. I disagree with John Wheeler's cosmology but I treat him with respect. I don't think V should be treated with respect. And I think it was disgraceful that the *NYT Magazine* last Sunday treated a Washington lady astrologer with respect.

I am a great admirer of Mencken. His reports on the Scopes trial seem to me models of the kind of rhetoric the trial deserved. He treated Bryan as a fool, which Bryan was. I would like to see the *The Zetetic* treat V the same way.

I am, of course, arguing for a tone and style. I see the *The Zetetic* as designed to *combat* the current wave of crap, not to *analyze* it. The suggestion by Carl Sagan, that the AAAS have a big symposium on astrology, strikes me as naive. What would it accomplish? The young people who are into astrology already know that the scientific establishment has no interest in astrology, and they couldn't care less. It would be on a par with a symposium to consider whether Jesus rose from the dead, or, to get closer to science, on whether all living things were created in six literal days. Certain battles are over and obsolete. The creationist and the astrologer, from my perspective, should be laughed at, and hit over the head with bladders and abusive rhetoric—not treated the way one school of linguistics treats a rival school. Believers are never unconverted by rational argument. They *are* affected by ridicule, especially if they are young.

I would like to [see] the *The Zetetic* designed as a nonscholarly, nonacademic, bad tempered magazine— calling the fool a fool and the quack a quack. I would like to see it perpetually skirting libel laws. The claims of Scientology about embryos should be treated like one would treat a flat earth theory. I would like to see astrology attacked the way Mark Twain attacked Christian Science, and V attacked the way H. G. Wells attacked H. Belloc—gloves off, no holds barred, and gobs of purple rhetoric. I would like to see it reproduce cartoons from magazines that ridicule a cult, and humorous poems, and reproduction of crazy news stories from the press. A department of horrors could give crisp summaries of abominations like the recent issue of *NYT Magazine*, and thumbnail reviews of every crazy book just published. I'd like to see pictures of von Däniken and Panati and Charles Tart, etc., easily obtainable from the PR offices of publishers.

In brief, I think the magazine will be little read and publicized if it stays academic and scholarly. It should be wild, and pasted together in a style resembling the *Whole Earth Catalog*. It should be aimed at college freshmen, not professors. I could be wrong, but I think this is the only way to build circulation and reach the kids who are still reachable.

Best,
Martin

Chapter 2

The Demarcation Problem

By 1977 it was becoming clear that Marcello's vision of CSICOP was not aligned with the rest of the Executive Council. Originally the exchange centered on the policies of the Committee. It then extended to the general "demarcation problem," the problem of distinguishing good science from bad science. Later it became an effort to change the other correspondent's attitudes—without success.

Demarcation is traditionally a philosophical problem. Martin, a philosopher, appeals to that literature. However, he mostly focuses on the practical aspects of the problem; on how/when to dismiss a claimant. Marcello, a sociologist, appeals to "the scientific method" and the rules of the "game" of science. He focuses on adherence to fair play, preferring to not be held to an opinion on demarcation criteria.

The Philosophy of Science is a field in flux and we can at best take a vote on a definition of demarcation. On the other hand, the "scientific method" is not agreed upon either. Marcello and Stephen Jay Gould (a well-known scientist/historian) would have little agreement over the definition. Gould wrote hundreds of essays that circled around the thesis that science has never been done with strict adherence to any "scientific method." It is done by humans with emotions, not automata.

Also it is not clear how science operates as a "game" except that there are winners and losers in debates. Science, as Marcello has repeatedly stated, is a self-correcting system. Consensus is the mechanism that keeps the system from being chaotic, and consensus needs strong evidence to change. Marcello famously said "Extraordinary claims require extraordinary proof" (Sagan made the phrase popular and the thought was anticipated by Laplace and Hume). Martin and Marcello disagreed about how to treat extraordinary claims that had not yet reached this bar.

[The Zetetic]   20 January 1977
Dear Martin,

Re the Velikovsky correspondence, Greenberg was actually quite pleasant for the earlier letters. His sudden "reversal" in tone surprised me. I am anxious to see his response, if any, to my letter responding to his last one (I think I sent you copies of both).

Re the tone and style of the journal, I recognize your preference for something more bombastic, and I would not oppose the existence of such a periodical, but a number of factors make yours impossible as I see it. First of all, the nature of our financing is highly limited and necessitates a stiff price for the journal. We are not a newsstand item and don't have that kind of distribution. There is also some question as to whether or not anyone, in the mass market would want us since magic sells but skepticism may not. So far, we have around 650 subscribers-supporters. Despite sending copies all over the place and a *NY Times* story, we do not seem to be getting flooded with requests for the journal. I hope we can get subscriptions up to about 2000 including libraries but have never really expected much more than that.

The second factor, aside from the market problem (which I think exists, but I know several of you others do not), is my own motivation. Producing the journal takes a lot of my time, that I would normally be using for other far more lucrative things (at least in a monetary sense). I would certainly enjoy reading the kind of periodical you hoped we might be, but I don't want to spend my labors (at least my free labors) on that.

Along your lines, I would suggest that some science magazine would run a regular column or section on quackery in science.

I also part company from you somewhat re who is a merely deviant scientist and who is a pure quack. I would certainly relegate the vast majority of persons into astrology into the nut category; but I would not place Gauquelin in that category at all. I would place von Däniken into the nut camp, but I would see Velikovsky as between that and the legitimate deviationists camp. I just don't see how one can put Velikovsky into the same group as von Daniken or others who are clearly intellectually dishonest. I would be all for hitting the von Danikens and the regular astrologers with bladders, etc., but I can not see anything useful come from similar treatment of Gauquelin or Velikovsky–in fact, I think the historical facts have clearly shown a boomerang effect, I do not suggest that we show these mavericks *respect*, but I do think they deserve *courtesy*.

I do not agree with Sagan's suggestion (which you cite but which I had not earlier been aware of) that the AAAS have a special section on astrology. That would largely act to legitimate the quackery. However, I think it would be useful to have such a section dealing with what Gauquelin has tried to term astro-biology. If such a session were held, I think the basic ground rules should include a clear repudiation of classical astrology by all concerned, including Gauquelin (who has generally written against classical astrology in his books but has been in the awkward position of having them support him while scientists ignored his work). The same sort of thing applies to UFOs. If Hynek is seriously considered, he needs to clearly repudiate the groups of UFO advocates, like the contactees about whom he has privately said one thing, while publicly keeping largely quiet. I think, maybe naively, that given simple courtesy (which, again, I see as quite different from respect), some of these more serious what I have called proto-scientists will get proper consideration (and probably proper repudiation) in a way that will not back-fire on the sceptics.

I am primarily concerned about those who think themselves neutral about such claims. I hope to convince some of the people who already hold contrary views, but I am especially concerned about those who have not made up their minds. Ridicule when backed by clear supportive facts is fine, but ridicule based on mere authority is ultimately damaging. As much as I grew up with tremendous admiration for Darrow and joy in reading Mencken, I am surprised to now learn that Bryant was not really as he was portrayed by them, even at the Monkey trial.

Of course, all of this does not mean that ridicule will be absent from *The Zetetic*. I think Omohundro's piece has some re von Daniken, and Rawlins piece on Velikovsky will certainly be nasty at times. But I want the ridicule to be of the crackpot ideas rather than of the men (unless they are clearly intellectually dishonest, like Hans Holzer or von Däniken).

Part of the problem is where to draw the line. I gather that your review of T&P (from what Ray Hyman tells me) will largely ridicule their work. I don't think that inappropriate for you or the *NY Review of Books*, but it would be inappropriate for *The Zetetic* or for us (Ray, Persi and I) when we go to Washington to argue against the Defense Department's giving them $3 million or so. T&P are taken seriously by others and the SRI. I think that is unfortunate but it is also a fact that we must accept. We therefore must deal with their claims seriously to be credible ourselves.

Individually, the Committee Fellows can be as extreme as they might like. But as a group, we must above all show fairness and courtesy to our opposition. I hope we can include humor (in fact, I will be running a cute item by Phil Klass on fortune cookies in the next issue), and I expect individual authors to sometimes get quite nasty about things. But the editorial position must remain above that if at all possible.

Christ, this was supposed to be a brief note!

Best,
Marcello

Dear Marcello

Our exchange is beginning, I think, to get focused. You don't want to call Velikovsky and Gauquelin "cranks." I do. It's a semantic argument, of course, but arguments over words betray important attitudes.

I do not believe that the word "crank," as it is commonly used, applies to a charlatan who writes books to make money. Von Daniken is a charlatan. The writers of books about contacts with UFOs belong in the same category. Martin Ebon is an interesting example of a semi-charlatan. He "fence sits" in his anthologies only because he wants to sell books. Privately he regards pyramid power as insane. In his book on pyramid power his attitude is "there may be something to it." Ebon is not a crank. Neither is he an out-and-out charlatan. He's a half-charlatan. Mencken's famous "you may be right" letters are a different matter entirely: this response was designed to keep cranks off his back. I do the same thing when I return angle trisections. I know the trisection is false, but if I say so, the writer then telephones me, and sometimes pays me unexpected visits.

The typical "crank" is a man who passionately believes in his system, but for one or more reasons is blind to the evidence against it. Sometimes, as in the case of George M. Price, the Seventh Day Adventist creationist, and also Velikovsky, the cause of the blindness is a passionately held religious faith. Sometimes, as in the case of W. Reich, it is paranoia. We cannot put V in the same class with Von Daniken but otherwise he is the very model of a "nut." If he is not a nut, who do you consider one?

Would you call Horbinger a nut? His Germanic tomes are as carefully reasoned as V's book. Indeed, he attracted more intelligent followers in his day than V does today. If H was a nut, so is V. I wonder if you fully realize the magnitude of V's claims. There are a hundred well established facts that sharply contradict him. I will cite only two. The formations in Carlsbad Cavern testify to the absence of a catastrophe, such as V demands, a few thousand years ago. The evidence for the organic origin of petroleum is overwhelming. To avoid the first fact, V has to invent whole new theories of limestone cave formations. To avoid the second, he has to rewrite historical geology and evolution. (In his view, you recall, the oil dripped out of the comet that had erupted from Jupiter and which eventually became Venus.)

I consider it hilarious to treat V with the *same kind* of courtesy one would extend to a genuine maverick such as John Wheeler. The difference is one of degree, of course, but the difference is enormous enough to make it qualitative. V is much closer to Price, or even to Voliva. Let's consider

Voliva for a moment. I used to hear him lecture in Chicago. He was passionately sincere about his flat earth views, and, so far as I know, intellectually honest. Should one treat a Voliva with the same kind of "courtesy" one extends to a John Wheeler? Clearly not. I am not saying that Velikovsky is as ignorant a man as Voliva, but I am saying that his views are considered by astronomers to be *almost* as crazy, and in this they are right. He is *not* a "scientist" with offbeat views. He is the best living example I know of the pseudoscientific nut who is rationalizing a belief in the literal accuracy of Old Testament stories.

Reputable scientists with eccentric views are all over the landscape. Feinberg and his tachyons, for instance, or McConnell and his flatworm experiments. These men deserve courtesy. But Velikovsky and Gauquelin? Holy Kepler! I read your words, but I can't believe you are saying them. Of course they deserve "courtesy" of a sort—nothing is gained by calling them fools and idiots, but neither do they deserve the "courtesy" one extends to a Wheeler.

It is true, as you say, that sometimes a too vigorous attack "backfires." But there is another side of the coin. To argue against a crank with respect and courtesy has the effect of *legitimatizing* his views. The stupid public says: "Well, if the scientists are debating this seriously, there must be something to it."

That's why I think an AAAS symposium on astrology would have the effect of strengthening the public's astrological obsession. (By the way, this suggestion of Sagan's appears in the current *Zetetic*, so I'm surprised you write that you hadn't heard of it before.) The general public can't follow careful arguments, and they are too ignorant of elementary statistics. The fact that the AAAS discussed astrology would mean, to them, that astrology had become a legitimate theory that the establishment just happened to oppose today. But tomorrow? After all, the establishment opposed Galileo, etc, etc.

I think the major issue between us is the degree to which we find it horrendous that in this day there are intelligent people around who (1) take the Old Testament tales as literal fact (Velikovsky) and (2) believe there is "something to" astrology (Gauquelin). Both V and G deserve courtesy *of a sort*, but not much.

Best,
Martin

[Euclid Avenue]   27 February 1977
Dear Marcello

NY Review of Books finally got to my blast at P and T. Copy enclosed.

Harper and Row are bringing out an interesting book by an old friend of mine, Raymond Smullyan, a professor of mathematics at City University, in NYC. It's called *The Tao is Silent*. It's a collection of essays about Taoism and Zen but from a responsible point of view—filled with Carrollian humor and references to magic. Smullyan worked his way through grad school (Princeton) by doing professional magic, and his book contains some amusing jibes at ESP, fad diets, etc. But I mention it mainly because there is a chapter on astrology that contains what he believes (and I agree) is the best possible defense of astrology. Not that Ray and I believe in astrology, but we both believe this is the best way to defend it! You'll find the book very amusing. Harper and Row have uncorrected bound proofs they can send you: Write to Clayton Carlson (editor) H and Row [address withheld].

Did you know that *Quest 1977*, the slick new magazine (just out), edited by a former Harper's editor, is backed by money from Garner Ted Armstrong? I predict it won't be long until articles with a Protestant Adventist angle start to turn up in it, and articles attacking evolution. It's amazing how much of the Armstrong gospel is taken from the Seventh Day Adventist movement: the Saturday worship, the doctrine of soul-sleeping (or conditional immortality), the annihilation of the wicked, the imminent Second Coming and Millennium, and the creationist arguments which are, without exception, straight out of George M. Price, the famous SDA anti-evolutionist.

The SDA's, by the way, have the best attack on the Armstrongs in print (a paperback), and you can't blame them for being annoyed. The old man was formerly a member of an Adventist splinter church before he decided to start his own denomination.

Ronald L. Numbers, who wrote the fine book about Mrs. White (*Prophetess of Healing: A Study of Ellen Gould White*, Harper and Row, 1976) tells me his next book is going to be about SDA science, with special emphasis on the role of Price. Maybe you could get a good article from him. He's at the University of Wisconsin Center for Health Sciences, Madison, 53706.

Martin Ebon knows which way the wind blows. His next anthology is, on the Apocalypse! He insists that the occult revolution peaked last year and is starting to decline, at least in the media.

                                    Pip pip,
                                    Martin

[Euclid Avenue]   [no date]
Dear Marcello

Sherman Stein, a distinguished mathematician, sent me this, and asked
for suggestions in case *NY Review of Books* doesn't publish. I suggested,
of course, the *Zetetic*.

If you can use the letter, you might drop him a note about it. I told
him I was sending you a copy.

I find it hilarious, and so typical of the amateurishness of Tart. An
undergraduate would have noticed the absence of pairs in the data. Then,
too, that high score of one subject is certainly suspicious. It would be
interesting to know the exact circumstances of his (or her) testing, but
I'll wager Tart won't even reveal who it was. Naturally there weren't any
magicians consulted at the time.

Best,
Martin

[Euclid Avenue]   27 March 1977
Dear Paul

I would strongly recommend no cooperation with the testing proposals
of Gunther Cohn. I know very little about either Mohrbach or Cohn, but I
do know about Evan Walker who is the central figure behind the proposed
tests.

Walker is the leading rival to Sarfatti as the theoretician of the paraphysicists. After I described Sarfatti's views on how quantum mechanics
can explain psi (in my *Technology Review* article), Walker wrote me a letter to say that Sarfatti had stolen ideas from him. "...he borrowed many
of these ideas from my works without referencing me." Walker feels that
Sarfatti has corrupted his theory. As to Geller, he says he does not know
yet if he is genuine or not, but "I can say that psi phenomena now falls
within the understanding of physics."

Walker is the author of many articles in which he takes for granted the
validity of ESP, PK and precognition. He is a great admirer of Puthoff and
Targ, and his writings praise their work. At the close of his long paper on
"Foundations of Paraphysical and Parapsychological Phenomena" (a copy
of which Mohrbach sent to you), he writes: "This result is quite adequate
to account for the observed PK effects in the data Puthoff and Targ give.
We find there is an entire new physics here!"

Outside paraphysical circles, I know of no experts on quantum mechanics who take Walker's theory seriously, any more than they take seriously Sarfatti's similar theory. Cohn obviously thinks it is a great theory. If you read his document carefully you will see that he is not at all concerned with testing psychics to see if, they possess psi powers; he is concerned only with testing those powers to see if they support Walker's theory! He wants a magician to "minimize" possible "adulteration" of the experiment by fraud. I would bet a large sum that Cohn is firmly persuaded that both Geller and Swann have genuine powers, but may occasionally cheat (now the official view). In short, he is proposing an experiment to test a dubious theory of *how psi works*. There is no indication that he would participate in an experiment that would permit him to conclude, if the experiment failed, that the tested psychics had *no* psi powers. Of course the experiment might fail to show evidence of psi if we could get a trained psychologist and a magician in on the testing as observers. But then Cohn could easily fall back on the old Catch 22 and say that the sceptics prevented the operation of psi, and we would be blamed for having "blocked" adequate testing.

But my main point is this. We have here once more a case of men trained in physics and engineering who are seeking funds for experimental work that calls for trained experimental psychologists. These men have no training for psychological testing. Note how proudly Cohn speaks of "20 years experience with designing electronic instruments..." as though that qualified them for psi testing! He feels, of course, that psi power has a physical basis, Walker has provided an excellent theory based on quantum mechanics, so physicists should be the people to test it. This of course is hogwash.

One possible course of action would be to go along with them to see how far they would go in allowing a sceptical psychologist and a magician to be present throughout the testing. Obviously it is not enough to have them as "consultants" (the word Cohn uses) and then keep them off the premises! The psychologist and magician have to be participants in the experiment, or then assistance is worthless. Of course if they agreed to this, it is certain that Geller would refuse to come, and I suspect Swann would refuse also.

I favor making no attempt of any sort to go along with them. My guess is that they feel that if our names were involved it would give respectability to the plans and they would be more successful in obtaining funding. I think they hope to use the Committee in this way. I recommend that we not waste time on such nonsense. Our position should be: Attempts at replicating psi phenomena should be in the hands of trained experimental psychologists who are not firm believers in the very phenomena that they are supposed to be testing. These men take for granted that psi forces exist, therefore nothing would be gained by giving them funds to test whether psi fits a questionable theory. We should support funding by any reputable group of psychologists who are not true believers, and who are trained in the kind of experimental design needed for such testing. Our position should be that such testing should not be in the hands of physicists, regardless of their beliefs, but certainly not in the, hands of physicists already strongly committed to affirm results that would support the reality of psi.

Best,
Martin

CONFIDENTIAL - FOR EXECUTIVE COUNCIL ONLY
[Clipping on an SF/Occult fair from NYT January 20, 1977]

[Euclid Avenue]   23 June 1977
Dear Marcello

Thanks for the copy of the excellent article on Sai Baba. I certainly hope you print it. He is the Uri Geller of India, at the moment. I enclose a couple of recent clips about him you may not have seen. Mantra is the Indian magic periodical.

Amusing letters coming my way from Sarfatti. His physics consciousness group was originally funded by Werner Erhard, of est, with whom Sarfatti (according to his recent letter in the MIT journal, *Technology Review*) has a "warm personal relationship." Something must have happened between them, because now Sarfatti is convinced that Erhard is a fascist at heart, and a potential danger to America! He has been digging up evidence that est was backed by Birch Society money, that est trainees are now getting paramilitary training, etc., etc. I'd suggest you ask Sarfatti to write a piece about est, now that he's disenchanted by it, except it probably would be libelous.

Ray Smullyan's *The Tao is Silent* is now published (both hardcover and paperback) by Harper and Row. Might ask for a review copy and review it because of the astrology chapter. I enclose a review I did for *New Leader*.

I have a "tart note on Tart" coming up in *NY Review of Books*, pointing out the most probable loop hole in the experimental design of tests that provide the data in his University of Chicago Press book. Not knowing anything about magic, the poor fellow overlooked the possibility of a time-delay code! But I now find the religious scene much funnier than the psi scene – especially the Armstrongs who are almost as funny as Moon.

Best,
Martin

[No letterhead]   10 October 1977
Dear Paul

I thought Marcello's points were well taken in the letter he's just sent around, and thought I'd write again to urge that we adopt a new, name for the periodical, which I take will be quite different in format, approach, etc. from the Zetetic. It is Marcello's title, and I for one would certainly welcome its continuance under the suggested title of *The New Zetetic.* There is certainly a need for a scholarly quarterly put out by the sceptical side as distinct from the Journal of Occultism, and I can't see any real conflict between such a journal and the periodical the committee hopes to get off the ground.

                              Best,
                              Martin

Marcello
[handwritten addendum] Marcello: An editor at Basic Books told me yesterday that they will be publishing a book by Ehrenwald on the psychiatric approach to psi! Ehrenwald is opposed to all statistical studies, lab work by the physicists, etc, and wants to get back to basics—dreams, hallucinations, premonitions, hauntings, OOBEs, etc! So it goes.

[Euclid Avenue]   20 October 1977
Dear Marcello

1. Note from one of the mathematicians at Davis says that Tart is about to replicate his experiments, with the mathematicians supervising, and with controls to rule out time delay code, etc. I urged him to get Persi or someone who knows mentalism in on the actual test, as an observer, but I doubt if this will happen. The mathematicians are not, of course, equipped to detect secret signaling, so if Tart gets good results, we can expect a lot of noise from him. I proposed picking the two students (sender and receiver) at random from a pool, but I doubt if he'll do that either. What happens, of course, is that preliminary screening identifies the "good" subjects, which means (in many cases) that it identifies the clever subjects who, for one obscure reason or another, want to cheat.

2. Thanks for copy of the Feynman piece. Persi had also sent me one, but it's good to have another copy.

3. Haven't yet seen *Future Science.*

4. I enclose a paper that is clearly in your domain, rather than Frazier's.

I told the author I was sending it to you, and that he would hear from you on it. Shall I continue such practice if other things like this come my way?

5. Don't know how much you discussed with Baywood, but you might now consider a possible tie-in with them. They seem to have done well with the *Journal of Recreational Mathematics*, as well as other periodicals they handle.

Best,
Martin

[Euclid Avenue]   8 January 1978
Dear Marcello

It's okay with me if you want to put my name on the list of advisors. However, if you find that this means you can't get anyone like, say, Tart to serve, then I won't mind in the least if you take my name off.

Ray sent me a copy of his very strong letter to Paul, which of course I was happy to see; and by now you've probably seen Paul's apologetic letter. It now appears to me as if Dennis is probably the main cause of all the furor, so perhaps his letter to Rockwell will, turn out to be fortunate in that it may silence him. I do wish that you would rejoin our committee, even on just a token basis, but I can understand why you wouldn't want to.

Re: Gardner Murphy. A little known paper by him is a 21-page entry on "parapsychology" in *The Encyclopedia of Psychology*, ed. by P. L. Harriman, Citadel Press, 1946. It is the best summary I know of his opinions, as well as an excellent history, with a good bibliography. Perhaps you know it.

Have you seen the new Crowell book on occultism and Nazi Germany? Lots of information that was new to me, especially about Hess. I have no way of knowing how reliable the book is in any overall way. I must try to curtail the amount of time that I am spending on the occult revolution. It is diverting my energies from more important tasks; it's like spending time attacking Roman Catholicism or Moslemism. This is why my style is to treat the whole thing humorously. Considering the current cultural scene, I actually think it might be *easier* to convince a Catholic that Jesus was an illegitimate child of Mary than to make any dent whatever in Hynek's ufology the central convictions of any top parapsychologist. Here, I think, is perhaps the heart of our disagreement as to rhetoric. You want to approach a parapsychologist in the same spirit as an establishment physicist might approach Weber and his gravity waves. I think there is a qualitative difference, and that such an approach toward parapsychology isn't worth the trouble. The Webers of orthodox science are no fools, but the Velikovskys really and truly are.

Everything is, of course, grist for the sociologist. It's all a matter of tone. I'm suggesting that your best approach is that of a sociologist studying an *aberration* (I belong to the school that thinks sociologists should make value judgments—partly because they can't avoid making them), not a sociologist studying the kind of fringe science that is *healthy* for science, and which does indeed introduce new "paradigms". It takes only a glance at any issue of *Reviews of Physics* to see how wild are the suggestions being made (e.g., the "many-worlds" interpretation of quantum mechanics!). From my point of view, this healthy genuine "wild" science has nothing in common with Gauquelin, Velikovsky or Puthoff and Targ. When I say "nothing in common" I don't mean this literally, since *any* two things have something in common; I just mean that they are so far separated on the continuum as to constitute a different phenomenon. I think you and I have a different evaluation of this "distance," and that it is this, more than anything else, that accounts for our difference in strategy.

Best,
Martin

[Euclid Avenue]   11 January 1978
Dear Marcello:

Here's my Popper review—very nontechnical. In my youth I used to do reviews for the *New Leader* (I'm a democratic socialist), and still occasionally do one for them in return for a copy of an expensive book.

Thanks for telling me about *The Case of the Philosopher's Ring* (all new to me) and the outline of your book. The outline makes for exciting reading; you should have little trouble finding a publisher. Sorry I can't give you any information on R. Targ. Several years ago she wrote to me, about something, I can't recall now what, but I had no idea she was related to Russell Targ.

The big news in parapsychology circles is John Wheeler's blast at the recent AAAS meeting in Texas. I saw only a brief story in *NY Times*, but a friend is sending me a copy of his speech with the appendix in which he urges the AAAS to reconsider its decision about having a parapsychology branch. One of Wheeler's subheads in his paper (I was told) I found marvelous. It's "Where There's Smoke There's Smoke." Almost as good as your Brother Can you Paradigm? Wheeler was the ideal scientist to make his suggestion, because the paraphysicists have been fond of quoting him. His subjective approach to QM (like that of Wigner's) has made them think of him as on their side, and now it turns out to be quite the contrary! As for Wigner, I still don't know. I just received a clipping showing him sitting beside Rev. Sun Moon at Moon's latest conference on the unity of science. After Moon spoke, Wigner is quoted as praising Moon for his eloquent address (in Korean, of course). I keep getting literature for big conferences on "holistic healing" at which Wigner is a featured speaker (along such people as iridologists—who diagnose ills by specks in the iris), so he may be more sympathetic to parapsychology than Wheeler. Wheeler (again, I'm told) said that had he known he was to share the platform with Puthoff and Honorton, he might not have been willing to speak.

Cheers,
Martin

[Zetetic Scholar]   12 January 1978
Dear Martin,

Delighted that you have agreed to let me list you as a Consulting Editor (so far I have had agreements from Ray Hyman, Richard de Mille, William F. Powers—an aerospace engineer friend who works with NASA regularly and is up on UFO-related stuff and I presume I will shortly get an affirmative reply from Randi, since he said "I will if Martin will" earlier). I appreciate your comment that if getting some of the pro-psi boys on board is impossible "with" you on I can drop you; but I feel very strongly that that is their problem and I have no intention of disowning you or Randi. I do not expect to get anyone one like Puthoff (whom I suspect of intellectual dishonesty). I have asked McConnell (who has a very interesting piece on differences between the proponents and the skeptics on psi in the current issue of the *Journal of Parapsychology*) and also Martin Ebon (whom I see as basically honest but an opportunist and fence straddler since he seems to rely for his income on psi survey books). I will, of course, let you know the full list when it emerges.

Thanks for the good reference to the Gardner Murphy paper, which I will get hold of shortly.

I just got hold of the *Who's Who In Science* and *Who's Who In America* listings on Theodore Rockwell, but that's all I know about him. Do you have further information? I strongly suspect that his work on nuclear submarines and his Johns Hopkins connection may be instrumental in Targ and Puthoff's alleged remote viewing work with the Dept of the Navy. Do you have more on that?

I agree with your portrait of our differences. I do not see the big qualitative gap between some of the wilder "conventional" people and the less acceptable wild unconventionals. Orthodox science is full of commonly accepted pseudoscience pockets. Sociology and psychology are particularly full of them, but they exist in physics, etc., too. Oddly enough, I think that part of our difference is due to the fact that you may (in my view) not appreciate how much stuff passes for science inside the normal scientific community. Thus, I see Velikovsky (as crazy as he might be) as little different in terms of acceptable science from most psychiatrists, lots of economists, etc. Bad science is (analytically) bad science whether or not it is practiced by socially respected "scientists" or outsider mavericks. In fact, I think many aspects of maverick science are sometimes truer to acceptable scientific method than some of the science that is institutionally accepted. We criticize astrologers, for example, because most of them present an unfalsifiable conceptual system. But when economists, or sociologists do it we somehow think that is qualitatively different. I don't agree. From my standpoint, most clinical psychology today still bases itself upon archaic notions like "self," "personality," etc., not to mention unconscious and subconscious processes of motivation, etc. Even the basic construct "motivation" is technically dubious. Much of this stuff posits a kind of homunculus that is really no more explanatory (in a truly scientific way) than older notions of "soul" and "character." Yet the latter would be considered pseudoscientific by many while the former are not. While you and some others of our Committee associates probably perceive me as less skeptical (debunking) in my outlook, I am, I think, actually far more skeptical than any of you. For I am skeptical about much of what passes for science in the orthodox science areas. It is because I see psychiatrists as little different from witchdoctors, for example, that I am more sympathetic about the practices of the witchdoctor. I see neither the witchdoctor nor the psychiatrist as true scientists, but I see little reason to be more tolerant of the latter than the former. And since I recognize that we know little about the area that both are working in (in a strictly scientific sense), I am inclined to tolerate both rather than denounce both (since there is no ready replacement in society for their functions). Compared to some of the Committee members, I am very willing to emphasize how much we simply do not know yet (scientifically). And if a Gauquelin or a Hynek (or anyone intellectually honest and willing to play by the scientific rules of evidence for judgment by the historical court of science, wants to play in the search-game for the truth (which is simply trying to discover what the hell is going on "out there" in the empirical world), I welcome them into the search party. I recognize that some may not be intellectually honest (will lie, fudge their data, etc.), but this will eventually come out through replications; and I find similar dishonesty among the orthodox (where it is usually far less visible). I also think you tend to underestimate the importance of anomalies in determining the growth of science. Of course, I would strongly agree with you that the reality of any anomaly must be established first. But finding anomalies and reconceptualizing our views to include them in our theories is the life blood of science. Science has to be concerned about Type II as well as Type I errors. You seem overly concerned with Type I (thinking there is an anomaly present when none actually exists) to the neglect of the Type II error (not noticing an important departure from the norm when one exists). I am growingly convinced that different attitudes, towards what is a serious error (Type I or Type II) are to be found in the skeptic and paranormalist camps. But the fact is that both are potential types of real error and both need to be guarded against.

Did I miss the coverage of the TM claims in *Scientific American* that I thought were to appear therein? If so, could you tell me the issue it came out in (the stuff on levitation, etc.). I checked through back issues but couldn't find it. Or is it still forthcoming.

Re TM, I saw the Maharishi on the Dick Cavett show the other night. He on the other hand claimed that levitation by TMers did contradict the laws of physics and also mentioned that he was trained as a physicist before becoming a guru. This juxtaposition was particularly funny. Do you know anything about his background as a physicist (prior to becoming a guru)?

Finally a comment re Paul and Dennis. I am inclined to disagree with your interpretation. Dennis is, of course, a bit cracked in his letter, but I think it came from the information he got from Paul. You mentioned Paul's apologetic letter. But I don't see it as such. Paul has never apologized to me either for his own distortions (earlier) of my requests or for Dennis' nonsense. I gather he has apologized to you fellows, but it seems to me that that is just an act of conciliation and his recognition of his error that he went too far. Thus, he wrote you fellows about how I was angry and vindictive and causing problems but then wrote me saying he was sure I had nothing to do with the attempts by outsiders to use my resignation against the Committee. I suspect he has written Dennis an even different sort of appraisal of the situation. Since I am the guy he did the damage to, I think it rather perverse that he apologized to all but me. He may just be stupid, but I think it more likely that he is simply two-faced. I appreciate your (and Randi's) stated desire that I might rejoin the Committee. But I hope you also see that Paul as leader and controller of things makes that impossible for me. Until I got the surprising letter from Dennis, I thought my problems were only with Paul. Unfortunately, they seem to have gone beyond him to include Dennis (and possibly Phil Klass, too, to some extent). I truly wish the rest of you success in living with Paul as your leader; but I am afraid that is now impossible for me barring some very extraordinary actions on Paul's part. And I remain a skeptic about matters extraordinary. So that's unlikely. In the long run, it may all be for the best. The Committee is becoming an advocating body rather than a scientific arbitrator of controversies re the paranormal. This is a useful role and needed. I think it complements the kind of role I want to play. You guys can act as lawyers for the defense of orthodoxy while the opposition advocates their own side. The general scientific community can act as judge, and I hope to act as a kind of *amicus curiae*, a friend of the court who recognizes the rules of evidence and the adjudication procedure and tries to help the process work more efficiently and fairly.

Best,
Marcello

[Euclid Avenue]   17 January 1978
Dear Marcello

I was pleased to get a copy of the paper you sent on "Pseudo-Effects in Experimental Physics...". It deals with the kinds of controversy that are constantly arising in *every* field of science.

I know I am repeating myself, but I write to say once more that *this* kind of controversy is *qualitatively* different from the controversy between the science "establishment" and the views of Velikovsky, Gauquelin, E. H. Walker, Puthoff and Targ, etc. In view of this difference (which of course lies on a spectrum), there is no justification for treating the far-out cranks in the same way that one treats the "responsible" deviants from orthodoxy. A physicist, who finds fault with Weber, writes against Weber with restraint and compassion. An astronomer who finds fault with Velikovsky is under *no* such restraint.

I think Wigner is wrong in his interpretation of quantum mechanics, but I would never attack him as a fool. I think Velikovsky is far enough out on the spectrum of pseudo-science to be called a fool. Is this a genuine difference between us?

Best,
Martin

[Euclid Avenue] 18 January 1978
Dear Marcello

First to answer specific questions in yours of January 12. The *Scientific American* TM story got pushed out for more important items, and now I don't think it will appear. . . I know even less about Rockwell than you do. I didn't even know of his work on submarines. . . Don't know anything about the Maharishi and physics!

Now to our friendly differences. Like all arguments of emphasis, and on how to define parts of spectrums, it is difficult to focus things sharply. You compare Velikovsky with psychiatry. I would not for the following reason. There are areas of science about which we have very little firm knowledge. One of them is surely psychiatry. It is inevitable, therefore, that there will be widely differing schools. Let's take Freud as a type. I long ago came to the conclusion that Freud was probably 90 percent wrong, but I cannot see him as coming even close to Velikovsky. Had V stuck to his Freudianism, it would have been different. But V decided to theorize about astronomy and physics. His wild assertions in *this* area have a qualitative difference from Freud's speculations. Outside of psychiatry (the area where there was in Freud's time, and still is, very little hard knowledge) Freud was very respectful of the organized body of knowledge—i.e., of physics, physiology, neurology, chemistry, and so on. There is nothing in Freud's theories comparable to such assertions as that oil is not of organic origin but came from a comet that rocketed out of Jupiter, and finally became Venus. There is nothing remotely like this in the writings of Freud, Adler, Jung, Sullivan, Horney, etc., Take the case of Reich. His early work, in an area where there is little solid knowledge, was quite respectable. When he became paranoid and fancied himself a physicist, he wrote utter nonsense.

You say I underestimate the importance of "anomalies" in science! This I vigorously deny. Again, it is our varying interpretations of what an "anomaly" is. The Michelson-Morley experiment in physics was a whopping anomaly, and Einstein's relativity did indeed strike many establishment physicists as mad. But the difference between *this* kind of anomaly, on which scientific progress does indeed depend, and the "anomaly" of claiming a correlation between a person's profession and the position of planets in the sky when he was born, is like the difference between night and day. (Night and day, of course, are parts of a spectrum, with no sharp division.)

Even more revolutionary were the anomalies that led to quantum mechanics. The two-slit experiment with electrons is *absolutely* impossible to explain in classical space-time terms. I am at the moment working on an article about the EPR paradox (Einstein-Podolsky-Rosen paradox), which is another whopping anomaly that is now at the forefront of enormous debate among physicists. Einstein argued (and I agree with him) that the EPR proves that either Q theory is incomplete, or there is instantaneous transfer of information across vast distances. It goes to the very heart of the mystery of Q mechanics. I myself am extremely fond of paradoxes (anomalies). Where we differ is in our belief that a rational man, knowledgeable about a science, does not find it difficult to distinguish genuine anomalies, essential to scientific progress, from the fake ones that are constantly raised by the cranks.

Now if you could see yourself as stimulating an adjudication procedure between orthodoxy (whatever that is) in a particular science, and the visionaries who are responsibly raising anomalies, fine. But the danger is that you may not distinguish very well between the responsible visionaries and the irresponsible. Naturally there are borderline cases, and it is not always easy to decide. But there is such a thing as a crackpot. I mean this in the same sense that I can say there is such a thing as a chair. Velikovsky and Gauquelin are examples. Wheeler and Wigner (to mention two famous scientists who hold extremely far-out, far from orthodox views) are not. One does science a service by trying to adjudicate between, say, Wigner's solipsistic view of QM and the orthodox Copenhagen interpretation. You don't do anybody a service by treating Gauquelin and Velikovsky as a responsible opposition to orthodoxy. In short, I applaud your motives, but fear your judgment.

Best,
Martin

[Eastern Michigan University]   24 January 1978
Dear Martin,

Sorry to hear *Scientific American*, decided not to publish the TM expose. I hope *The Zetetic* might do so.

As regards our differences, I think you make a very strong case as to calling Velikovsky a crackpot, while being more tolerant of Freud, et al. Let me state at the outset that I am inclined to agree with you in regard to flagrant ways in which Velikovsky is wrong. If being wildly wrong in that fashion, is to be called a crackpot, however, I am not entirely in agreement. Privately, I would label V a crackpot, but I might do that for a number of fellow social scientists, too. Publicly, it seems to me that calling anyone who, (a) seems to have honest intentions (as opposed to a charlatan like von Daniken or a hoaxer like George Adamski), (b) wants to see his theories seriously discussed by the scientific community, in terms of its ground rules, (c) is willing to respond to criticisms made, and (d) is unlikely to cause anyone physical harm while such consideration is going on (as opposed to some medical "theorists" whose suggested cures may immediately damage some members of the public). A crackpot is in the long run detrimental to the scientific process. Even if correct, it is likely to produce a backlash of sympathy for the claimant who then becomes a martyr to science's authority system. Velikovsky's case is a good one in point. Please note, I do not say that such wild theorists must be confronted, they can be simply ignored. But once confronted, I think the arguments against them be rigorous and without character attacks. I do not say that no one should attack those like Velikovsky on unscientific grounds. I resent the comfort that V's theories gave the fundamentalists and frankly some pleasure in seeing some of the debunking of him that were beyond pure reason. This is to me in the area of publication of opinion and viewpoints that may rest on authority, rather than reason but which speak for individuals editorializing. But I think it imperative that such arguments not be presented as scientific actions even if they are meant to in some sense, defend science. Thus, I have usually approved of your writings in popular periodicals like the *NY Review of Books* even when I felt you might be somewhat unfair (in a strict sense) because you were speaking for yourself and not some scientific body (such as the Committee for SICOP was originally intended to be, at least as I had intended for it to be).

You are correct in seeing that our differences about anomalies concerns their definition. Mine is far broader than yours. When you speak of liking anomalies, you seem to mean (in some sense) "true" anomalies. You essentially seem to mean anomalies that you find "plausible" (in the way I think Polanyi means it in the article I sent you with my last letter). I would in large part agree with you even here. It is necessary, it seems to me, that an alleged anomaly must be validated before we construct a theory to explain it (although even this is no simple matter since the history of science shows that we often need a theory to cover an anomaly before we are willing to admit it exists). Thus, I think most Forteans simply care more about the alleged anomaly than they do in validating its existence.

The more I have gotten into these matters, the more complex they seem to be. For example, I often have stated that extraordinary claims require extraordinary proof. But extraordinary is always a matter of degree. Velikovsky's claims contradict so much established knowledge that it seems inconceivable that he could provide proof so extraordinary that it would convince either of us. The claims for psi vary in their degree, of plausibility. Thus telepathy seems within the confines of possibility (at least in some sense); but precognition would play havoc with most of what we think possible since it fundamentally contradicts our notions of time. A sasquatch is comparatively hardly extraordinary at all. The same is true for Unidentified Aerial Phenomena (which may, as Phil Klass argued in his first book, simply be strange plasma phenomena). Yet, our notions of plausibility and strangeness often have little to do with such theoretic consideration. So, many people would think that a bigfoot is far less plausible than precognition. People laugh at the idea of a Loch Ness monster while taking clairvoyance rather seriously; yet the former is hardly likely to alter ichthyology in any basic way while the latter would revolutionize psychology. Which new ideas "offend" our sense of the plausible varies a great deal with our other values. Thus, *I suspect* Velikovsky's support for Fundamentalists or Gauquelin's adoration by the astrologers is a serious factor in what you see as the danger in their ideas. (Of course, I may be totally wrong in this appraisal of your motives.)

The case of Gauquelin is a bit different from that of Velikovsky. Gauquelin is trained in psychology and statistics. His wife is a statistician. His early work debunked astrology (classical astrology) entirely. He claims to have found consistent correlations in the process of this debunking that (a) do not fit the predictions of astrology so do not support it directly at all and (b) do not have any current explanation. He suggested the idea of "cosmic clocks" and tried to relate his findings to some of the similar work done by some others on animal studies. He does not claim to understand or even have much of a theory for why these relations (which he claims to have found and which he claims can and should be replicated) seem to exist. He calls his approach (for the most part, in most places) astro-biology. He has sought analysis of his data by others and has encouraged replications. Now I find all this straightforward. I can see—following Polanyi's reasoning—that the plausibility of these claims plus the lack of theory (mechanism) to explain them should result in their having a low priority for science to investigate further. I can not condemn anyone for simply ignoring Gauquelin's work. I think it unfortunate that the astrologers find kinship in his claims even though he repudiates them. What I do object to is anyone viewing him as irrational or a crackpot. He claims to have stumbled on a finding. He found on replication that it persisted. He does not think it is due to chance and can not see any other thing that would cause such a correlation. Although I give his findings a low priority for investigation, I certainly can not condemn anyone who gives it a higher priority and wants to replicate his stuff. My only concern is over the integrity of the study conducted. I have no reason to think Gauquelin is a charlatan. I have much reason to think he is not. (Along these lines, let me call your attention—if you are not already aware of it—to Charles Sanders Pierce's most interesting comments on the character of psychical researchers whom he respected but with whom he disagreed. I have recently been reading much of Pierce and am very impressed by his observations on research into the paranormal As a logician yourself, I presume you are well aware of Pierce's writings in general, but you might have overlooked the stuff on psychical research in his Collected Works.)

Because of my own biases, I am emotionally inclined to agree with you that Velikovsky is a crackpot in the way you characterize that. But I do not at all feel that this is so of Gauquelin. This shows how hard the line is to draw. You state that you are in favor of those who "responsibly" raise anomalies. But how are we to judge such responsibility? I can see that there could be some ground rules that are easy enough: the absence of fraud in the presentation of data; willingness to answer criticism; etc. But these probably would not apply to either Velikovsky or Gauquelin. It seems to me that the question of responsibility-irresponsibility is something we determine after we consider the merits of the cases presented. I believe the truth *will* out. I believe along with Pierce that the first rule of science is to *do nothing that will block inquiry*. That is why I am willing to tolerate ignoring those we think are "too far out" in their claims; but I am not willing to attack their ideas in any way that will block inquiry into those ideas by any that otherwise might want to pursue such inquiry. (One tangential point: I am not absolutely opposed to ad hominem arguments if such arguments show a past history of deceit; this simply warns us to be that much more skeptical towards the claims made but does not in itself deny such claims; it means look further into this matter and do not take the claimants' word for matters.)

Finally, I am sending you an article by Michael Scriven that you may not have seen that is especially relevant to our exchange and your "defense" of psychoanalysis as non-crackpot. Scriven compares psychoanalysis with parapsychology and argues that the former is pseudo-science (close to what you might call crackpot) while the latter is not. I think he makes an interesting case. The point, however, for me is to demonstrate how difficult it is to draw the lines you want to draw when you say there is as surely such a thing as a crackpot as there is a chair. Scriven is a first-rate philosopher of science who apparently disagrees about where to draw the line with both of us.

Best,
Marcello

[No letterhead]   24 January 1978
Dear Ray

The problem of distinguishing good from bad science is, of course, incredibly difficult and in many cases hopeless. Carnap devoted the last ten
years of his life to laying the foundations for what he hoped could become a
logic of induction, and I have followed with enormous interest the conflicts
between the Carnapians and the Popperians, and other schools of thought
concerned about the question.

I maintain that all this is quite distinct from the task of identifying a
"crank," and that in most cases there is high unanimity among working
scientists as to who the cranks are in their field. This is especially true
in the physical and biological sciences, less so in the social sciences and
psychology, and least true in psychiatry where there is the least amount
of agreement among the experts. I think it is important that we make
a distinction between the person who is commonly called a "crank" or
"crackpot" by his peers, and the scientist who is called "unorthodox."

You remark that I included Rhine in my *Fads and Fallacies.* Yes, but I
went very much out of my way to emphasize that I considered him unorthodox but not a crank. Let me quote from page 299: "It should be stated
immediately that Rhine is clearly not a pseudoscientist to a degree even remotely comparable to that of most of the men discussed in this book. He is
an intensely sincere man, whose work has been undertaken with a care and
competence that cannot be dismissed easily . . . he is an excellent example
of a borderline scientist whose work cannot be called crank, yet who is far
on the outskirts of orthodox science . . . " I have similar qualifying remarks
about Count Korzybski.

You are absolutely right in pointing out that many continuums are
involved in defining "crank." But this is a difficulty encountered, in trying
to define most nouns. Like the words "game" or "chair," the word "crank"
stands for a family cluster of things. The cluster is held together only by
a given culture at a give time. Different people may have slightly different
meanings in mind. The best we can do, to avoid confusion, is try to keep
as close as possible to common usages. Fortunately in most discourse the
fuzziness of definition is not a major problem. Consider the word "chair."
It fades off along a dozen continuums into other meanings. Along, one
dimension it fades into a sofa, and pieces of furniture actually are made of
such in-between status that it is arbitrary whether you call them chairs or
sofas. This fact, however, doesn't destroy the usefulness of the word. If I
ask someone to bring another chair into a room for a guest, it is unlikely
that he'll come back lugging a sofa.

My disagreements with Marcello are partly a disagreement over how to use words, and partly a disagreement over how to treat people—the latter being in part dependent on how the words are used. Being a sociologist, Marcello is naturally interested in the "sociology of knowledge" literature, from Karl Mannheim to Thomas Kuhn He is impressed by the way science progresses by "revolutions," the rise of new "paradigms," the role of "anomalies" in forcing change, and so on. I have no quarrel with this approach. Anomalies are the life blood of science. But—and here is my main point—this has nothing whatever to do with persons like Gauquelin and Velikovsky. Cranks, as scientists use the word, are not by anybody's standards (except their own) the sort of people who produce useful anomalies, or who issue unorthodox challenges that lead to new paradigms. Marcello, in my view, constantly confuses the "crank" with the "maverick." I thought Freud's way of distinguishing the two, which you quoted, was excellent. You point out that in your opinion Rhine is on the carbolic acid side, not the jam. As my quotes from myself indicate, I have always believed that, and still do. I do not consider Rhine, or Tart, or McConnell, or Gardner Murphy "cranks." I may think that they are poor scientists, but that has nothing to do with crankery.

The genuine anomalies and unorthodox scientists are all over the place. They are reported every month in the journal. Black holes are anomalies. Nobody calls Oppenheimer a crank because he first described them. David Bohm is an authentic maverick, and in my opinion a great physicist. In math, "catastrophe theory" is the latest unorthodoxy. About 80 percent of the top mathematicians think little of it, but I've never met a mathematician who would call Thom, who started it all, a crank. I think the Michigan McConnell is dead wrong about his flatworms, but I would never call him a crank. I always believed Weber was wrong about his gravity waves, but never considered him a crackpot. I predict Roger Penrose will get nowhere with his wild-theory about "twistors" and their role on the microlevel, but he certainly is not a crank. I believe Wigner and Wheeler are off base in arguing that quantum mechanics makes it necessary to think of the universe as not "real" unless it is observed. But they are *not* cranks! I think psychoanalysis is a poor form of therapy, but I don't confuse Freud with, say, the "psychiatry" of Jimmy Carter's sister. But people like Velikovsky, or Wilhelm Reich in his "orgone" phase, or Jule Eisenbud in his Ted Serios work, are clearly cranks. By "clearly" I do not mean anything absolute, since nothing is absolute. I mean "clearly" in the sense that I can say "clearly" the earth is not hollow and open at the poles. I mean "clearly" in the sense that you cannot find a single professional astronomer, anywhere in the world, who thinks it remotely possible that Venus was a comet that erupted from Jupiter, and is responsible for all the earth's petroleum, not to mention altering the earth's spin a few thousand years ago, and several hundred other things. If V is not a crackpot, then the word has no meaning. And if he is a crackpot, as all astronomers believe, then he should be "treated as a crackpot, and not as an "unorthodox" scientist.

It is precisely over these questions of judgment that Marcello and I differ. Marcello is willing to use the word "anomaly" for the claims of Gauquelin. I believe this is a serious misuse of the word. Marcello seems to think that the current rise in occultism and pseudoscience should be viewed as serious challenges to the orthodoxy of science. I believe this is a misperception. I think Marcello would be on sounder sociological ground if he viewed the occult wave as a social aberration, comparable to, say, the wave of excitement about the Second Coming that swept over America in the 1840's, and which was tied to all sorts of physical phenomena such as the "falling of the stars" in 1933. I have never asked Marcello about his attitude toward "values," but I have always assumed that he is not a cultural relativist, so that he and I both agree that there is a difference between good and bad science. If Marcello doesn't agree to that, then it's a different ball game. If he does agree, I would put our differences this way. Good and bad science are hopelessly intertwined. Science makes progress by continually meeting challenges to orthodoxy (witness relativity, and quantum mechanics, and evolution!). But totally outside this ongoing process are persons so far removed from science e.g., Velikovsky that they are legitimately dismissed as "cranks." The cranks do not aid science. And if their views are widely believed, they damage science by diverting energies into blind alleys, and sometimes (as in Nazi Germany) by diverting state funds in unproductive channels. From my point of view, attempts to establish "dialog" with the genuine crank are foredoomed to failure, and a waste of time unless one does it for laughs. A dialog with mavericks, yes. In fact, this goes on constantly in the science journals. You can have a meaningful exchange of ideas with the maverick. You can't with, say, the promoters of biorhythm or pyramid power. That is why I believe in writing about crackpottery with humor and scorn. To write about biorhythm in the same tone one would use in attacking, say, Wheeler's cosmology, is to misperceive distinctions that scientists take for granted.

Now to less philosophical matters. Yes, it would be good to make some definitive tests and pin down just what goes goes on in that finger lift bit. Thanks for references on liar dice. It is a dice version of the "I doubt it" card game, or maybe it's the other way around.

Your syllabi look great! The references are all carefully chosen (I'm honored to be included), and I love your informal comments on them. The Uri bibliography is enormously valuable, and I couldn't think of a single important reference you missed. Ditto on the deception biblio. Should

Erdnase be added? Not so much for the card methods, but for his revealing remarks about how the card hustler is unique in having a great skill that he has to perpetually conceal, etc.

On fake medium, maybe Houdini's *Magician Among the Spirits* is worth listing, since there is (I'm told) a reprint edition available. Mulholland's *Beware Familiar Spirits* is an excellent book. Houdini's *Miracle Mongers and their Methods* contains a wealth of material by no means dated. Has anyone done a book yet on computer cheating? (I have a vague memory of seeing a reference to such a book recently, but I'm not sure.)

What do you think of the idea of our committee giving an annual "Uri" (little statue of a bent spoon?) each year to the year's most gullible scientist? Like the Harvard Lampoon's annual Kirk Douglas award for the worst picture. Might get us some amusing publicity. I've dropped a note on the idea to Paul and Ken.

Best,
Martin

CC: Marcello

[Euclid Avenue]   27 January 1978
Dear Marcello

I hadn't read this particular essay by Polyani before, but I am familiar with his writings. I am not one of his admirers, but have no time at the moment to go into detailed reasons why. I suppose you know that he is admired by sociologists much more than by philosophers. I enclose a typical review by the English philosopher Mary Warnock that you at least will find amusing. It expresses my views exactly. I had my first run-in with Polyani about 35 years ago! He had published an article in the *International Journal of Ethics* on the replication of science experiments. (I was then reviewing books occasionally for the journal, and even did a paper for it called "Beyond Cultural Relativism".) In it he argued that it was nonsense for physicists to talk about experimental replication confirming anything, because there were always scientists who couldn't replicate. As an example, he cited the case of Professor Miller who was forever getting positive results when he repeated the Michelson-Morley test. Now here was a case where the MM experiment had been replicated ten thousand times, in hundreds of labs, using widely different equipment, at all heights above sea-level, and under varying conditions, and all confirming the original test. As you know, it is a cornerstone of relativity theory. And here was a young Polanyi, making the identical mistake that you yourself are making: which is to say, recognizing (correctly) that science never provides anything but *probable* truth, but failing to realize that differences in probability are all important, and provide ways of distinguishing good science from bad. I wrote a strong counter article to Polanyi's piece, but the editor of *Ethics* (a friend) felt it would have been impolite to print, so it never appeared. I notice that Polanyi, in writing about V, is making his old mistake all over again—playing up V's correct prediction about temperature on Venus, as though that somehow balances several thousand ridiculous aspects of his "theories." (See Sagan on this.)

Marcello—there is no reasonable sense in which V is playing the role of presenting science with anomalies, new paradigms, etc. On this point the agreement of astronomers is well nigh universal. As I have said before, the views of George McCready Price (see my *Fads and Fallacies*) are a much stronger challenge—and indeed, Price still has an even bigger following than V. But Price, too, is outside the pale of the kind of challenges that Kuhn has in mind, and which indeed are responsible for the progress of science. That you can believe a *useful* "dialog" can be established between Hynek and Klass, or Hynek and *any* UFO unbeliever, is beyond my comprehension. It would be easier to establish a useful dialog between Sun-Moon and Harvey Cox.

Marjorie Grene and her husband David, were around Chicago when I was there. She shares many of Polanyi's metaphysical views, and has written quite a bit about new theories of evolution, which challenge the mutation theory, and are (in effect) restorations of Lamarck. I could list 20 such major areas of science where there are genuine anomalies; genuine challenges, that are well worth debating, and a magazine that established dialog between the opponents would be enormously useful. But Velikovsky and his critics! Or Hynek and his critics! Or Gauquelin and everybody else! Marcello —where is your sense of *proportion*? If it's a game you're playing, for your own amusement, okay. But if you think it serves *science* in any way—well, what can I say that I haven't said before?

Best,
Martin

[*The Tacit Dimension*, by Michael Polyani—This book is divided into three lectures, entitled, Tacit Knowing, Emergence, and A Society of Explorers, In the first Polyani introduces the concept of tacit knowledge, by means of the assertion that we always know more than we can say. In the second, as he explains, he enlarges the concept by suggesting that evolution, our gradual adaptation to our environment, depends, on this iceberg quality of knowledge. In the third, he uses the same theory of knowledge to introduce some reflections on the problem or problems, deriving the form of this problem from Plato's Meno. To those interested in an account of the development of Professor Polanyi's thought over the last nine years this book will be welcome. It contains, for example, an unusual bibliography with no items except the author's own writings. To those without this particular interest the book has little to offer.—Mary Warnock *Mind* (July 1968)]

[Euclid Avenue] 31 January 1978
Dear Marcello

Yours of January 24 just received. Yes, our basic argument is over how to use words. I am prepared to argue that your definitions are too far away from those used by most scientists, as well as by most people in general. For example, at the start of your letter you say one shouldn't call a person a crackpot if he is honest, as opposed to a charlatan. I am convinced that if you checked this with scientists picked at random you would find that they do not consider charlatans cranks at all, but just plain charlatans. It is the very *essence* of the "crank" that he passionately believes what he says. We have discussed this so often before that I cannot understand why you bring it up again. Voliva believed firmly that the earth was flat. Gardner (no relation to me) was convinced that the earth was hollow. The angle-trisectors who write to me couldn't be more honest. Nobody doubts Velikovsky's sincerity. So lets forget about dishonesty—it only clouds the issue. Crackpots are a separate category from charlatans.

First Polanyi. Now Scriven! I know the paper you sent because I have the book *Frontiers of Science*, in which it appeared. Scriven has a minor reputation as a philosopher—I mean minor compared to Carnap, Popper, Reichenbach, Russell, etc. He is virtually alone among his colleagues in buying almost every aspect of the psi scene. I have exchanged letters with him on this for years. He is well known for his violent dislike of psychoanalysis (much stronger than Hook's or Popper's) and for his gullibility with respect to psi. I enclose a copy of his endorsement of Mishlove's *Roots of Consciousness*, the wildest psi book of the past 50 years. Mishlove, by the way, is working for his Ph.D. under Scriven.

Yes, I know about Peirce's views, and own his *Collected Papers*. William James, whom I adore, was even more tolerant of psychic research. All of which has nothing whatever to do with the question of how to define "anomaly". You state that Gauquelin, because he is honest, and knows statistics, has presented science with a genuine anomaly. I do not believe that you are using the word "anomaly" in the way Kuhn intended. Can you name a single astronomer who would describe Gauquelin's results as

an "anomaly" in Kuhn's sense. Would Kuhn himself so describe it? I think
not. When words are used with unusual private definitions, the result is
not light but confusion. We have to have some perspective about where
to draw lines on continuums. Otherwise, we couldn't talk at all. I have
an amusing idea. Why not publish in your journal a questionnaire, and
ask your readers what they think the words "crank" and "anomaly" mean.
Use Gauquelin as the touchstone. Inform your readers that Gauquelin
maintains that there is a correlation between a person's profession and the
position of certain planets on his or her day of birth. Are they willing to
call this an "anomaly" in Kuhn's sense and do they, think that the word
"crank" should be applied to Gauquelin? This is the only way I know for
deciding in an argument such as ours, which is essentially one of deciding
on how words should be used.

There are, by the way, precedents for using such questionnaires for deciding verbal disputes. One of the reasons why pragmatism eventually died
was that James and Dewey (not Peirce!) wanted to redefine the meaning of "truth", in a way that contradicted its traditional correspondence
meaning—a meaning accepted by scientists, philosopher's, and bartenders.
The result was an enormous epistemological confusion until Tarski cleared
the air, by giving such a precise definition of "truth," in the correspondence
sense, that pragmatism quickly expired. It is not that James and Dewey
were "wrong," so much as that they wanted to change, the meaning of a
word in a way that was "not necessary." I mention all this because actual
tests were made, to see what people *meant* by "truth" (as distinct from the
criteria for identifying it), and the results were an overwhelming support
for Tarski.

So, how about a survey of your readers to see how many of them are
unwilling to call Gauquelin a "crank, and who view his "discovery" as an
"anomaly" in Kuhn's sense of the word.

                                Best,
                                Martin
CC: Ray

PS: You suggest that I call Velikovsky a "crank" mainly because *his
views* support fundamentalism, against which I have a strong animadver-
sion. I, of course, deny this. I can list 100 cranks who have nothing to do
with fundamentalism. Let me make a counter suggestion. I believe that
you are unwilling to call Gauquelin a "crank" because you have met him
and find him to be a charming, reasonable sort of fellow. In my experience,
almost *all* cranks are charming, reasonable fellows. It is, indeed, one the
traits of the kind of crank who develops a following. I could cite dozens of

instances, but don't want to make this letter too long. I think you have in your mind the notion that a "crank" is some sort of psychotic with a wild look in his eyes, who rants and raves about his beliefs. Nothing could be further from the truth!

[Eastern Michigan University]   5 February 1978
Dear Martin,

Sorry if you feel I am being redundant in some of my arguments. I am getting a great deal out of your letters to me, so I really hope I am not just wasting your time.

I am not sure you got my main intended point re Scriven's piece. I merely sent it because it seems to me to demonstrate that someone of reasonable intellect has pictured psychoanalysis as crackpottery and para-psychology as respectable scientific deviance. I don't agree with Scriven, particularly, but since his position here seems to me the exact opposite of your own labeling, it was intended to show the confusion present. And such confusion, I argue, reinforces my position that terms like crackpot just don't belong in scientific discourse at all.

Now re such terms as these, I think I can make myself a little bit clearer to you. First of all, the terms crank and crackpot are not at all equal in my own mind. The connotations for me are quite different. "Crank" mainly means someone stubborn and obstinate, someone eccentric possibly, but not necessarily irrational about it. The term is not that offensive to me. "Crackpot," however, suggests to me that the person is nuts, insane, irrational. Now in both cases the person may be infatuated with an idea or claim, but I presume the "crank" can be argued with and possibly won over whereas the "crackpot" is probably hopeless. Velikovsky may in fact be a crackpot by this meaning. Gauquelin probably is not. Gauquelin is in my view, however, not even quite a crank. Let me see if I can explain why.

As I understand his writings, Gauquelin, a trained psychologist, was in-terested in some astrological claims. He thought they could be easily tested and looked at some of the stuff purporting to do so. He found it completely fraudulent. He conducted some correlational studies which tested the claims and found them all negative as far as astrology was concerned. But in the course of doing these tests he found some curious correlations present. His curiosity was aroused and he found the same effects present upon replica-tion. He has not called these "findings" supportive of astrology. He does not claim to understand why these may be present. He has sought to find other factors that might account for their presence but has not been suc-cessful. He simply says that these "findings" should be double-checked by others, investigated further, and, if others agree about his findings being valid, argues that they constitute something of interest. Something that needs explaining. I call this, therefore, a *claimed* anomaly by Gauquelin. Now I note that you don't think it is an anomaly in Kuhn's sense. As far as astronomy goes, that is in part true. Or if it is an anomaly for astronomy it may be a largely inconsequential one, certainly not likely to have revolutionary impact. But Gauquelin (and I) see it an anomaly for psychology (and sociology) that is possibly of greater importance than its anomalous status in astronomy. *If* the correlations Gauquelin claims are present are indeed present, they pose an important riddle for personality theories. I don't have Kuhn's book handy here at home (it's at the office) but I feel confident that he would define Gauquelin's alleged findings as anomalous. Certainly Willard C. Humphreys' *Anomalies And Scientific Theories* would give it that label. Now I must emphasize that Gauquelin is merely alleging the existence of an anomaly. It is not yet an accepted anomaly by the scientific community. It can only have impact in Kuhn's way after it is accepted as a true anomaly.

This brings me to why I don't see Gauquelin as a crank. If he were claiming that he had established his claimed anomaly as a true anomaly, that would be one thing. But he has not. It is not merely that. I found Gauquelin a charming and reasonable fellow in our interactions. The important thing to me is that he wants his stuff checked by others. He recognizes that his claim is whacky. He does not claim he understands why the correlations he found are present. He sees no particular implications for his finding once accepted. He merely thinks he has stumbled on a set of phenomena that don't fit others expectations and that did not fit his own. He's not trying to sell any new theory of the universe. He's not trying to deny our materialistic views. He just thinks he's found something very strange indeed and that it is quite replicable. He seems to me to be willing to discard his claims if replication fails. He is, of course, somewhat interested in seeing his claims supported. He has spent much time and money in conducting his research in the face of opposition. But I don't see him as irrationally wedded to his ideas any more than any other scientist that thinks he's found something new and exciting about the way the world operates. Now, he may in fact tenaciously cling to his claims after others falsify them. At that point he may turn into a crackpot. I don't know. But right now he seems to me to be honestly seeking to get others to test his claims. He would be the first to admit the claims are surprising and implausible. He offers no theory to explain the alleged findings but wants others to help him come up with an explanation. This kind of persistence when others want to ignore you may be what you'd call a crank. If so, thank god there are some cranks in science. But this doesn't make him a crackpot. More important, what do we gain by labeling him a crank? The important thing is the arguments he brings to the claim and the testability of these. We agree he seems not to be a charlatan. We both would agree the claim he makes is "far out." But if he is willing to play by the rules of science and simply seeks a hearing by the court of science, why condemn him by the use of such terms as crackpot? I can, following Polanyi's argument, see that you may find his claim so implausible and so insignificant even (assuming it were true) that you might want to ignore him. I'm not saying we all should leap into spending our time and money in checking out G's claims. He and his claims may deserve very low priority for our investigation. But why imply by terms such as crank and crackpot that he is somehow outside of science? When someone like Jung can come along with archetypes, etc., that are not even falsifiable, why consider Jung acceptable and Gauquelin beyond the pale?

I brought up Peirce because of his concern for doing *nothing* that would block inquiry. My whole point is that labeling some protoscientists (whom I have defined as those willing to play by the rules of science in having their claims accepted or rejected but who have not yet been accepted as scientists by the general science community) as cranks or crackpots can do nothing but block inquiry. When we use such labels about proto-scientists (as opposed to those demonstrably irrational or fraudulent), we are simply telling others, that their ideas are not worthy of investigation. I think all ideas are worthy of investigation but some are more worthy than others. I think we would agree on the low priority that should be given research into G's claims. I would not recommend that NSF fund such work. But if someone is interested in testing his claims, I don't want to erect any obstacles for them by making them fear rejection, from their more orthodox brethren.

Gauquelin's case to me is quite different from Velikovsky's. I must admit that the evidence to date would seem to indicate that Velikovsky is a crank and possibly even a crackpot. But V's case has been obscured by so much irresponsible activity by his critics. A good case in point may be the recent book containing the AAAS papers. I do not think V is correct. But if one can at all believe C. J. Ransom's description of the events at the AAAS meetings in his book *The Age Of Velikovsky*, matters are not procedurally that clear. I think that V's claims re astronomy are outrageous, as you do. But his new chronologies have found some support from others. I think V would have one hell of a hard time answering Sagan's and others criticisms. But I think it is significant that he intends to try to answer them in his forthcoming book. I look forward to seeing that book. It is not that I think he could be right; he is almost certainly wrong. But it is important that he is willing to answer the criticisms. He has not retreated into a shell expecting "faith" to prove him right. I would be willing to see him labeled a "crank" (just as Linus Pauling on vitamin C may be so labeled), but that term would have to be used without the connotation of irrationality that "crackpot" carries with it. V is clearly stubborn and almost certainly wrong, but he still belongs, in my view, in the scientific arena. What I object to is throwing, anybody out of court (the scientific arena) who is willing to abide by the court's rules and decision. But like a courtroom, some things have to get on the docket before others, and V may have to await his turn for quite some time.

I am not sure where our differences are re the definition of an anomaly. I hope I have it correct that you mean "accepted and significant anomaly" where I use the term anomaly. I agree with you that most astronomers would not consider Gauquelin's correlations an anomaly. They would deny the anomaly exists; they would not deny it was an anomaly if it were proven true. Part of the problem with so much of this concerns the role of the claimant going outside of science to a lay audience for support. Velikovsky is a good example. Parapsychology is another. A key question here to me is why did they go outside the science arena? If they could get no hearing at all within science, they may have gone for support outside science to get support and/or to put pressure on science for the hearing. The charlatans are normally not forced to go to the public; they choose to go there first. From what I can see, Gauquelin (and to a large degree Velikovsky) went to the public after failing to get a hearing from the science community. This is very different.

A final comment. In the case of Velikovsky you may be right that he is indeed a crackpot. I think you may know his case more fully than I do. But I think I know Gauquelin's 'case quite well, and I do not think the pejorative labels really apply to him. In addition to the above factors, an important difference between Velikovsky and Gauquelin is that Velikovsky may actually be a terribly rotten and biased researcher. That is, in addition to being wrong and contradicting known physics, he may be sloppy and terrible in his reading of history and geology, etc. So, even if what he does is science, it may be terribly bad science. But I think that Gauquelin has not misrepresented other claims, etc. His own work seems respectably conducted. In fact, if his stuff does not get replicated by others, his error could even be due to charlatanry. But this seems most unlikely since he really has not profited from his claims to any substantial degree (and has probably lost from making such claims). I think it is very significant that Gauquelin, unlike Velikovsky, is not really pushing his claim as evidence for anything more. G does not seek to revolutionize astronomy. G's position is more like that of a cryptozoologist who is trying to tell us where to look to see a strange animal without knowing what the hell the animal is, or what it signifies for us. Unlike the fellow who simply claims to have seen such a beast, G is telling us where we can look and presumably see one too. We may not want to look, or we may doubt it is worth looking, but we can not accuse him of being a crackpot if all he wants us to do is look and come to our own conclusion. My conversations with G indicate that is all he wants. Not for us to believe him, but to look for ourselves and confirm what he says. Surely that does not make him a crackpot if he really means it—and we have no reason at this time to suppose he does not mean it.

Well, I see I have rambled on quite a bit. Hope you don't feel it just repeats what I have already said.

To other things: Have now read the new issue of *The Zetetic*. I very much liked your piece and Kusche's. I detested Rawlins' piece as I am sure you can imagine. I am frankly most surprised that Ken published it without severe editing or without editorial comment. But I guess the general tone of this issue (debunking, primarily of the mass media and lightweight claims rejected by most in the occult community) will make the differences between *The Zetetic: Skeptical Inquirer* and my *Zetetic Scholar* that much more clear. Since the book reviews in this issue were the ones I turned over to Ken along with a couple of the articles, I must wait until the next issue to make a real judgment of where Ken is going with *The Zetetic*. But it looks like he is following Paul Kurtz's direction quicker than I thought likely. My only real complaint about this issue is the listing of all my Consulting Editors as though they were still with the journal when (1) I dismissed them all with thanks when I left, (2) I so informed Ken, and (3) Ken does not even know who some of them are since they are close friends of mine in a few cases and unknown to Paul entirely. My superficial impression is that they just kept them listed to make it look like nothing had changed except my departure (which also went pretty much unnoticed, though I expected that). Oh well...

Best,
Marcello

CC: Ray

P.S. If you haven't read it, you would probably enjoy Harriet Zuckerman's "Deviant Behaviour and Social Control in Science" (In *Sage Annual Reviews Of Studies In Deviance*, 1977), which deals with many of the issues we are discussing. She makes many useful distinctions. The section on etiquette in science, particularly on ad hominems, is particularly apt.

[Euclid Avenue]   11 February 1978
Dear Marcello

"Crackpot" is a bit stronger than "crank." The Collegiate Webster defines "crackpot" as a person given to eccentric or lunatic ideas. It calls a crank an eccentric, or a person who is "overly enthusiastic about a particular subject or activity."

You write: "terms like crackpot just don't belong in scientific discourse at all." If by "scientific discourse" you mean the language of technical papers, then of course you are right. I doubt if you could find the word crank or crackpot in any technical journal. By the way, there are other kinds of scientific discourse. In the informal language of scientists talking to one another, both terms are used often. And more to the point, they are often used in popular writing about science. Bertrand Russell, for example, has some hilarious passages about his technique of arguing with cranks (see his *Outline of Scientific Rubbish*); you and I have never disagreed over the avoidance of terms like "crackpot" in scientific papers. We disagree only over their use in popular rhetoric.

I have the feeling that you don't want to call anybody a crank even in popular writing, but surely that must not be the case. Consider Cyrus Teed, whose book puts forth the conjectured "anomaly" that the earth is hollow and we are all inside. Teed actually made empirical tests using a team of surveyors. His book is filled with charts showing how light rays curve so that one cannot see across the inside of the globe. His theory was taken seriously by some Nazi officials, and an expedition was actually funded to see if infrared light might not photograph across the interior. I cannot imagine that you would hesitate in calling Teed a crackpot. If so, then our disagreement shifts a bit, and it is only a question of when the word should be applied to certain contemporary figures that I call crank and you do not.

Of course anyone, by Humpty Dumpty's principle, can use words any way he likes. But the peril of going against common usage is that more confusion than light results. Now I cannot think of any scientist—astronomer, psychologist or sociologist (excepting yourself)—who would not in informal discourse, consider Gauquelin a crank. Indeed, the notion that there is a correlation between positions of planets in the sky at the time one is born, and one's later choice of a profession, would be regarded by most scientists as on the same level of absurdity as Teed's hollow earth theory.

Let's take a look at your distinction between a "claimed anomaly" and a "true anomaly." In a strict sense, nothing is ever proved true in science—there are only degrees of credibility—but aside from that, all the great anomalies in the history of science were put forth as conjectures. What we are concerned with is not the difference between a conjectured anomaly and a "true" one (by which I assume you mean one that is finally accepted by almost everybody), but between sensible conjectures and absurd conjectures. If the conjecture is too absurd, no one calls it an "anomaly." It is called a crazy idea. Teed's conjecture is not an anomaly in Kuhn's sense. The results of the Michelson-Morley experiment provide the anomaly par excellence because it ushered in the relativity revolution. Remember that Michelson himself did not believe it to be true and died with that conviction. The black hole is another excellent and legitimate conjectured anomaly. I would guess that about 1/4 of today's cosmologists doubt that black holes exist. Phil Morrison is an example of a distinguished sceptic.

The question before us can now be put simply. Should Gauquelin's conjecture be dignified by the name of "scientific anomaly" or dismissed as a crazy notion. Here we cannot decide the matter between us, since it is one of language choice, but we can make a statistical survey to see how the science community uses words. Suppose we mailed a questionnaire to 100 psychologists and sociologists and asked them to check one choice:

1. Gauquelin's conjecture is an anomaly in Kuhn's sense, worthy of serious consideration by scientists.

2. Gauquelin's conjecture is a crank notion, not worthy of any serious attention by scientists.

How do you think the 100 would respond? I am willing to bet that the majority would check 2. If so, it follows that you are using words in an eccentric way, and I am using them in the usual way. It is not a trivial debate because it underlies a disagreement over the kind of rhetoric that is justified in popular writing.

I would be very much surprised if Scriven, for example, would be unwilling to call Gauquelin's conjecture a crazy hypothesis not deserving of serious rebuttal. More than that, most of the people I consider "cranks" would, I believe, take the same position with respect to almost all the other "cranks" except of course themselves. Hynek, for example, often uses the word for ufologists who disagree with him. I'll bet you ten bucks Hynek would be willing to call Gauquelin a crank, and that both Hynek and Gauquelin would call Velikovsky a crank, and that Velikovsky would call both of them cranks. If the word didn't exist, someone would have to invent it. Neither you nor I would use it in a technical paper, but at popular writing it would be, in my view, absurd to write about the hollow earth theory in the same tone of respect one would use in debating, say, the steady-state theory.

When we talk about Gauquelin we are not talking about twilight figures like Rhine and Freud, but about someone who is, in every scientist's opinion except your own, outside the pale of serious science. In my view your language blurs a useful distinction, like that between day and night. The result is that the language itself becomes an impediment in understanding the peculiar phase that our culture is going through.

Best,
Martin

CC: Ray

[Eastern Michigan University]   19 February 1978
Dear Martin,

Thanks for your most recent (11 Feb.) comments on our dispute. I find myself still in disagreement, but I am frankly not entirely sure about the source of our divergence in the comments you made. You seem to understand what I mean about a claimed versus a confirmed anomaly. Obviously, the distinction is not an absolute one; again we agree. But we very much disagree about where to place Gauquelin's claims. I would agree, with you that the Flat Earth theory should be called crackpot if anything deserves that name. And I agree with you that a distinction should be made between writing in a popular periodical and a journal of technical character; thus, "crackpot" would not be used at all in a technical, scholarly journal; in fact, a totally implausible' theory like the hollow or flat earth stuff would not even find its way into discussion in a scholarly-technical science journal but would probably be ignored. But I don't at all see that Gauquelin fits that categorization. And, contrary to your suggestion, that even Hynek would call him a crackpot, Hynek respects (though is skeptical about) Gauquelin's claims (Hynek even wrote a nice Foreword to Gauquelin's book *Cosmic Influences On Human Behavior*).

You seem in your letter to confuse, in my view, conjecture and anomalies. Let me clarify. A conjecture to me is a speculative picture given as an hypothesis. These can be wild or plausible. The conjecture that the earth is flat is a wild conjecture. To make it fit our observations one must reinterpret just about everything we believe we know (light rays bending, etc.). An anomaly is something that is supposed to exist (a unicorn or a strange correlation) based upon some kind of experiment or sighting. Thus, we conjecture that a unicorn might exist without anyone claiming to see one. But if we have a report of a sighted unicorn, that is a claim of an anomaly. If we, catch a unicorn for everyone to see, that is no longer a claimed but now a verified anomaly.

The point is, you want to refer to Gauquelin's stuff as conjectures. But they are not, really. They are claims of anomalies. Let me remind you of what Gauquelin is claiming. Gauquelin has no theory at all. He started out by examining the empirical, statistical claims of the astrologers, some of whom claimed to have proof of their claims. He found that Krafft and the rest had no really valid evidence; they misused statistics, etc. He then did some independent correlational studies. He found that they totally debunked classical astrology's claims (their conjectures). But in the process of data gathering he found some strong correlations that he could not understand which had nothing to do with astrology's claimed patterns. He then replicated and expanded his investigations and was surprised to find the same sorts of correlations present. He then proposed the study of such relations by others, found some similar strange correlations in the works of some biologists studying animal behavior and relation to lunar and similar phenomena, and proposed the name astrobiology for the study of such processes. Now he definitely thinks that he has discovered some implausible but demonstrable processes. He would agree with you and that they are very strange indeed and make no sense in terms of what we would expect. He does not offer any theory for their explanation. He does not think they support astrology. Unlike parapsychology or Velikovsky, he does not see his "findings" as significant for our view of basic philosophical matters like "spirit" or Old Testament tales, etc. He is not trying to establish anything except that he thinks he has discovered some relationships between cosmic and terrestrial events. He seems to have stumbled on his finding and did not come to it hoping to find it. He insists that others need to replicate his work before it can be accepted or established. He has sought counter-explanations but can not come up with any. He is not opposed to explanations others have offered and has tried to test for these (controlling for various factors). In my view, this is a perfect case of claimed anomaly. His attitude about his work is not excessively defensive. He welcomes debate and replication studies. What else can you ask for from him? What would he have to do to make you not label him a crackpot? Let us assume for a moment that he has been careful and has indeed found what he thinks he has found. Surely just because the finding is implausible does not make it crackpottery? From what I have learned of his work, Gauquelin has a scientific attitude about his work. Like any scientist, he has some investment in his finding and wants to defend it. What's wrong with that? Look at his comments in *The Humanist*. Are those the comments of a crackpot? I don't think Abell considers Gauquelin a crackpot, and I don't think Paul Kurtz does either (since Paul is publishing one of Gauquelin's books and spent so much space and money on the Gauquelin-Abell test). They believe him to be incorrect and offering a claim that is quite extraordinary, but that's not the same thing as thinking him a crackpot. Gauquelin has clearly stated what it would take to falsify his claims. The Abell study in *The Humanist* simply did not result in such a thorough falsification. It raised some questions but it did not settle matters on either side. I think Abell would admit this. (I might add that from our brief comments on Gauquelin, I don't think Persi would call him a crackpot either.) Again, I must stress that Gauquelin is just about as amazed by the correlations he claims to find as we are. He tells me he would welcome a conventional explanation. He is not positing some sort of strange paranormal forces. He thinks it is due to some normal forces we don't fully understand somehow.

Now it may be that G's claim is so implausible that *even if true* it does not deserve serious attention by scientists (Ray has argued somewhat the same way about psi). It may simply be too strange a phenomenon with too little theoretical or practical relevance to deserve high priority for our attention. But that does not make it non-scientific or crackpottery. And it may be that G's data have been erroneously handled by him somehow. But that would make it bad rather than "mad" science. As I have always said, it is different to ignore strange claims as opposed to defaming the claimants by pejorative labels.

In a very real sense, I think you are actually saying that the claim being made by Gauquelin is so implausible and far out that you want to call it crackpot. But I must presume that persons and not their ideas are crackpots. I certainly would agree with you that the alleged anomaly claimed by Gauquelin is wild and far out. So was Wegener's claim for continental drift. It is mainly far out because we have no known mechanism that could produce the result claimed. This was the case with Wegener. But I think Gauquelin is going about trying to establish his claimed finding the way the most reasonable scientist might do it with any unconventional finding. (I realize, by the way, that the Wegener parallel is a poor one since Wegener's was a theory to account for many known anomalous facts, whereas G's "facts" —the correlations—are not really a theory at all.)

I note in looking over your letter that you also refer to G's claims as an "hypothesis" as well as a "conjecture." I don't really think it is a hypothesis in the sense we often mean that either. Gauquelin claims to have found a correlation. That is like finding a unicorn. He says look "here" and you will consistently find the same correlation. He says that like one might say you can look in woods X and consistently find a unicorn. He also says that the correlation is not spurious just as someone might say is not a horse that was mistaken for a unicorn. I normally think of an hypothesis as something derived from a theory. G's claim is not at all derived from a theory. He just stumbled on what he thinks consistently exists. To me it is like going looking for a Yeti in the woods, and happening to find a strange species of animal. The Yeti (astrology) may have led you to go looking but what you found does not support that. How then can you convince others that you found the strange beast? You have to get them to go with you or separately into the woods and see for themselves. I may be too busy to go into the woods and look, or I may think the probability of finding the beast is small for me to take the trouble, but I would not be correct in calling the claimant a liar or a crackpot, especially before I looked for myself or examined the evidence offered.

****

Have you seen the newest issue of *Kronos*, which attempts to answer Sagan et al.? I was very upset to also get the previous issue last week to find they had done an article around my correspondence with them and Sprague de Camp. I considered that a private correspondence in which I was trying to mediate between Sprague and Greenberg. Greenberg's article (mainly on de Camp's letters) is really playing foul with what I felt was an attempt on my part to be courteous with them. This breach of science etiquette on their part certainly argues strongly for labeling them cranks (though I still would like to avoid the term crackpot, at least for Velikovsky himself—though maybe not for Greenberg).

Cordially,
Marcello

CC: Ray

P.S. Enclosed is a flyer I think I sent you last week. Three others have since offered their support: Ellic Howe, Harry Collins, and Robert L. Morris.

[Zetetic Scholar]   19 February 1978
Dear Ray,

1) Enclosed is a copy of my last letter to Martin re our continuing discussion.

2) My discussion at the AAAS went over very well, I think. It looks like our whole session will appear as a book. I enclose a copy of my comments as presented (sans the notes which I am still completing). I intend to extend this discussion to specifically deal with each of the four papers after they are revised (there is indirect reference to them in the paper here but most of it is of a more general character since I did not get the papers in time to really deal with them in detail). Any comments would, of course be appreciated.

3) Harry Collins, Ellic Howe and Robert L. Morris have now (joined the editorial consultants group. I enclose a copy of the flyer I am sending out. You might want to send it to your library for consideration.

4) Collins was at the AAAS meeting as were numerous parapsychologists (Honorton, etc.). I am very impressed with Collins' work.

Best,
Marcello

DISCUSSION COMMENTS AT "THE RECEPTION OF UNCON-VENTIONAL SCIENCE" SECTION FOR THE ADVANCEMENT OF SCIENCE MEETINGS, WASHINGTON D.C., FEBRUARY 16, 1978, PRE-SENTED BY MARCELLO TRUZZI

In speaking of what he termed the "essential tension" in science, Thomas Kuhn has noted. that: "the successful scientist must simultaneously display the characteristics of the traditionalist and of the iconoclast." It is this problem of equilibrium that faces the scientific community in its collective reception of unconventional theories. The balance is a difficult one to put into operation, and the history of science is replete with examples of failure. In general however, institutionalized science has tended to be conservative and protective of its existing bodies of currently accepted facts and theories. Michael Polanyi has similarly noted that:

The professional standards of science must impose a framework of dis-cipline and at the same time encourage rebellion against it. They must demand that in order to be taken seriously, an investigation should largely conform to the currently predominant beliefs about the nature of things, while allowing that in order to be original it may to some extent go against these.

But Polanyi has defended the position of a strong orthodoxy for science arguing as follows:

Journals are bombarded with contributions offering fundamental dis-coveries in physics, chemistry, biology or medicine, most of which are non-sensical. Science cannot survive unless it, can keep out such contributions and safeguard the basic soundness of its publications. This, may lead to the neglect or even suppression of valuable contributions, but I think this risk is unavoidable. If it turned out that scientific discipline was keeping out a large number of important ideas, a relaxation of its severity might become necessary. But if this would lead to the intrusion of a great many bogus contributions, the situation could indeed become desperate. The pursuit of science can go on only so long as scientific judgments of plausibility are not too often badly mistaken.

These sentiments by Polanyi are probably shared by many critics of unconventional theories. But such an extreme defensive stance runs counter to the fundamental openness of science to new data and theories that most of us would value. Charles S. Peirce wrote that "Do not block the way of inquiry" should be the first rule of reason. Many of us would agree that this must apply to all reason in science. In addition, many contemporary philosophers and historians of science have emphasized the role of anomalies and puzzle solving as at the heart of science. Thus it has been argued, not only should we not suppress anomalies, we should actively seek them out. For anomalies not only offer puzzles to solve, they many constitute falsifications that force us to expand and redevelop our cognitive maps of the world.

Ron Westrum has suggested that the problem can often be to statisticians' concerns with Type I and Type II errors. The orthodox concern, is with avoiding, a Type I error: thinking there is variation when in fact there is not. The proponent of the unconventional claim, however, is often concerned more concerned about making a Type II error: missing an important source of variation in the world by mistakenly thinking nothing special is happening. Most scientists (as with most early statisticians) are concerned with Type I errors. They are more interested in the general pattern they already have and don't want to mistakenly acknowledge "unimportant" exceptions. But the unconventional theorist commonly feels the "exception" is highly important and fears that it will be missed. Thus, the proponent of acupuncture or psi thinks that the implausibility of such claims is secondary to their importance if they are true. These questions of significance may be related to broader philosophical concerns and may be part of what Polanyi calls the "tacit knowledge" of science but which may also be viewed as part of what Harriet Zuckerman has referred to as a pre-public or "private phase of scientific inquiry."

The problems surrounding the reception of unconventional science can perhaps be better understood by unpacking the central issues for each of these interlocked elements: *reception, unconventional,* and *science.* Let me consider these in reverse order.

The Definition of Science.

The basic issues surrounding the definition of what constitutes a science continue to be debated. In at least a rough sense, made much rougher by recent work in the sociology of science we can separate the social institution called science from the basic method of science (i.e., the attempt to intersubjectively validate or falsify systematic conjectures about the empirical world). "If one takes this separation seriously, scientific knowledge is defined not so much by its content as by its form. This perspective has special meaning for our common categorization of much unpopular thought as pseudoscience. As I have noted elsewhere, this definition of pseudoscience as *methodologically flawed* science strongly suggests that there is some pseudoscience accepted as legitimate within the scientific community while some methodologically proper work may be considered illegitimate by that community and thus gets labeled "pseudoscientific." It should be noted that some forms of pseudoscience are simple charlatanry, as noted by P.H. Abelson in his complaints in an editorial in *Science*, but many publicly labeled pseudosciences are honest attempts to gain scientific acceptance of alleged extraordinary events'. I have suggested the term *proto-sciences* for such esoteric views, and I have argued that belief systems can actually be taxonomized along a continuum from normal science at one end to mystical occultism at the other.

The Unconventional Aspect.

The recent scholarship in the history of science has revealed a far more discontinuous process to us than is generally portrayed in our textbooks. Our notions of plausibility and importance are often relative to our membership in particular science subcultures. This insulation may have healthy consequences. Thus, most biologists ignorance of contemporary physicists arguments against the plausibility of evolution had the happy result that; physics later changed its estimate of the age of the sun so that the potential conflict never substantially emerged to inhibit the development of evolutionary theory. The point is, however, that different scientific, subcultures may have very different views of just how extraordinary or anomalous a new claim is. Seymour Mauskopf's description of the statistician's view of parapsychology versus that of the psychologists (especially those in the specialty of perception) is a good case in point. Acupuncture's reception by dentistry and veterinary medicine has been far more favorable than that by general medicine.

In addition to some relativity as to *how* extraordinary or plausible a new idea is, we should also note that new claims can be highly specific or very general, even to the point of claiming a need for a whole new scientific specialty for its study. This is well exemplified by the parapsychologists, many of whom view themselves not as a specialty group within psychology but as a distinct science. I have elsewhere argued that we can differentiate *crypto*scientific claims (ones positing the existence of an extraordinary variable, e.g., a unicorn) from *para*scientific claims (ones positing the existence of extraordinary relationships between ordinary variables, e.g., astrobiological correlations). Both crypto- and para-scientific claims occur separately within all sciences. Most of these constitute what Isaac Asimov recently termed "endoheresies" or deviant, perspectives from within science. These are usually treated with some respect and courtesy by the orthodox scientist. But there are also "exoheresies" proposed by those outside of science, and these are commonly received with far less courtesy. One should probably distinguish two forms of exoheresy. One form consists of a scientist from one specialty area invading the domain of another. (This may also be viewed as a form of endoheresy relative to the total science community.) Wegener's background as a meteorologist made him a somewhat unwelcome intruder into the domain of geographers, and this probably had negative effects on the reception of his theory of continental drift. It should also be noted, however, that the prestige structure of science is such that cross-overs of research interest may be greeted differently depending upon the direction of the entry. Thus, some physicists' recent entrance into parapsychology (despite psychology's general rejection of psi claims) has probably enhanced the plausibility of psi's standing in the general scientific community. Another version of such exoheresy may be cross-national. Thus, Chinese medicine's acceptance of acupuncture was widely viewed as an implausible treatment seeking entry from outside Western medicine. But exoheresy from total or near total outsiders to the scientific community (a notable example being psychoanalyst-physician Immanuel Velikovsky's astronomical and historical theories) have commonly been greeted with strong negative sanctions or ostracism, sometimes to our later embarrassment. All of this strongly suggests that we should concern ourselves more with the methods employed by a proto-science and concentrate less on its substantive claims.

Unfortunately, recent work in the philosophy and history of science has further complicated our view of science's method. Facts do not simply speak for themselves; the presence of theories facilitates our acceptance of implausible facts, and Joseph Agassi has even pointed out that there may be times when we should ignore evidence in favor of an hypothesis. Maier has even put forward a satirical law (following in the footsteps of Parkinson) based on his analysis of what is actually taking place in psychology, stating that "if facts do not conform to the theory, they must be disposed of." These problems are clearly interrelated, and I can only hope future unpacking may lead us to clarify matters to the point of a solution

Factors in Reception.

By Scientists. A recent study by psychologist Michael Mahoney has confirmed what many of us practicing science strongly suspected. Scientists are not the paragons of rationality, objectivity, open-mindedness, and humility that many of might like others to believe. Though the American tradition in the sociology of science surrounding the work of Robert K. Merton emphasizes the existence of scientific norms that should promote such virtues, the recent empirical work on scientists and their behavior, particularly by the British sociologists of science, emphasizes the frequent absence of sanctions and the negotiable aspects of many of these alleged norms. Much of scientific knowledge is perceived by these critics as socially negotiated. They have argued that this is well demonstrated by examination of the reception of deviant theories and ideas within science.

By the Public. A confounding of our problem is due to the possibility of direct appeal by some claimants of esoteric views to the mass media and the general public, often prior to submitting these to the science community. Zuckerman has characterized this as "publicity seeking" that violates part of the etiquette within science. This has been a special problem for parapsychology whose funding has frequently come from extra-scientific community sources. Yet we too often neglect to remember that the exclusion of such theory groups from science may leave them little choice if they wish to continue. This has sometimes resulted in public demand for the science community to respond to such publicized claims. Unfortunately, the scientific community's response has sometimes been most imperfect, sometimes even been irresponsible, and seldom has been systematic.

Seeking to institutionalize a responsive mechanism for orthodox science to the more extreme and often publicly headlined claims of paranormal phenomena, a Committee was formed in 1976 to help the lay public better understand the debates surrounding such claims. This Committee of critics and its publication (originally called *The Zetetic* but now the *Skeptical Inquirer*) have generated considerable media attention and may have had some positive effects towards balancing the usually one-sided publicity given claims of the paranormal.

But there remains no balanced, systematic, and institutionalized forum for fully debating issues between the proponents and critics of claims of anomalies, initiation of such a publication, the *Zetetic Scholar*, is now underway, but its success remains problematic. Its own reception by the scientific community may tell us much about the problem. The social and intellectual matrix is complex, but responsibility demands an attempt be made to produce a more rational and efficient means for adjudicating unconventional claims. Science has little to gain by simply labeling its proponents of unconventional ideas as "crackpots," "pathological," or "pseudoscientists" without responsible examination of the evidence. Such authoritarian labeling is sometimes justified by its alleged effect on a gullible public. But as C.S. Peirce noted:

> the general public is no fool in judging of human nature; and
> the general public is decidedly of the opinion that there is such a
> thing as scientific pedantry that swells with complaisance when
> it can sneer at popular observations, not always wisely.

Like any form of deviance within a social group, unconventional ideas in science are seldom positively greeted by those benefiting from conformity. But science's basic dependence upon such innovations for its growth should remind us of a special need for tolerance often absent in the rest of society.

The "essential tension" remains with us, and we need to find better ways to live with it.

[43 footnotes omitted.]

[Euclid Avenue]   23 February 1978

Dear Marcello:

Yours of February 19 to hand. Since we are still arguing about how to use words, it obviously can't be settled by each of us repeating remarks about our language preferences.

First, a subtle point. I opened my last letter by saying that crackpot was a stronger word than crank, and cited dictionary definitions in support. Thereafter I was careful never to call Gauquelin a crackpot—only a crank. Yet repeatedly in your letter you attribute to me the word "crackpot" with reference to G. Let's be fair. Crackpot, as I granted, suggests lunacy. Crank does not. It means, to most scientists, a scientist with an obsessive interest in a notion too wild to be taken seriously. Let me quote you: "I think, you were actually saying that the claim being made by G is so implausible and far out that you want to call it crackpot." Change the word to crank and you are exactly right. Now the only way to settle such a dispute is to query a substantial number of scientists and see which of us is using words in an eccentric way.

Why don't you try this on your sociology colleagues and grad students? Explain to them what G claims—a correlation between positions of planets at the time of birth, and a person's later choice of profession. Then ask them two questions:

1. Is the word crank legitimate, in informal discourse, for G?

2. Should G's claim be called an anomaly in the sense popularized by Kuhn?

From what you tell me, I now no longer think Hynek would call G a crank, but why don't you ask him? If he will write to you, "In my opinion, G is *not* a crank," then I'll send you the ten dollars I bet. I must add, however, that the loss of this bet would not in the least suggest to me that the word is not right for G. It would only cause me to think that Hynek is a much bigger crank than I had suspected.

In your description of G's attitude toward his finding, I think you are being too trusting, of what G personally told you. After all, his discovery is his chief claim to fame, his books are in print all over the world. I have yet to meet a scientist who ever heard of him, or a person who was not in some way into astrology. He is surely aware of why his books are selling so well. I cannot believe he is the honest, simple soul you make him out to be. But—and here is a major point—to the degree that he is like what you say, to that degree he is even more a crank than if he were partly a fraud. As I have said so many times, it is the essence of a crank that he passionately believes in his work. In G's case, it is a belief that he has found a genuine correlation. The correlation is an "hypothesis" because we don't yet know if the raw data is accurate—i.e., if there is a correlation or not.

You close by saying that the "breach of science etiquette" on the part of the editors of *Kronos* "certainly argues strongly for labeling them cranks."

Marcello!—I thought we long ago settled the notion that unethical behavior has a role in the definition of crank. You seem here still to be clinging to the notion that unethical practices somehow make a person a crank. The great cranks of the past have been noted for their morality. Surely one of the greatest of modern geology cranks was George H. Price, but he was a devout Seventh Day Adventist and one of the most moral men I ever encountered. Crankiness is defined by how far out a scientist is on the spectrum of plausibility, and by nothing else.

I can only repeat myself. The correlation G says he has found is so far out that the probability (in the minds of the majority of scientists) is enormous that his raw data is biased. It is the highness of this probability that justifies calling G a crank, and justifies not dignifying his claim with the name of anomaly. But now we are back, to our debate over how scientists use words—a debate that can be settled only by a questionnaire.

Best,
Martin

CC: Ray

[Eastern Michigan University]   28 February 1978
Dear Martin,

In response to your letter of Feb. 23rd, you are correct that I misattributed the term crackpot as your label for Gauquelin. You did indeed refer to him as a crank and did make a distinction earlier between crank and crackpot. I apologize for that mistake. But I think I can show why my mistake is a reasonable one in light of the comments in your Feb. 11th letter. You indicated that crackpot was defined as one believes in "lunatic ideas." A crank you defined as merely one eccentric and "overly enthusiastic about a particular subject or activity." Although you do refer to Gauquelin as a crank when commenting about Gauquelin's claims as a "crazy notion." Now I saw "crazy" related to the term "lunatic" in the definition of crackpot given earlier. However, your position that Gauquelin is merely a crank is what I take you to mean now that you have called my attention to your exact words. I accept that as your consistent position, which I may have misinterpreted.

Obviously, defining a crank as someone "overly enthusiastic about a particular subject or activity" would make that apply to Gauquelin. I will even grant that his view is "eccentric." But this definition also obviously applies to a great many people within orthodox science as well. Anyone obsessed by a particular area (subject or activity) who is seen by others as someone overly enthusiastic or eccentric in his perseverance about it would thus be properly called a crank. But you obviously want to go beyond that from what you say in your letters. At least so it seems to me. It seems to me that you want to dismiss such crank ideas and figures as simply unworthy of scientific discourse. That is the issue between us if I understand you correctly. The issue is not whether or not to call Gauquelin a crank. As you define the term crank, I would accept that label for him as I would accept it to characterize many other scientists (B. F. Skinner on his form of behaviorism which almost everyone in psychology considers an extreme view, Phil Klass on his attempt to explain the best UFO cases as examples of plasmas while most plasma experts would disagree, T. X. Barber on the view that suggestibility explains' acupuncture effects, and many other minority viewpoints held persistently by several of the Committee fellows). Using the term crank this way is not offensive to me at all. The way you seem, to be defining it, using the denotative dictionary definition, it says nothing at all about the truth value of the claim of the cranky, only that he is overly enthusiastic and eccentric relative to his scientific peers. As long as you do not imply anything about the truth value of the claims, the label is certainly acceptable by me as designating Gauquelin (just as it would fit Linus Pauling re his views on vitamin C, etc.). I do not and never have disagreed with you that most scientists probably think Gauquelin is an eccentric and overly enthusiastic. But my impressions from our past conversations and what may be my improper inferences from your letters (I must admit I can not find a direct statement by you on Gauquelin in your letters in which I can truly pin you down on this) is that you also imply connotative aspects, to your definition of a crank. Thus, throughout your letters you speak of Gauquelin and other cranks' positions as absurd. You obviously seem to think such cranks are *wrong* as well as merely eccentric or overly enthusiastic. This is where we seem to differ—if I read you correctly. The question of whether Gauquelin is correct or not is largely irrelevant to your definition of him as a crank; that is, it should be irrelevant, but I don't think it is for you. You seem at times to think he is eccentric and overly enthusiastic because he clings to an absurd (wrong) idea, not merely an idea that others think unlikely to be true. To the degree that you do this, and I must emphasize that I might be wrong in thinking you do this (I hope you will correct me here if I am wrong for it would straighten matters out greatly for, I think, both my and Ray's understanding of your position), you would be guilty of prejudging Gauquelin's claims based on their plausibility rather than the actual empirical evidence involved.

Now let me turn to your second question re whether or not Gauquelin's claim is an anomaly. It appears that I was not somehow clear enough in my last letter on thin point. Neither I nor Gauquelin would call his claim an anomaly. It is at this time a conjecture. Gauquelin alleges the existence of a correlation which *if validated by others* would constitute an anomaly. At this point it is merely an alleged anomaly. Gauquelin is most clear on this in both his conversations with me (which you think I overemphasize) and in his writings. Gauquelin believes that he has discovered an anomaly, but he does not believe that he has established its existence for science. He insists that others need to replicate his work. He has as far as I can determine always and consistently insisted on that. He has consistently found this alleged anomaly present in his own researches. So *of course* he believes it exists. But this conviction is not dogma with him. He says it is not and I simply see no reason to disbelieve him in light of what he has publicly said in his writings. Since he believes others will find the same thing, he persists in seeking to get others to investigate independently despite the manifest implausibility of the claim. This may make him a crank, of course, by your definition (which I admit is the common dictionary one), but it does not make him wrong. The thing that I think is so important in his claims is that he admits that others should find the claim implausible. He does not argue for its plausibility, only its empirical presence. He says "I see it every time I look, and I can not explain it away." He asks others to look for themselves. They say "nonsense" and refuse to take the trouble to look. (I don't object to anyone's not looking—we all have limited time and priorities—but I do object to calling it nonsense without looking.) Since he is suggesting something implausible and is therefore seen as eccentric just by doing that, and is therefore defined by others as "overly enthusiastic," he is called a crank. But I guess what I am suggesting is that he may be eccentric but is not really *overly* enthusiastic. Given the extraordinary character of his observations, and given that these observations have been correctly made (which he thinks is the case), he would quite naturally be enthusiastic to get others to check out his claims and give him credit for his discovery. He really should be called overly enthusiastic only if his evidence does not sufficiently warrant his degree of enthusiasm. Yet it is his evidence that has not been properly examined by his critics who call him overly enthusiastic.

If and only if Gauquelin's claims are confirmed through replication and examination of his methods of data collection, etc., these conjectures of his would constitute an anomaly in the sense Kuhn means it. Neither I nor

Gauquelin claim such corroboration yet exists. So, I would agree with you that Gauquelin's claim should not be called an anomaly at this time. But I think you continue to confuse his conjecture with anomaly. He conjectures that an anomaly is present (that is, he alleges an anomaly); but this is not the same thing as his establishing the existence of the anomaly as truly present.

Re Hynek, I never accepted your bet as such so forget the $10 part. But I think Hynek would answer your question differently depending upon how you phrased it. If you asked Hynek "Do you think Gauquelin is a crank?" Hynek probably would say "Not really." But if you asked him if Gauquelin posited eccentric views, which most scientists would find implausible, I am confident Hynek would say "yes." As with me, the central thing is whether or not you separate the truth value of G's claims from the social reality of his position in comparison to the dominant scientific viewpoint.

I have now reread your letters several times. The more I do so the more I think we really don't disagree very much on all this. You mention that the high probability of his being in error justifies calling him a crank. As mentioned above, I can agree with you on that given your definition of crank as not implying anything about the truth of his position. You also say it justifies not dignifying his claim with the name of anomaly. Here I disagree. The probability of his being wrong has little to do with choosing whether or not to call, his claim an anomaly. The probability you refer to here concerns the prior plausibility of his claim. This probability makes his claim unlikely to be true. But something called an anomaly depends not on its implausibility but upon whether or not it is corroborated or falsified empirically. I agree with you that G's claim is not an anomaly. He presents some data that alleges an anomaly exists. That data needs to be examined critically. If found to be superficially proper, replication is still necessary. Following that we may, if results are positive, claim G's findings constitute an anomaly.

Finally, I think you misunderstood me re the *Kronos* reference to etiquette. The reason I said that Greenberg's actions argued strongly for labeling them cranks is because this act of bad faith on his part indicated to me that there was an essentially irrational and even somewhat paranoid aspect to his actions. When I wrote this I was concentrating on what I thought then you meant by crank, someone not merely tenacious or eccentric but dogmatic and somehow blind to valid criticism. It is not that unethical behavior on Greenberg's part makes him a crank, as you impute to me. I was referring here to what I thought reflected a kind of blindness to my role of trying to actually be helpful to him and de Camp in discussing their differences. Greenberg's action seems to put people into camps of those who are either with him and Velikovsky or against him. There seems to be no middle position possible for Greenberg. In light of the terms as you and I now use them, I should have said that his action strongly argues for calling him a crackpot (since this action was somewhat nuts as far as I am concerned, not merely an act of over-enthusiasm or eccentricity).

Whether or not Gauquelin is the "honest, simple soul" I made him out to be, I am not completely certain. I can only go by his writings and his brief conversation with me. But I have also spoken at great length about G's views with Ron Westrum who knows G very well and speaks French fluently with him while in Paris. Ron shares my impression of G. I will be seeing G in Toronto in a couple of weeks at a conference on astrology held by the U. of Toronto. I should know much more about him after our exchanges there.

***

This afternoon Gordon and Breach called me to offer me the editorship of *Psychoenergetic Systems*. I tentatively accepted. We need to work a lot out including changes in format, editorial board, etc. Things look promising however. Apparently Chris Evans was offered the job, initially showed interest, but then backed out. Any advice you can give me would be most appreciated. I know nothing about Gordon and Breach. I hope I am doing the right thing.

Best,
Marcello

CC: Ray Hyman

[Euclid Avenue]   5 March 1978
Dear Marcello:

When I cited, the brief definitions in my Collegiate Webster, it was merely to indicate that common usage implies that a crackpot is mentally deranged. I didn't intend the Webster reference to express what I think scientists mean when they call an idea "crank" or the person who defends it a crank. You are absolutely correct in assuming that I believe the probable truth value of an idea is essential to such a definition. We are concerned now with the informal language of scientists, as when a physicist speaks of a "perpetual-motion crank," or a mathematician talks about an "angle-trisection crank," or a geologist or astronomer about a "flat-earth crank."

When a scientist calls an idea crank I think he means the following. The idea is so eccentric, so far out on the continuum of plausibility, so unlikely to be confirmed by investigation, that it is not worth taking seriously. By "take seriously" he means serious enough to spend time, energy or funds in trying to evaluate it. The person who advances such an idea is called a crank. If he shows signs of being psychotic, he may be called a crackpot or crazy. However, "crazy" is often used as a euphemism for a crank idea without suggesting that the proponent is literally crazy.

Let's consider some typical examples. Reich claimed in his later years to have discovered a new type of physical energy. It is blue, causes stars to twinkle, dowsing rods to turn, can be trapped in orgone boxes to improve someone's orgasms and cure cancer, can be used to produce rain, and so on. He thought one could even run a motor with it. Now no physicist, to my knowledge, took this claim seriously. It was universally considered "crank." Lots of intelligent people bought the idea, including men with medical degrees, but physicists are busy people, and there are thousands of claims they considered more worth testing.

Paul Edwards, in his article on Reich in the Encyclopedia of Philosophy, writes: "most professional physicists who have heard of orgone theory have dismissed it as nonsense [another synonym for crank]." Then Edwards adds something extremely naive: "In fairness to Reich it should be added that a really unbiased investigation of his physical theories remains to be undertaken."

I once discussed this article with Phil Morrison, who reviewed the encyclopedia in *Scientific American*. We both found Edward's second remark hilarious. Why? Because no serious physicist would consider it worthwhile to test Reich's orgone energy claims. His ideas were too implausible.

Second example. Hubbard claimed that immediately after conception the embryo makes recordings of everything said near the mother, and these "engrams" cause later mental ills. Here again is a hypothesis that can be tested. Did any biophysicist do so? Not one. Why? Because the theory is too crankish, too unlikely to be true. (We are not considering here the interesting question of whether Hubbard is a crank, crackpot, or charlatan.) Moreover, biophysicists made this judgment without bothering to read Dianetics and find out precisely what Hubbard thought the evidence was. Life is just too short to waste time on every crank book that comes along.

It is my contention that the majority of scientists (including sociologists and psychologists), if told, of G's claim, would consider it crank and not worth testing. Indeed, I believe most scientists would regard it as *less* plausible than the claims of either Hubbard or Reich. After all, some distinguished scientists (e.g., Frederick Schuman, professor of political science and author of a widely used text on international relations) bought Hubbard's claim. Ditto for Reich. I know of no comparable converts to G's claim.

I contend that to suggest that G's claim is worthy of serious investigation is to misconceive the nature of the scientific enterprise. It would require a whopping grant to persuade any statistician to go to Paris and check G's raw data. Who will be so foolish as to provide such funds? It would be easier to try to replicate, but this too is costly and time-consuming. Most statisticians, if approached on such a project, would consider it beneath their dignity to engage in such a task. I believe that they are right to feel this way, after being told G's claim, and without any effort to read a single book by G.

If you want to call this "prejudging" a claim by its implausibility, before making further tests, I have no objection. But surely you must know that science does this constantly, and rightly so. If scientists didn't react this way to highly implausible claims, they would never get any work done. I made this point so often in my old *Fads and Fallacies* that I am almost embarrassed to bring it up again, but it is fundamental to our disagreement. Thousands of implausible claims, less implausible than G's, are being made every year by cranks. The science journals are deluged with such papers, many of which they actually print. Hundreds of other such claims are trumpeted by various cults, and argued in best-selling books. It is absolutely necessary for the health of science that the more outrageous claims be "prejudged." Scientists simply do not have the time, inclination, or funds for investigating such claims.

When you write that G's claim should be "corroborated or falsified empirically," you seem to be suggesting that some reputable statistician drop what he is doing and to work on G's claim to see if he can confirm it or not. Here is the nub of our argument. I believe that the claim is too outrageous to justify trying to test it. I am not just saying that this is my opinion. I am saying that this is the opinion of the scientific community. Of course G wants someone to test his claim. Perhaps here and there a statistician will be sufficiently amused by it, or curious, to try to test it. From what I know about cranks, I would predict that no amount of negative results would alter G's opinion in the least. So far as I know, you are the only person outside of G, who can call himself a scientist, who thinks G's claim is worthy of any serious attempt to evaluate it.

Let me cite a personal instance to support my claim that prejudging a crank theory is essential to the health of science. About 15 years ago I got in the mail a bulky manuscript from a man who said he had been working on relativity theory for 20 years and found that it was all wrong. I had just had published, a popular book on relativity. He wanted me to read his mss and help him get it published. He added that he had previously sent it to Gamow. Gamow had the gall to return it *unread*, with a curt note saying he had no time to read it and please don't write him about it. The man was distressed by this prejudgment of what he considered, a serious challenge to physics. I can't recall his profession, but I think he was probably an engineer with some background in science. I repeated Gamow's prejudgment. I returned the mss with a curt note saying I had no time to read it.

Now this is how science operates, indeed, how it *must* operate, with respect to far-out claims. When you insist that G's claims are sufficiently plausible to deserve serious testing, to find out if it's an anomaly or not, you are writing in a manner characteristic of Paul Edwards, and not of the science community, I find this sad because it blunts your effectiveness as a writer on such questions, and damages your reputation.

As I see it, we have two major differences: one over how a scientist should use the word "crank," and another over what action should be taken toward claims such as G's. Once again, I propose an empirical test to see which of us represents a majority point of view. Query 100 scientists. First outline G's precise claim. Then ask two simple questions:

1. Do you think this claim deserves the term "crank"?

2. Do you think it has a sufficiently high probability of being confirmed to justify funding a serious scientific effort to test it?

Perhaps you will agree with my prediction of a yes/no answer, but wish to argue that most scientists are "wrong" to have such an attitude, that it betrays an "establishment" opposition to new ideas that should be opposed by right-thinking scientists. If this is the case, then our argument is no longer over what scientists believe, and how they use language in informal discourse, but shifts to a deeper level in which we debate the most effective procedures by which science can progress.

As for Psychoenergetic Systems I too know nothing of Gordon and Breech, nor why Chris Evans turned down the job. Isn't this the magazine Stanley Krippner once edited? If G and B will give you a free hand, it would be a good way to finance the magazine you wish to produce. My only advice at this point is: Don't weaken your standing in the science community by suggesting in the magazine that claims such as G's deserve more consideration by scientists than they have given to the claims of Reich and Hubbard.

Best,
Martin

CC: Ray Hyman

PS: An afterthought that also focuses our argument nicely. You are fond of saying, and I have picked up the phrase from you and repeated it, that an extraordinary claim requires extraordinary evidence to be taken seriously by scientists—that is, to be regarded as a genuine anomaly in Kuhn's sense.

Where is the extraordinary evidence for G's extraordinary—fantastically extraordinary—claim? He presents nothing but one man's analysis of one's man accumulation of French statistics. This is extraordinarily *weak* evidence. The claim is much more extraordinary than the remote viewing claims of P and T, and in my opinion, supported by much *less* plausible evidence. Claims of statistical correlations, to support wild theories, are a dime a dozen. Witness the vast literature of books correlating economic depressions with sun-spot activity. I can dig up hundreds of such instances, the topics of large books now forgotten. G's claim is certainly more extraordinary than, the claim of statistical correlations put forth by the biorhythm people, to give another example. Obviously the only reason G is getting any attention at all, and selling books, is the widespread public interest in astrology. That his claim doesn't support traditional astrology is beside the point. In fact, I find your continued interest in G to be almost as extraordinary as G's claim.

[Zetetic Scholar]   8 March 1978
Dear Ray,

Here's another installment on the Gardner-Truzzi pseudoscience exchange. I think the issues may be getting clearer between us.

I still find it hard to believe that Kurtz is really going to debate Rhine at the Smithsonian. Can you confirm this? I can see his asking you or someone else qualified from the Committee to do so, but it takes a lot of gall for him to do it himself (I am presuming Bergson's column information is right since his correspondent-informant is in Buffalo).

I still have heard nothing from Randi. Maybe his silence is really for the best.

Your suggestions and comments re *Psychoenergetic Systems* would be appreciated.

Best,
Marcello

[Euclid Avenue]   8 March 1978
Dear Marcello:

I find myself agreeing with about everything in your AAAS speech, but reading through it I perceive once again the heart of our differences. On general questions you seem to me on firm ground. But when it comes to exercising judgment in distinguishing "unconventional science" (of the sort meant by Kuhn, Peirce, and all the others who have written on the topic) from the crank claims that are too implausible to be placed in this category, we differ sharply.

Gauquelin focuses the problem sharply for us. You seem to regard his conjectured anomaly as roughly on, the same level as those historic examples of "unconventional science" of the past that proved to be true: stones falling from the sky, germ theory of disease, and so on. Or to take current anomalies: black holes, acupuncture as an anesthetic, the claims of Charles Tart, the correlations Zeeman is claiming for catastrophe theory, and so on. I find G's claim very close to Freud's example (which I learned from Ray) of the center of the earth being made of jello. I am not alone in this. G's claim is simply too far out on the continuum of plausibility to be treated like the others. You have to exercise *some* kind of judgment on these matters in order not to dissipate energies, on irrelevancies.

I note for the first time that G is listed on the panel of those you describe in your prospectus as "distinguished consulting editors." Gauquelin distinguished? Marcello—I read the words but I can't believe them! That's no way to run a magazine of the type, you envisage!

> Best,
> Martin

CC: Ray, Persi

[Zetetic Scholar]   8 March 1978
Dear Martin,

Just received your latest (March 5) letter and am responding immediately here at the office (where I don't have our past correspondence handy so am herein referring exclusively to this latest letter).

By now you have, probably read my AAAS remarks and this may already have clarified some points taking up our differences in your letter. You will, I hope, have noted that your letter here is almost an exact rephrasing of Polanyi's position which I sent you earlier and which I quote in my

AAAS remarks. (Oddly enough, your comment on Polanyi was a negative one in a past letter even though I see you and him in basic agreement on the points I am raising here.)

You seem to me to consistently misunderstand my main point re Gauquelin. I do *not* argue that Gauquelin has proven his claim. I do *not* argue that the science community should leap into careful analysis of his claims. I have said over and over again in my past letters that I think it perfectly appropriate that you and others should be allowed to simply *ignore* Gauquelin's claims if you think they are too implausible to justify your time being spent on them. (Of course, Gauquelin would likely disagree with me on this point). Thus, I think it perfectly proper that both you and Gamow returned manuscripts alleging revolutionary scientific evidence as simply not worth your time. *Every* scientist must constantly make judgments about the probability of pay-offs in various areas in order to make judgments about where to put in his own efforts and resources. That is perfectly proper. I do *not* consider returning the manuscripts unread to such esoteric thinkers as demonstrating a prejudgment of the truth of their claims. It is a judgment of the probability of any successful results emerging from careful reading and analysis of their alleged case for revolutionizing our views. This is an important distinction that must be made, and I fear that you may not fully appreciate it yourself. The fact that you failed to make this distinction yourself—the fact that you seem willing to call it a prejudgment of the evidence as the author accused you of—suggests that in a sense you may actually be doing prejudging. I get crazy manuscripts for comment myself just as you do. I normally return these with a courteous comment indicating that my time is limited and though their idea may have merit, I simply must assign consideration of that idea too low a priority for my limited time. You, on the other, hand, would probably not include a phrase like "though your idea may have merit" because I think you would prefer to stress in your mind the likelihood that it does not. I, on the other hand, would not want to imply that the idea had a strong likelihood of being correct, but I would not want to say anything that would indicate to the author my prejudgment of his conclusions. This difference also comes out in your reference to giving a "curt" reply. To me (and to my Webster's) the term "curt" means "brief to the point of rudeness." I simply see no good reason for any rudeness in this case at all. Now (as Nixon would say) let me be perfectly clear. I am not against rudeness per se. If I took the trouble to read the manuscript and thought the author was nuts, I would not be opposed to telling him so (though I probably would not go out of my way to be insulting unless he offended me somehow). My point is that prior to consideration of the evidence being offered, I really can make no serious judgment of the claims at all. I can only make the judgment as to whether or not I want to spend my time examining the claims. This is quite separate. The history of science, is full of people making both Type I and Type II errors (as defined in my AAAS commentary) and I think both are important types of errors I wish to avoid if possible.

In regard to your two questions (1) Do you think the claim (G's) deserves the term crank? and (2) Do you think it has a sufficiently high probability of being confirmed to justify funding a serious scientific effort to test it? you misguess my answer. I agree with you that the scientific community would answer "yes" and "no" respectively. Our clarification of your use of the term "crank" in a way that would eliminate its pejorative meaning, whereby we were commenting mainly on the character of the claimant rather than on the truth value of his claim, would make me also respond "yes" to #1. But I would also give a qualified "no" response to question #2. You seem to imagine that I would recommend NSF funding for Gauquelin's work. I would not. I think it too unlikely to be fruitful for me to assign a high priority to that kind of study (though I would say the same of much stuff that does get funded by NSF, especially in the social sciences) But I would recommend that studies like Gauquelin's should have a proper right to *apply* for NSF funds; and be ranked for priority just like other scientific studies. You and others come dangerously close to wanting to exclude him from science entirely. That is where I think we disagree. He must be considered seriously at least to the point of allowing him to compete with other claimants of a more conventional kind. I don't expect him to get funding from normal science channels, and I would probably rank him low in priority for such funds myself; but he has a perfect right to ask to be ranked with the others. He is still a member of the scientific community, *not* some sort of *anti*-science advocate as some would portray him. But these negative comments mainly apply to public funding of his work. I feel my ranking of priorities is relevant to government spending. But if Gauquelin can find some individuals willing to fund his strange claims, I see absolutely nothing wrong with that. I think it enough that the general scientific community indicate that his views are deviant from the general views held by most scientists, that he is a "maverick" investigator that many think is unduly wedded to an implausible claim. If this discourages someone from giving him funding, so be it. That is an honest statement from the general science community. But stating that he is a crank (in the pejorative sense of suggesting his ideas are wrong as opposed to merely implausible), or going further and calling him a "crackpot" or "nut" is going further than is reasonable since it may block inquiry on ultimately non-rational (not irrational) grounds.

I don't want to discourage anyone from trying to conduct proper experiments on the empirical world. Let me take a parallel case that might clarify. Let's say a Catholic priest wants to experimentally investigate transubstantiation. I would think it silly to do so since I don't believe in transubstantiation. If NSF funded the priest I would be among those writing to Senator Proxmire to complain. I would think it a waste of time. But if an eccentric millionaire gave money for such investigations, I would not disapprove. If the priest claimed positive results, I would like to see some follow up through further private funding. Again, I would not want public money used. But I should remind you that the priest could get negative results (just as investigating G's claims could get negative results) and this would corroborate my view and be considered an important falsification of transubstantiation (or G's claim). Negative knowledge is valuable too. (In fact, I am amazed that so little credit is given Gauquelin for his significant debunking efforts on classical astrology, which he alone has acted, to almost exclusively falsify. His own positive claims have next to nothing to do with astrology—even Paul Kurtz used to insist that Gauquelin was not an astrologer and has published Gauquelin's book.) In the final analysis, I think science and its method is important for what it allows us to learn about the empirical world. I consider our conceptual scaffoldings and theories essentially secondary in some ways to the spirit that science expresses in letting nature itself tell us about nature. Ken Frazier once wrote that "science is the opposite of dogma." I largely agree with this. Our preconceptions (including those derived from our scientific expectations) must play second fiddle to the way the universe really operates. Our notions of plausibility change. I understand why science needs to be conservative. I am in favor of that. But we must only be as conservative as we have to be to successfully operate. There is a point of diminishing returns. Just because a crazy-sounding idea is likely to be crazy, there is no need to prejudge it as being crazy. We can simply ignore it or give it low priority for examination. To do otherwise soon leads us from an essentially conservative position to what amounts to a reactionary one (like Dennis Rawlins') that sounds more like religious dogma than open minded (but still conservative) science. I am not asking you to open the doors to floodgates of crazy ideas. I am simply asking you not to gratuitously shut open doors. That, it seems, is the difference between us—if it is really is a difference between us, for I really think that in the final analysis we may be in substantial agreement.

Thanks for the comments re Gordon and Breach. I will be careful in dealing with them. I see in *The Linking Ring* (Bergson's column) that Paul Kurtz will be debating Dr. Rhine at the Smithsonian in April. If so, couldn't Paul find someone better from the Committee to take on Rhine? I would expect Rhine to make mincemeat of Kurtz. I look forward to reading the transcript if available.

Cheers,
Marcello

CC: Ray Hyman

P.S. I see I forgot to comment on your P.S. comments. I certainly agree that G's claims are extraordinary and require extraordinary proof. However, I do not agree with you about his claim being more extraordinary than remote viewing, and I do not think his evidence offered is less extraordinary than Targ and Puthoff's. Again, I think you misunderstand what G is claiming. He claims merely to have found a consistent correlation. He does not infer from this alleged data any form of causality such as in astrology. He does not claim that the planets cause occupational choices. He simply claims a correlation exists that needs explaining. He has seriously looked for alternative variables that link these two phenomena and has tried to control for such. But he claims the correlation persists despite whatever controls he has been able to come up with. He still thinks there may be some common factor linking the two things. He has also suggested that the causality may be the reverse of astrology's claimed direction; that is, there may be something genetically present in the fetus that gets triggered somehow by cosmic events and somehow signals that it is time for the fetus to emerge. He admits this is not very plausible either. He claims a consistently observed phenomenon. He would like to have others confirm it; but in the meantime will further investigate it himself in the hopes of finding the causes involved. If his data is correct, he may have found something important. Maybe not. Compare this to Targ and Puthoff. Their claim is far more extraordinary and their evidence far weaker. The claim is that the human mind can transcend space and time in ways completely alien to modern physics, physiology and psychology. When the subject remotely identifies a target, they consider this a direct causal relationship via clairvoyance/precognition. If they simply pointed to a curious correlation between a subject's guess and a target, said they didn't know what produced this remarkably consistent coincidence, and were willing to consider serious alternative explanations (alternative variables), their position would be more like Gauquelin's. But their claim is far more broad and extraordinary. They insist on a process that we would label magical. Gauquelin does not do this. He would accept an explanation that was quite natural (e.g., athletes are mesomorphs, mesomorphs tend to come from mesomorph mothers, mesomorph mothers tend to be more erotically inclined when weather is colder so have more sex then, and this tends to produce mesomorph babies more frequently during months when Mars is more commonly in a certain position). As for evidence offered, Targ and Puthoff offer some dubious experimental data which really can not be checked (as can G's data) and the data they offer is not the evidence itself; the evidence is certain inferences they make from the data. Gauquelin's data is itself his evidence. Both Gauquelin and Targ and Puthoff cite non-chance patterns. But the non-chance pattern itself is Gauquelin's claimed anomaly. For Targ and Puthoff the inference they make as to the causes of this non-chance pattern (clairvoyance/precognition) is their anomaly, not just the non-chance pattern itself. So, you see, G's evidence is really much stronger than Targ and Puthoff's to my mind. But most important, his claim is itself not so extraordinary since I can concoct numerous possible explanations which might explain his data should his data be confirmed. And these explanations need not be very implausible at all.

[Zetetic Scholar]   13 March 1978
Dear Martin,

First, I note that you sent a copy of the last letter from you (8 March) to Persi as well as to Ray. I wonder if it will make much sense to Persi without the earlier letters. Or have you sent him copies earlier of our exchanges? If he wants to enter in, I can do so if you haven't.

Glad you liked the AAAS comments though I am a bit surprised since I make the same points in it I made in our correspondence earlier. You say in your latest note that you think the difference between us concerns specifics rather than the general points. Maybe, but really think it goes deeper. Our consistent difference has been over the distinction you keep making between unconventional science and crank claims. My central point is that the only difference that can reasonably be made between unconventional claims and crank claims must concern *not* the substantive issues claimed but the methodological differences in approach. I have consistently tried to argue with you that as far as I know, and far as you have thus demonstrated, the only thing about Gauquelin's claims that seems to make you want to consider them "crank" rather than "unconventional" is their implausibility to you. I insist that plausibility should have little to do with defining a claim as scientific. Plausibility should determine the priority we are willing to give to some claim in determining whether or not we want to bother to investigate it. What makes something scientific as a claim is the approach one takes towards its verification (and presuming the claim is an empirical one). I don't understand why you cannot simply agree with me that for most scientists Gauquelin's claims probably should have a low priority for investigation-funding-etc. If I conjectured that a mermaid can be found at the top of a mountain based on a picture I claim to have taken last summer, and if I state that I don't expect anyone to believe me but I hope someone will go look for themselves, that is an empirical claim amenable to investigation and basically a scientific claim. No one would blame you for not going to look for my mermaid, stating you think the likelihood of validation of my claim is extremely slight, etc. But you should not mock someone else who chooses to go look for himself. I see the guy who goes looking, assuming he will honestly report his finding or non-finding, as a scientist playing a long shot—a damned long shot. But I would never decry anyone going to empirically test out any claim that can be tested. I wouldn't fund him, but I would not put him down for looking, for that, to me, is basically what science is really all about.

Now let's look again at Gauquelin. Briefly since I've done so at great length with our past exchanges. I do *not* as you suggest think that Gauquelin's claims are on the same level as the great historical example as meteors. But the reason they are not on the same level to me is because I really think G's claims are unlikely to reveal anything that important. This, however, is a prejudgment based on my biases—and yours. In an analytic sense, they are, on the same level in that if you were a contemporary of Lavoisier, you would have argued with all the sensible scientists of the period that extraterrestrial origins for meteors was as implausible as you know seem to think G's claims are.

If I read your letter correctly, you seem to place acupuncture, black holes, and Tart's psi claims in the category of "unconventional science" while placing G's claims in the crank category. Yet I, *Even though I seriously doubt Gauquelin's claimed findings*, consider Gauquelin's claimed finding far less implausible than Tart's psi claims. All G claims is a correlation that needs explanation without offering any explanation. *Even if G's claim was completely true it would not seriously alter our view of the universe because Gauquelin does not claim a cause and effect relationship directly between the planets and occupations; he claims only a correlation.* You just don't seem to want to acknowledge this in our exchange. You seem to persist in equating his views with the causality views of the astronomers whom he has consistently and openly rejected. I can hypothesize dozens of possible causal chains that might account for G's findings even if they are validated. In addition, as I have continuously pointed out, G's claims can independently be checked by skeptics and non-believers. There are no Catch 22, 23, 24, etc. as in parapsychology. Extraordinary claims require extraordinary proof, but G's claims (as I argued in my last letter) are simply not nearly so extraordinary as are Tart's.

The fact that you and I seem unable to agree about what is more or less plausible should clearly demonstrate why plausibility alone can not differentiate well between what you want to call unconventional science and crank ideas. Obviously, some science is more unconventional than other science, but if both cases are amenable to empirical test, falsification, etc., that is what is important and makes both of them science. Why do you insist on a sharp line between the more and less conventional forms? We would both agree that low priority should be given to the less conventional claims. Why write those unconventional inquirers off the science map? In fact, don't you force such protoscientists into the welcoming arms of the true anti- and non-scientists and occultists when you thus reject their honest inquiries? Isn't modern science committed enough to its method as final arbiter to tolerate such mavericks? Can't you see that your wanting to throw them out of the science community (motives aside, that is the result of such categories as you use) is like branding them heretics, something I think you find objectionable when done in the past by religious orthodoxies? Why stifle any kind of proper inquiry by others. You have yet to demonstrate anything improper in the attitude of Gauquelin. He has taken criticism in the proper spirit. His attitude has not been vituperative or unreasonable. Compare his reaction to the Zelen-Kurtz analysis rejecting his claims to, say, that of Velikovsky's advocates.

Finally, you seem amazed at my referring to Gauquelin as "distinguished." Of course he is. He has a good reputation as a psychologist in France quite aside from his astrobiological claims. And even his astrobiological claims in the several books he has written have received generally favorable reviews. I remind you again that Kurtz is publishing, his *Dreams And Illusions Of Astrology* with Prometheus Press, and this was initiated long before the current Gauquelin-Zelen controversy. If you take the term "distinguished" as simply meaning "conspicuous" the term certainly applies. If you mean "noted or eminent" I think it does, too. The whole point of my editorial board is that I want the best known critics and advocates on it. Surely you would agree that Gauquelin is the best of the even remotely pro-astrological crop. Even as a critical writer on astrology, Gauquelin leads the pack and is far superior to Jerome or Bok. By any reasonable standards, Gauquelin is an expert witness. From what I have read of his work, Gauquelin's knowledge of the would-be scientific writing on astrology is second to none. If you object to the use of "distinguished" because you think that term has strong connotations of respectability, the term could not be used about most of those on the list in the sense that few of us are past-presidents of our scientific associations or Nobel Prize winners.

Best,
Marcello Truzzi

CC: Ray, Persi

[Euclid Avenue]   13 March 1978
Dear Marcello:

Re: yours of March 8. I am in full agreement with all your *general*
remarks. I have never assumed that you think G has proved his claim.
How could you suppose I had such a preposterous view, especially since I
repeatedly stress that no scientist ever "proves" anything. I have always
assumed that you give G's claim a very low probability of being true.

As always, our disagreement is over how to evaluate a specific claim,
in this case G's. You write in your PS that you consider his claim to
be "stronger" (i.e., more likely to be confirmed) than the remote viewing claims. To me this passeth all understanding. Claims of clairvoyance,
and experimental results confirming clairvoyance, involve hundreds of researchers over a period of many decades. And there are many quite respectable theories to account for clairvoyance. Hundreds of distinguished
scientists (including Einstein), who are not convinced of clairvoyance, nev-
ertheless consider it a highly respectable and not at all extraordinary claim.
As before, our difference is one of language. I know of no way to settle the
matter except to ask other scientists: Which do you consider the most
"extraordinary"—the claim that the mind can receive impressions of remote objects, or the claim that there is a correlation between professions
and the positions of planets in the sky at the moment of birth? Surely
you must realize that you are virtually alone in considering the clairvoyant
claim to be the most extraordinary.

And this difference in our evaluations has practical consequences. It
leads you to accept G as a "distinguished, consulting editor." I am arguing
that he is not distinguished, that his claim deserves to be called "crank,"
and that evidence for his claim is close to nonexistent.

You write, "His claim in itself is not extraordinary since I can concoct
numerous possible explanations which might explain his data *should his
data be confirmed*." What on earth can you have in mind? It is my view
(I assume yours also) that the simplest explanation of G's result is that his
raw data is biased—i.e., that there *is* no correlation. Publishing statistics
doesn't establish anything unless it is carefully confirmed by responsible
others. Will you outline for me a "plausible" theory to account for G's
conjecture—should it be confirmed—that is less extraordinary than, say,
any of the many theories put forth to explain psi?

As for G's disconfirmations of conventional astrology, I consider this
irrelevant to our argument. Scientology falsifies Freudian theories of neu-
roses, the evidence published by Price for his flood theory of fossils falsifies
other creationist theories. Forwald, in his book defending his gravity theory

of psi, reports on all kinds of tests that he himself made which falsify other theories of psi. E. H. Walker reports on similar falsifications of theories other than his own. I am totally unimpressed by negative astrology data presented by a man who, working solo (in contrast to the tens of thousands of past astrologers), presents positive data for a correlation fully as extraordinary as the correlations claimed by traditional astrology.

Let me try to sum up. We both agree on the generalities. *Of course* doors should be kept open to any claim, however absurd. *Of course* G has every right to *apply* for grants. To assume I would think otherwise is to caricature my position. We also both agree that responsible scientists have to make value judgments about how to spend time and money. On this point I do indeed agree with Polanyi. I made the point very clearly in my old *Fads and Fallacies*—it is a point that is obvious and hardly worth making. Where then do we differ? We differ over whether a man like G should be called "distinguished" and made the consulting editor of a magazine that aims at a serious consideration of the claims of unorthodox science. The point is that there are thousands of unorthodox science claims floating around, many of them published in establishment journals, that are far more likely to be confirmed, that are much less crankish, and deserve far more consideration in your magazine. It is over how to evaluate a claim, where to put it on the relevant continuums, that is at the heart of our differences.

Best,
Martin

CC: Ray

[Eastern Michigan University]   19 March 1978
Dear Martin,

I get the impression that: your last letter to me was overlapped with the last one I sent you, so I am going to wait a week or so to make sure you have all my last comments.

Let me comment on a related matter that has emerged. A basic problem (not necessarily between us but re the whole paranormal debate) concerns what constitutes something's being extraordinary and the problem of degree of extraordinariness. This is intertwined with the second question (which we may have been confusing with it), of plausibility/implausibility. You seem to be using extraordinary in the sense of "strange" or "bizarre" which is only one kind of extraordinary. Thus, when you speak of most people probably thinking astrology is more extraordinary than psi, you are probably correct in that most people think astrology is more "far out", than psi. But my meaning of extraordinary is meant specifically in relation to existing theory. In an earlier paper of mine, I spoke of ordinary versus, theoretical anomalies. The problem here is related, but I think I earlier oversimplified things. Thinking about, the matter now, I believe one can start with distinguishing observations from explanations. Obviously, the observed X can be ordinary or extraordinary, as often seen (expected) or rarely or never seen (unexpected). If I saw a green hairy creature resembling a man, that would be an extraordinary observation. It might be explained by an extraordinary theory (a Martian) or an ordinary explanation (a movie creature made up for the occasion). I can also find extraordinary explanations for ordinary observations, e.g., my normal cold is the result of a hex by Q. In the case of Gauquelin's alleged observation (a correlation, in data gathered), you seem, to me, to confuse the extraordinariness of the observation with the extraordinariness of the explanation (astrology) that some (but not Gauquelin) would give for it. In the case of psi claims, the explanation for the observations is usually what is extraordinary. In general, I think that the extraordinariness of observations is not, or should not be, a real problem. Replication of the observation is usually easy enough to check for observation error. Thus, if you are right about Gauquelin's simply making a mistake in his correlation computations or data gathering, that can be easily checked through replication, or examining his data and computations. In a very fundamental sense, however, extraordinary observations should be more acceptable than extraordinary explanations; for our theories (explanations) are supposed to fit our observations, not vice versa. It seems to me that we start with a world that is by its character unexpected. We create maps, both observational and explanatory, that make things more "expected." Even in our observation (apart from our theories and explanations), we come to expect various things to empirically occur. Some of this stuff is quite direct (people have two arms, one head, etc.) but some of it is more inferential (e.g., statistical rules). My point is that since Gauquelin is speaking of rather direct observations, for the most part, confirmation of claim (that a statistical relation is present that is not due to chance), if some of these claimed observations are confirmed, it really does not have very serious consequences. We might be able to account for such unexpected (extraordinary) correlations through rather ordinary third variables that account for the co-presences. In psi claims, however, what is extraordinary is the explanation offered to account for the ordinary event (that two people should think the same thing at the same time or one before the other). In the psi claim, the extraordinary explanation is based on the idea that all ordinary explanations have been eliminated through controls. That is not the case for Gauquelin who does not claim any solid explanations at all. If you told Gauquelin that he failed to control for variable X that might link the planetary cycle with the occupational choice pattern, Gauquelin would welcome that mechanism. For Gauquelin, an ordinary explanation would, be acceptable for his extraordinary observation. The reason all this is so important to me is that my view of science (which is essentially pretty positivistic and not relativistic) insists that I take the actual state of the empirical world as my "given" (as much as I can ascertain that) no matter, how extraordinary the facts may seem to me. Of course, if the facts are extraordinary in the sense of running counter to either my past observations or my explanations, I should be skeptical about such claimed facts. But if they are really present, I am obligated to give them priority over my theory that says they/are not there. I can ignore them until I get a theory that fits them, but I must be very careful not to ever deny them.

All of this, I think, is very relevant to the Gauquelin case; for unlike the parapsychologists, Gauquelin really is claiming to see the, equivalent of a "white crow." The number of alleged replications of psi experiments is simply not comparable to the sturdiness of Gauquelin's claim. First because the psi claims are offers of extraordinary explanations rather than extraordinary observations. Thus you can say a clear yes or no to Gauquelin's claim by checking his work or doing it over independently *once*. Secondly, the psi claimants do not guarantee a positive result, for any replication. For Gauquelin's claim to be true, you would have to get a positive result with no real excuses (Catch 23, etc.). All of these comments of mine do not, of course, argue for any greater likelihood that G is correct. I am merely arguing that his claim is not as extraordinary as Tart's claims. This question of extraordinariness is separate from the issue of plausibility which seems to concern the subjective likelihood you would attribute to the claim but seems to extend further than mere probability to include general reasonableness.

I don't know if the above really adds anything new to our exchange, but I am inclined to think that we really need to unpack such terms as "extraordinary" and "implausible" before we can really come up with guidelines for the specific case studies we seem to disagree upon. As far as I know, this has not been done in the philosophy of science literature but needs doing. Corresponding with you on all this has been very helpful to me in sorting out many of my own thoughts. So, even if we don't end in full agreement, I have appreciated your comments and the stimulation they have provided me.

***

I spoke with Ray earlier today and he informs me that the whole Rawlins matter re the U. of Toronto may be far more complex than I had thought.

On the one hand Dennis' actions may indeed have led to the cancellation of the conference. On the other hand, Dennis was apparently originally invited to the conference and then cancelled, and this also may have happened to Abell. I might add that I too was initially invited and then cancelled as a presenter because they got too many positive responses to preliminary invitations to participate. I later rejoined merely as a panel member when I told them I planned to be there anyway. I suspect the same thing happened to Dennis and possibly Abell being cancelled. A critical question seems to be whether or not anyone got invited after the cancellations. But whether or not there was such a screw-up, the conference, directors had a right to invite anyone they wanted without attempted censorship by Rawlins. As it happens, the conference was indeed going to be balanced since Jerome and I and a couple of astronomers were going to be participating. In any case, my concern is not with Rawlins getting it cancelled, so much as whether or not he was getting it cancelled in the name of the Committee. His letter to the President of the U. of Toronto was on Committee stationery. Many people are now angry about the cancellation and believe the Committee caused the cancellation. If Rawlins was indeed invited and cancelled and if this was in any sense a breach of contract, he could merely have sued. I intend to write Abell to get the facts from that end and will write the President of the U. of Toronto for his side of the matter.

Cheers,
Marcello

CC: Ray Hyman, Persi Diaconis

[Euclid Avenue]   21 March 1978

Dear Marcello:

I haven't been sending copies to Persi. It's just that it occurred to me that since he is our statistician, he might be interested from now on—though let's hope that the "now on" will soon end!

True, G has no cause-and-effect theory. It is the claim of a correlation that is so extraordinary. Bizarre correlations of this sort are constantly being claimed by cranks and quickly forgotten. G's claim is in the news for only one reason—the great popular revival of astrology among the scientific illiterates. Let's try to divorce his claim from this strange social phenomenon, and compare it with something similar.

Back in the 40's a man named J. H. Kenneth began to publish statistical data that supported his claimed correlation between the gestation periods of hundreds of different kinds of animals and multiples of pi. He had an article in *Nature* (1940) and later did a book about it that ran into at least three editions.

How did biologists, world over, react to this extraordinary claim? Total disinterest. The claimed correlation was so far out on the plausibility spectrum, so unsupported by a reasonable theory, that nobody was the least interested, in trying to find out how K's raw data got biased. His work was dismissed as nonsense, and in my opinion rightly so. Hundreds of similar examples could be cited. Since his extraordinary claim (Note: of a correlation in data only) was not supported by extraordinary evidence, the science community reacted rationally in not wasting time on it.

I maintain that the science community, for exactly the same reason, has no interest in G's correlation claim. I think it was bad judgment for Paul Kurtz to publish G's book, just as I think it was bad judgment for Nature to publish Kenneth's paper. We seem to agree on the low probability of G's claim being accurate (i.e., actually supported by the birth records), but we differ in how one should treat G. I believe it was poor judgment for you to put him on your of advisors.

You write that even if G's claim turned out to be true, it would not alter our view of the universe. Do you really mean this? If so, I am so astonished that I hardly know how to reply. If there is in fact a correlation between one's choice of a profession, and the positions of planets in the sky at the time of one's birth, that is a fact for more extraordinary than Rhine's claim that there is a correlation between a willed digit from 1 through 6, and the falls of dice. Rhine, too, has no theory that explains PK. He claims only the correlation. I find his claim far less unreasonable that G's. As you know, I think it has low probability of being true, but at least his correlation claim is supported by hundreds of experiments published all over the world by scores of parapsychologists. G is virtually a loner. Yes, Eysenck says his statistics are sound. But of course Eysenck has not checked the data against the actual birth records, nor has anyone else.

You still write as if astrology, or at least G's curious variant, has something in common with Kuhn's anomalies. I maintain that astrology is not a challenge to science, but the survival of an ancient superstition, no more deserving of respect than palmistry (which has had as long a history). That astrology is having a revival today no more raises its status as a challenge to science than the revival of the flood theory of fossils raises its status as a challenge to geology. Have you thought about seeking a flood-theorist as an advisor? I actually believe that there are more people who are creationists around the US than there are people who take astrology seriously, and what is more, they actually have a testable theory with considerable "evidence."

By all means let the fundamentalists hold a conference to debate "Evolution: Science or Pseudoscience," but let them hold it at Wheaton Colleges not at Harvard. The very notion of a university as distinguished as Toronto allowing itself to be the site of a debate on "Astrology: Science or Pseudo-science," strikes me as hilarious. That you would bother to speak at such a conference impresses me as even funnier. But quite aside from my personal reaction (which I believe to be typical of the science community, and especially your colleagues in sociology), I believe that your manner of treating G with the, kind of respect that one would use in treating, say, Wigner with his unorthodox views about quantum mechanics, is damaging your reputation except among the promoters of pseudoscience who naturally welcome any academic who will bother to listen to them.

Best,
Martin

CC: Ray, Persi

[Euclid Avenue]   24 April 1978
Dear Marcello

Here's all I have of Beauregard manuscripts. The paper you mention was the one I sent to you, a year or so back, saying you might be able to use it, and you wrote me that you had passed it on to Ken Frazier. You may keep the enclosed.

I enclose a copy of my review of the first Webb book. You're welcome to print it if you like. I had only a set of badly printed galleys (some so faint that I couldn't read the print), and without chapter headings, index or notes. I mention this because I really ought to look over the book itself to make sure I didn't misjudge something. But if you've read the book, perhaps you can let me know if you think I made any errors. If Webb is no longer living in London, you should correct that line. And I've not seen his second book. All I know about it is from a brief review I clipped from *Fate*.

I'm reviewing a spate of books about catastrophe theory for the *NY Review of Books*. (I don't think much of CT, even though *Scientific American* published Zeeman's article on it.)

Best,
Martin

[Cornell University, Laboratory for Planetary Studies]   27 April 1978
Dear Martin

I wonder if you could give me the benefit of your opinion on the division between the *Zetetic Scholar* and the journal being edited by Kendrick Frazier.

With best wishes.

                                    Cordially,
                                    Carl Sagan

[Euclid Avenue]   2 May 1978
Dear Carl

The editorial board members of our magazine, and the editor Ken Frazier, probably all have slightly different ideas about what the magazine should be like. Best I can do is give you my own perception of what I think are the main differences between our journal and the one Marcello intends to publish.

As you know, Marcello resigned from the Committee. From the start there had been a personality clash between Paul Kurtz and Marcello, as well as disagreements ever the kind of magazine desired. Marcello wanted a scholarly magazine, without pictures or cartoons, that would try to offer a dignified dialog between opposing sides, or at least a dignified opposition to the occult side. He did not want a magazine that would "debunk." Other members of the Committee, certainly the majority, wanted a more popular format, and one that openly presented itself as sceptical, and in strong opposition to the rising tide of irrationalism we saw sweeping the country. We saw our main task not as one of establishing dialog between, say, Velikovsky and his critics, but a crusading magazine, a debunking magazine if you will, that would go to media people and other molders of opinion—a magazine dedicated to the skeptical side, and giving it in an informal way that included humor and satire.

Marcello said he could not edit such a magazine, so he first resigned only as editor. Later, he felt he had bean unfairly treated, especially by Kurtz, he resigned from the committee and announced plans to set run up his own organization and issue his own journal.

Since then Marcello and I have exchanged many letters trying to focus our differences. The correspondence now runs to more than 50 pages. I could copy all of it for you, if you like, but it would probably tell you more about our differences than you care to know!

Let me try to summarize. Both Marcello and I agree that some claims are so outlandish—so unlikely to be true—that can at once be dismissed as "crank." For example, if someone sent in a paper arguing that the earth is flat, or that he had the power to walk across the surface of Lake Michigan, neither of us would hesitate to reject such a paper as not worthy of consideration. But when we get down to specific cases of individuals who are actively participating in the current occult wave, we react differently to them, and want to speak about them in different languages.

Most of our arguments have been over Gauquelin. I have argued that his claim—to have found a correlation between positions of planets in the sky at birth and one's later choice of a profession—is so implausible as to constitute what I, in informal discourse, would call a crank or crazy notion. I do not believe that scientists are obliged to treat G with the same respect as, say, they treat a respected scientist who advances an unorthodox theory.

Marcello argues that G *is* a serious psychologist, and a respected one, who should not be called a "crank" in any sense. He likes to say that G has found a "white crow"—a genuine "anomaly" in the Thomas Kuhn sense— that should not be treated with scorn. It should be taken as a serious challenge, and the sceptic must do one of two things: (1) show that G's raw data is biased, or (2) replicate his study to see if the correlation holds.

I counter: The claim is too far out to deserve such expenditures of time and energy and funds. No statistician will check the raw data (compare it with the thousands of actual birth records in Paris, etc.) without funding, and attempts to replicate are an equal waste of funds. Any negative results, I argue, will have no effect whatever on G's own views or those of his disciples. I view G as one among those hundreds of individuals who, from time to time, propose wild schemes of statistical correlations. I recently sent Marcello some information about one of my favorites, a man who wrote an entire book about the correlation of pi with the gestation periods of animals. In short, I do not think that a responsible scientist should treat G as a challenger to orthodoxy, or that his claim should be dignified by term "anomaly." I argue that there are too many really important "white crows" around that deserve far more to be taken seriously.

Our differences have practical consequences. Marcello has named G to his panel of what he calls "distinguished consulting editors." Although I have tentatively agreed to be on the same panel, I am not comfortable with having G as a co-panelist. I do not consider him "distinguished." Nor am I comfortable to be on a panel with Hynek, or with Martin Ebon, former secretary to the late medium Eileen Garrett, and a leading writer of trashy paperback books on the occult.

I hasten to say that Marcello has not the slightest personal belief in G's correlation claim, or any sympathy for the UFO views of Hynek, or the ESP views of Ebon. Our differences are over tactics and terminology. Marcello wants to establish dialog between Hynek and UFO sceptics, between G and his detractors, and so on. He does not want to put out a "debunking" magazine. He believes that his approach is the best way to combat irrationalism. My view is that meaningful dialog with cranks is usually impossible. Your criticism of Velikovsky, for example, has had not the slightest effect on V, and so far as I can see, about the same minimal effect on his admirers. In my opinion, our magazine is not obliged to treat the views of cranks as "anomalies." To me, a genuine "anomaly" is the black hole, or—also on the theory level—the twistor theory of Roger Penrose. I think it a tactical error to treat Gauquelin's claimed correlation as any kind of "white crow" to be taken seriously.

Since Marcello is likely to see our differences differently, I'll send him a copy of this letter so he can tell you where he disagrees.

Best,
Martin

[Zetetic Scholar]   6 May 1978
Dear Martin,

First, thanks for sending me a copy of your reply to Carl Sagan. I have written him some additional comments (as you suggested at least indirectly) and enclose a copy of those herein so you can "correct" anything I may have said to him that you think. might misrepresent your views. I hope I was accurate.

Re the Beauregard manuscript, thanks for sending me what you had, but it turned out to be the wrong thing. You sent me the *Antiscience* ms. and I wanted the draft of his paper applying the Bayesian formula to the criteria issue on judging the probability of a paranormal claim's being true. I will get the latter from Ray since he has a copy. He just met Beauregard last week in Oregon (B's in Portland and used to teach at Reed College). I think you will enjoy the paper when it is published (since you apparently have not yet seen it). I am hoping to get you and the other relevant editorial consultants to comment on the article for the 2nd issue.

The *ZS* is in full process and I hope to have it out by the end of this month. It will include a full bibliography on science and Geller, which amplifies Ray's earlier bib, and contains about 175 total items. I was surprised when I saw how much basic stuff there has been published on Geller (and this does not include the hundreds of newspaper articles around the country).

Re the review of Jamie Webb's first volume. I enclose a copy of your review, which I have slightly altered to up-date it (in red pencil). Only the first page is thus tampered with. However, I think Jamie's views come out more clearly in the second volume. I intend to publish this review of yours indicating it was intended for publication several years ago and is now being published belatedly because I think it should be of interest to the readership. Could you please tell me where it was supposed to be published originally and possibly why it was not. Also, what year was the review done? 1974? I should mention here that I rather disagree with you on one thing. I think Jamie's views are more clear than you took them to be insofar as the original title was *The Flight From Reason.* That is, Jamie sees the occult as unreasonable. His membership in the SPR is really irrelevant since one can belong to it without believing in psi, etc. In fact, the early presidents of the SPR were commonly skeptics. Further, a recent poll of the Parapsychological Association indicated about 30% of the members were unconvinced of the existence of psi. So, membership really doesn't mean that much in some of these groups. I think both Ray and I belong to the ASPR, for example. As for the "lurid" encyclopedias, Webb's pieces in them are strictly historical as are some of the other contributions in those books by anthropologists and other skeptics. So, you make a bit more of this than is necessary, I think. Of course, my views on this are influenced by my direct communications with and knowledge of Webb; you only had the book to go by. But for your information now, Jamie is a bit more hard line in his skepticism than I am. I think this comes out in his second volume.

Anyway, let me know if the red-pencil changes are ok, and let me know the answers to the questions I asked re the original intention for this review, etc.

I spoke to Ray last night and he told me that he had heard that Wilbur Franklin had died. Can you confirm or amplify on this rumor?

I am supposed to be writing a paper on roller coasters; I am going to a somewhat nutty conference on them at Cedar Point in, July (part of pop culture thing that Ray Browne has put together). I plan to write on the social psychology of why people go on the damned things despite the fear element (what I call "petty masochism"). Anyway, since you are an incredible storehouse of odd references, I'd appreciate knowing about anything you might be able to refer me to in this area. You may know of strange items to me.

The Magic Dealers Association meeting was here last weekend. Had a great time and met many interesting people. Spent a good bit of time with Jay Marshall whom I believe you know. I was particularly impressed by some brilliant mentalist effect handling by Jack London. Also talked to Micky Hades about getting involved with a survey analysis of mentalists he is going to conduct replicating and extending Bob Nelson's earlier poll (which I hope to get the data from Micky for reanalysis).

Best,
Marcello

P.S. Also enclose a new mentalism effect I just came up with.

[Euclid Avenue]   12 May 1978
Dear Marcello

I'll divide my reply into two letters, so I can send a copy of one to Sagan without burdening him with other matters. The changes on my review are fine; indeed, they are the changes I expected you to make (I didn't know the date of publication of Vol. 2.) Glad you can use it. The review was originally only half of a review that I was asked to write for the *NY Review of Books*. They actually paid me for it, and to this day I don't know why they didn't run it. Perhaps it was because the other book reviewed was Evans' *Cults of Unreason*, and it's possible the Scientologists sent them a threatening letter. (This not for publication, of course, since I am just speculating. Maybe they just didn't like my review.) The review was done before Webb's first book was actually printed, because I read only the bound galleys. It must have been some time in 1974.

Your idea for that ESP symbol trick is excellent. It uses the principle of the cut newspaper strip in a novel way. I wish I'd thought of it! Any magic periodical would be pleased to have it.

Yes, Wilbur Franklin died shortly before the appearance of the current issue of *New Realities* with his sad article defending Uri Geller. Franklin made three outright errors in one early paragraph (1) He said I'd written about the nitinol tests in *Scientific American* when it was the *Humanist*; (2) he said I offered no explanation as to how Geller could have used trickery(!), and finally, (3) that the wire is currently obtainable from Edmund Sc. Co.— they stopped carrying it many years ago. I sent a short letter on this to the editor, and Bolen replied by telling me Franklin had just died of a stroke. After "stroke" was an asterisk, with the note "perhaps diabetic coma." I have heard rumors of suicide, but they may be no more than that. Do you have any friends at Kent who could give you the facts?

No—nothing in the files on roller coasters! (Carnival games and rides, yes.)

I've seen Jack London do his prediction trick and it's very clever. Clue: the staples on the cardboard sheets are fake. The sheet goes together at one edge with rubber cement (and he uses a stooge from the audience).

Yes, Jay is an old friend, and married to an even older friend, Frances, whom I knew when she was about 18 (before she married Laurie Ireland). I used to hang out at Ireland's Magic Shop on North Clark Street, across the way from William Targ's old book shop (Targ is now president of Putnam's). Frances has written a charming autobiography that is worth reading. Jay used to be married to Al Baker's daughter. Everything, as a friend says is "intertwingled."

Best,
Martin

P.S. The Beauregard paper you asked for is the one I sent to you and you passed on to Ken Frazier. I didn't keep a copy. Do you recall? I wrote that it was *just* the sort of paper *you* could use.

[Euclid Avenue] 12 May 1978
Dear Marcello

Thanks for the copy of your letter to Carl Sagan. I won't attempt to argue against any of the views you express, and with which I have (as you know) disagreements more of emphasis than substance. I can't resist pointing out, though, that when you call yourself a "logical empiricist" and a "Popperian", you are saying something that would drive Popper up the wall. He has long taken credit for being the person who, almost single-handedly, slayed the dragon of logical empiricism. He and Carnap disagreed about almost everything, especially over "induction" (which Popper says doesn't exist), and over the question of "realism"—like B. Russell, Popper is a hard-nosed realist, even on the microlevel of quantum mechanics.

But more to the point. Naturally I deny that I've misrepresented Hynek. The only time I've written about him was in my review of *Close Encounters*, and everything I said about him there was taken straight from his own remarks in *Fate*. Since then he's become even more convinced that UFO's are "real" (by real I mean "out there" and not figments of the mind), but somehow projected into our minds. Ken Franklin told me a few months ago he heard Hynek say in a lecture that he doubted if a single "genuine" UFO sighting had registered on anyone's retina.

Our evaluations of Gauquelin's personality continue to be markedly at
odds. You believe he would change his mind if presented with negative
evidence. I predict that no matter what kind of negative evidence turns
up (either by comparing his raw data with birth records, or by attempted
replications), G will find a way to discount the evidence. The difficulty is
that "falsification" is as ambiguous as "confirmation." It is a vast over-
simplification to say that one observation of a "white crow" will falsify a
hypothesis. One simply patches up the hypothesis to "explain" the alleged
falsification. But enough...

                                Best,
                                Martin
CC: Sagan

[Euclid Avenue]   15 May 1978
Dear Marcello

Thanks for "Cranks"... [Bill Harvey, "Cranks—and Others," *New Sci-*
entist, March 16, 1978, pp. 739-741] I naturally disagree with the author's
viewpoint on so many details that it would take pages to discuss the article
adequately. As he himself says at the end, he writes from the standpoint
of cultural relativism, a position which seems to me to be exceedingly old-
fashioned, even in sociology. *Of course* science and scientific beliefs are
heavily influenced by culture—a point that is trivial—but to treat scientific beliefs as "true or false" relative only to a culture or subculture, as
Harvey puts it, strikes me as preposterous. Shades of Leslie White and
his "locus of mathematical reality"! I see he is invoking Kuhn, but his
arguments are caricatures of Kuhn.

There are many errors of detail. When he speaks of the von Neumann
(and others) proof there can be no hidden variables in QM, he should have
made clear that the "proofs" were so interpreted by popularizers of QM.
Von N (and the others) were well aware that their proof was no more than
that hidden variables cannot be reconciled with the *formalism of classical*
QM. This is quite a different thing. Naturally, if one revises QM, you can
add all sorts of hidden variables—as indeed, does Bohm and others.

Harvey's last sentence is his worst. He speaks of the world as "our
creation." Now again this is true in a trivial sense—i.e., that aspect of
reality that is mind-dependent is mind-dependent. Or, as I put it in my
old essay on White, it is like saying "that aspect of reality that is inside
this room is inside this room." But is the *world* our creation? This is the
road to subjective madness. No one would disagree with this more violently

than a realist like Popper, another person whose name Harvey invokes out of ignorance.

But enough of Harvey. I enjoyed your paper and you are quite correct in saying that I find Gauquelin's correlation (if true, of course) a bigger wrench to the whole of science than the truth of ESP. What you seem to me not to realize is that although G himself specifies no cause-or-effect modus operandi if the correlation is genuine, there has to be *some* kind of physical explanation of the correlation. You haven't yet told me what this might be. My claim is as follows: I believe I can think of many more plausible ways to explain ESP than you can think of plausible ways to explain G's correlation (if genuine). Hence I am as amazed by your attitude on this as you are by mine. Moreover, I think my amazement is more characteristic of even the majority of sociologists than yours. Here once more I know of only one way to settle the question of who has the most right to be amazed (i.e., a poll). Put in the terms of cultural relativism, I believe my amazement is typical of the subculture of science; yours is the anomaly. Or do you really think your amazement is typical?

See *NY Review of Books*, June 15, for my blast at Catastrophe Theory.

Best,
Martin

[Zetetic Scholar]   17 May 1978
Dear Martin,

A small word of defense. I was taught (possibly incorrectly) to distinguish logical empiricism from logical positivism. Popper was an antagonist of the latter (Vienna Circle, etc.). But he can be seen as within the former broader category of essentially scientifically oriented philosophers. I have quite frequently heard the term logical empiricist used to discuss those as different from one another as Carnap, Russell, Wittgenstein and Popper. I think we would both agree that these philosophers of science all exist within a broader tradition distinguishable easily from such Idealists as Broad, etc. Since recent philosophy of science seems to be trying to blend together the verificationist and falsificationist positions (as in Ayer, for example), some common term seems appropriate. That's what I meant by logical empiricist. Sorry for the confusion. Can you suggest a better label for what I mean?

As regards Hynek, we again disagree more about strategy than substance. In my dealings with Hynek (and I have attended several of his lectures now), I have been impressed by his awareness that UFOs remain essentially unidentified. You (and Klass) would seem to insist that they are really now identified. Because Hynek views them as still a puzzle, he considers the extraterrestrial hypothesis (and it is just that, an hypothesis) and some of the other hypotheses (like other realities, psychological phenomena, etc.) as more plausible or at least within the realm of reasonable discourse. I disagree with Hynek about the reasonableness of some of these hypotheses, but I agree with him that UFOs remain as unidentified phenomena in a great many cases. Because you (and Klass) think they have been adequately identified or have faith that they could be adequately identified, you give such low plausibility to these hypotheses entertained by Hynek that you are willing to label his public willingness to entertain such ideas as reasonable as a sign that he is a crank (if not a crackpot). At least, that's what I think you think. Tell me if I'm wrong. I should add that I think you have been generally far more reasonable than I think Klass has been about specific cases since he is far more willing to impute accusations of fraud with little evidence to support his case (and frequently a good bit of contrary evidence). My own view is simply that UFOs are frequently not identifiable in terms of known explanations, and I want ordinary explanations sought before leaping to extraordinary explanations. But a real problem exists here and science can benefit by further investigation. Let's just hope we find some new kinds of plasmas (as Phil suggested in his first book) rather than some of the proposed explanations Hynek is willing to entertain as hypotheses. From my vantage point, Klass and Hynek actually share more than they differ since both of them recognize that scientific investigation is warranted. What difference does it make if we call it Unidentified Plasma Phenomena or Unidentified Aerial Objects? If we properly follow Klass's ideas we look for plasmas: if we follow Hynek, we look for anything. Either way we agree it is worth looking. The important thing to me—as always—is not to block inquiry. By making fun of Hynek as you do, we may deny the existence of the problem itself. That's what I don't want to do. I realize that Hynek is "legitimating" a lot of nonsense by his presence, but would we be better off without his efforts? His Center is at least collecting a lot of data that will be useful to sociologists some day even if not to solving the UFO question. The nuts would remain without his presence and would gravitate to someone far more irresponsible than Hynek who is still basically agnostic about the UFO answer. Do you really see much menace in him or his work? Having now read everything I could find by both him and Phil Klass, I don't really see either as perfect, but I am really glad that we have the benefit of both of their views in the arena.

Finally, re Gauquelin. I remind you that he agreed to the original Zelen test. If it had been clearly negative he indicated that he would consider it a falsification of his hypothesis. The results were not clear cut. Even Abell and Kurtz agree that they were not and that is why they are replicating with an American sample right now. Michel has some concern about Rawlins being the one now responsible for the new test, and I don't blame him given my experiences with Rawlins lately, but Gauquelin has agreed to cooperate. Why not wait and see what Gauquelin's response is to negative results if they transpire that way? Why prejudge Gauquelin's behavior when it has been most reasonable to this point?

The first issue of *ZS* should be out within two weeks. It is quite a bit larger than I had anticipated (about 54 pages),and I think you will like it.

Cheers,
Marcello
CC: Carl Sagan

[Eastern Michigan University]   21 May 1978
Dear Martin,

I must agree with you about the too extreme relativism of Harvey and the other British sociologists of science like Harry Collins. However, they represent a very refreshing antidote to the, in my view, very culture-bound kind of sociology of science represented by Robert K. Merton and the American school of thought that has too long bought the scientific community's own appraisals of what they were actually doing. My own position is that taking things to a possibly still excessive relative standpoint may be very useful in a heuristic way and frequently leads to important insights. For example, Harry Collins' work on the actual different meanings of what constitutes "replication" within the sciences casts important light on what most of us previously took to be a non-problematic area.

Our differences over Gauquelin seem to become more rather than less clear, I fear. You say "there must be some kind of physical explanation of the correlation." Here I presume you mean "presuming the correlation is a real one and not just a goof." But here I think we seem to disagree. Let us assume there is no goof, that Gauquelin has done the work, computation, etc., correctly. If there were a claimed causality, there would have to be a physical linkage or chain of some sort. That may be what you mean. But since Gauquelin merely posits a correlation, not a causality, I don't agree that there has to be a physical connection at all. There may really be no connection of any sort, since a correlation establishes no connection, merely an association. For example: A planetary cycle may run X days. Some sort of human activity cycle may also run over X days. These could result in a correlation between them (because the two cycles happen to synchronize that was startlingly high. (Just as a group of women with a 28 day menstrual cycle, starting the same day, would be correlated highly with phases of the moon.) If your last letter meant what it said, you confuse a correlation with a connection. I argue that Gauquelin might have found merely a correlation. Period. It proves nothing much if so except two synchronous cycles. Sunspots may have a cycle similar to earth's warfare cycle. Again, no necessary connection at all. However, looking at one would result in a pretty good historical predictor of the other, and that might account for why one was used as a "sign" of the other in antiquity to the present. Now, obviously, Gauquelin does think that he may have more than a mere set of synchronous but disconnected cycles. He does not claim to have established that, but I think he hopes to find some connections. So, his picture of things may be more like your own of what he is claiming. But let me assume, now, that Gauquelin's claim is that of a correlation with some sort of connection somewhere. You then say, after commenting on there having to be a physical explanation of the correlation, that "You haven't yet told me what this might be." But I think I have done just that in my past letters. I specifically suggested, merely as an example of the kind of reasoning possible, that the hidden connecting variable might be the sexual copulation pattern of mesomorphs. Remember, all we have to account for is an above chance level of correlation between births of athletes and a planetary position at birth. If athletes are mesomorphs (most of the time) and more likely to come from mesomorph parents, and mesomorph-athletic parents have more energy so screw more frequently (say in the morning as well as in the evening), that would possibly create a difference between the birth pattern of such children and others, which in turn would result in a non-chance correlation being present. Imagination should allow one to come up with many other scenarios that might explain the relationship Gauquelin found. I am reminded of an old bit of nonsense when someone claims that the revolving doors in department stores are the motor force behind the escalators. If you look at the stores at night when the doors are not turning, you will see that the escalators are also still. This "correlation" is certainly a true one, but it says nothing about any direct physical explanation. In this case, the two events are linked through a third variable that causes both (the store hours). It is remotely possible that the planetary positions could somehow be linked through an involved chain of events to the birthing of the athletes. I doubt it, but it is by no means impossible to think of complicated chains that could produce the small, but above chance, correlation present according to Gauquelin. But I think it far more likely that we are here dealing with quite unconnected events that may simply have a cyclical pattern that coincidentally happens to be synchronous for the two cycles. It is just a case where no "explanation" is really involved at all but still one where prediction may be effective. If I am right, and if you agree with my comments here, you should be able to see why such a correlation is hardly of consequence compared to a claim of psi. You go on to say "I believe I can think of many more plausible ways to explain ESP than you can think of plausible way to explain G's correlation (if genuine)." By this I presume you mean that you can find non-psi explanations for ESP rather than find a way to keep psi from having important consequences for the rest of science. But if this is what you mean, you are really saying you can "explain away" rather than "explain" ESP. I say it is easier to explain away ESP statistical results (show possible non-ESP causes) than it is to explain away G's correlation. But you can equally easily explain away G's hoped for causal connection hypothesis. Please note that you said "if genuine" about G's correlation but did not refer to ESP in the same terms of "if genuine." You are not really making a direct comparison. Pro-ESP claimants are not simply claiming a statistical result (correlation or otherwise) between guessing scores and targets; they are claiming a causal link of psi forces that account for the relationship between guesses and targets. Gauquelin is not claiming any astrological or other forces between the births and the planets. If he did, you would be quite correct in your criticism. If he claimed strange forces between the planets and occupations, that are direct forces of some unknown kind (comparable to psi), I could understand your criticism and views. But he makes no such claims at all. That's what I really don't think you fully understand or appreciate. An astrological claim would indeed be more extraordinary than a telepathy claim, to me as well as to you, but that is not what G is making. He does *hope* there is a chain of connection which accounts for the correlation. He does not accept my argument that they might just be synchronous cycles without any connection at all. He accepts the logic of my view, I think, but he thinks there is a physical linkage (though probably a complex chain, not a simple force of any kind). Here we disagree, but he accepts the necessity of properly establishing any such linkage before anyone should accept his hypothesis of a real linkage. So, we continue to disagree but accept the same ground rules for what needs to be done to establish his mere hope (at this point) of going beyond his anomalous correlation to finding some sort of real connections. What's scientifically wrong with that posture? Martin, I really think you are blinded by the stigma of the association of G's correlation with the notions of that "superstition" astrology. I think you just want to see G as a kind of astrologer-in-sheep's-clothing. If I am right, that is really not fair to. Gauquelin, especially since he has personally done more to falsify astrological claims through his data studies than anyone else.

As to the issue of who has the right to be most amazed, let me again state my earlier point that I am sure a polling of scientists might very well—and probably would—agree with your view in the sense that more of them might be more amazed were G right than if the parapsychologists were right. But the issue is not who would have the most company in being amazed. It is who would have the right to be more amazed. I insist that the right to be amazed *as a scientist* (that is for scientific, which must mean theoretical rather than simply "empirical" reasons) must depend on the degree of extraordinariness of the claims in terms of scientific implications, not just the psychological strangeness to people who happen to be scientists. I must insist that G's simple claim of an unexplained correlation is almost inconsequential in its implications compared to the positing of a special mental force like psi. (Finally, I might add that one of my complaints about the parapsychologists is their willingness to jump from simple statistical results to positing strange "forces." If the parapsychologists were as cautious as Gauquelin, they would simply speak of residual and unexplained non-chance variations in guessing, which is a far cry from psi or telepathy, etc.) Now I would be very pleased if a polling of scientists did in fact reveal that they recognized the differences I am citing and resulted in their not being amazed like you are by G's claims. But I think you are probably right and from my standpoint, such a polling would merely state *who* was more amazed, not who had the *right* to be more amazed. My surprise at your amazement was due to my thinking you were far superior to the average scientist since you had vigorously thought about these issues before. I know the average sociologist would be more surprised at G's being right than at psi being in existence. But I think a lot more of you than I do of most of my colleagues. Unlike yourself, most sociologists simply don't deal with things very systematically, and certainly not very logically. You asked "do you really think your amazement is typical?" No, I do not. And I would hope not, for I don't think most scientists base their "amazement reaction levels" to anomalies on serious scientific-theoretical grounds but on more everyday psychological biases of "strangeness" which may have little to do with scientific-theoretic question of "fit" and implication.

I'm looking forward to the June 16 *NY Review of Books* issue for your blast at the catastrophe theorists.

ZS should be out from the printer next Friday. I look forward to your reaction.

Cheers,
Marcello Truzzi

[Euclid Avenue] 21 May 1978
Dear Marcello

Yours of May 17 to hand. I think we had best call a halt to debating the personalities of both Gauquelin and Hynek, since we simply repeat our previous arguments, I consider Gauquelin an almost clinically perfect example of the crank, obsessed by a wild notion that almost nobody considers rational, and I regard Hynek as a sincere man of enormous gullibility as you know, he is a firm believer not only in ESP but also in the "Geller effect" (or "was" a believe until—maybe—recently). That you would regard them both as "distinguished" scientists is beyond my comprehension.

Let me confine my attention to our minor debate over the most widely accepted use of the term "logical empiricism." "Logical positivism" was an early term, confined to the Vienna Circle. But members of the circle soon altered their own views, and were joined by others of similar persuasion. The term logical positivism was slowly replaced by logical empiricism. I quote from the first paragraph of the essay on logical positivism in the *Encyclopedia of Philosophy*, edited by Paul Edwards:

> Logical positivism is the name given in 1931 by A. E. Blumberg and Herbert Feigl to a set of philosophical ideas put forward by the Vienna Circle. Synonymous expressions include "consistent empiricism," "logical empiricism," "scientific empiricism" and "logical neo-positivism."

Of these synonyms, "logical empiricism" has been by far the most common in the English speaking world. You cite as three examples: Russell, Wittgenstein and Popper. Russell never called himself a logical empiricism. Indeed, his last great work, *Human Knowledge: Its Scope and Limits*, finally concludes that empiricism is not sufficient for a philosophy of science. Although Wittgenstein had some influence on logical empiricism, his major influence was on the analytical or "ordinary language" movement at Oxford, and he ended with a mystical approach in sharp conflict with logical empiricism.

Popper, like Russell, consistently battled with the logical empiricists. I'll cite one instance, picked virtually at random from *The Philosophy of Karl Popper*. Volume 2, page 973, he quotes what Hempel calls a central maxim of "logical empiricism," then explains why his own "commonsense realism" makes it impossible for him to accept the maxim.

You ask for a better label for what you mean. But I'm not really sure I know what you mean. Maybe the term "philosopher of science" is the best. All of them share a common respect for scientific method, and place it at the heart of their philosophy. But their metaphysical differences are subtle, multivarious and deep.

All of this is an aside, and irrelevant to our fundamental dispute. I believe that if you took a poll of all the philosophers of science in the US, as well as a poll of the 100 top scientists, you would find over 90 percent of them, regardless of their metaphysical differences, in full agreement with my characterizations of Gauquelin and Hynek. I suspect that a poll of sociologists would show a percentage almost as high. Among anthropologists, where cultural relativism is still a strong force, I would expect much more agreement with your way of looking at Gauquelin and Hynek.

But now we are back to our fundamental dispute. I believe it is essentially linguistic, but I also believe that I am using words that reflect common usage in the scientific community at large, whereas you are using words in a narrow and almost private sense that tend to cause more confusion than light.

Best,
Martin

CC: Carl Sagan

[Euclid Avenue]   24 May 1978
Dear Marcello:

When I said I could think of more plausible ways to explain ESP than to explain G's correlation, I did indeed assume that ESP was "true"' in the sense that the claimed correlations of ESP were genuine causal correlations, and not comparable to your metaphor of revolving doors and escalators.

Let me put it this way. A parapsychologist finds a correlation between the ordering of a psychic's guesses and the ordering of the targets. To say it is a "genuine" correlation is, to me, the same as saying that the correlation is explainable by a set of causal laws. Now parapsychologists have advanced numerous theories (sets of laws) to explain their correlations. In my opinion none of them are good, but at least they have a certain plausibility. I have just written a paper about the most fashionable current theory, Walker's theory based on QM. There is no reliable evidence to support this theory but neither is there any basis for calling it impossible.

Now consider G's claim. If the correlation is genuine (which I1. personally think it is not) it implies very strongly that it can be explained by causal laws. I now repeat myself. I find the theories advanced to explain psi more plausible than any theory I have heard to explain G's correlations. Of course the evaluation of a theory is not formalized, and is highly subjective.

All I can say is that my intuitions tell me that if science discovers someday that there is a psi force, not subject to the restraints of classical relativity, etc, I believe it would be less of a revolution, than the discovery of causal laws to explain G's correlation. Debate between us about this is clearly fruitless, since the probabilities are not well defined and all we can do is report our respective hunches.

But my main reason for having so little interest in G is that his claim is the work of one man, based on raw data not yet checked against source records, and unreplicated. By contrast, the claimed correlations of parapsychology rest on hundreds of experiments, made over decades by scores of sincere workers—and all claiming more or less the same correlations. The psi claims seem to me much more worth taking seriously for this reason, and also because they are so firmly entrenched in the cultures of so many nations. G's claims are comparable, to scores of similar wild claims of statistical correlations, advanced solo, that nobody bothers to worry about.

Now for your sample "explanation" of G's correlation via mesomorphic tendencies to copulate at certain times, and so on. I had not heard this before, or if I have, I had forgotten it. This strikes me as far more outlandish than Walker's QM theory of psi, Are you suggesting that the parents unconsciously are somehow aware of the positions of the planets about nine month's later? All the athletes I've met wouldn't be able to recognize Mars or Jupiter in the sky if their life depended on it, or even to tell you the phase of the moon nine months from now. If you don't suggest this unconscious awareness, what *do* you suggest? You haven't given me anything yet remotely resembling a plausible theory. In fact, I myself can think of two or three that are better than *that*. And what about the copulation habits of the parents of doctors-to-be, not to mention the musicians and painters and writers for whom G has also found correlations? I suppose you could work hard and patch up this theory, but by the time you patched it enough to account for all of G's correlation claims you'd have a theory that I would find enormously less plausible than half a dozen theories that have been advanced in the past to account for psi.

But let's call a halt to arguing about Gauquelin. If you prefer a final comment, write again and I will promise not to reply. To me, the G claim is totally uninteresting, and certainly not deserving the name of an anomaly or white crow. It is less interesting than seeking an explanation for the curious correlation found by the writer of the enclosed letter. Would you like to publish it in the *Zetetic Scholar* and ask readers for an explanation? It's a much whiter crow than G's ridiculous claim.

Best,
Martin

[Eastern Michigan University]   26 May 1978
Dear Martin,

I agree with you that our exchange has reached the point of repetition
so I will say no more for now re Gauquelin. We each think the other
is incorrect, of course, and I hate to see that be the case after so much
verbiage between us (that is, neither of us really shifting our views very
much). I guess that those who read our views (like Ray and maybe Sagan)
will have to make their own judgments. This is not to say the exchange
was wasted on my side at all. Not only do I think I understand your views
better, but I think I have clarified my own in the process and realize that
I may have been a bit harsh in my reaction to your use of the word crank.
Anyway, thanks for taking the time and trouble to have at it with me at
such length. I realize how busy you must be what with your column and
other writings. Since my own position is spelled out in my little piece on
evaluating the extraordinary in this issue of *ZS* (which I expect to have
from the printer tomorrow), it will be interesting to see reader reactions,
some of which, I presume, will take a position more like your own.

Enjoyed your piece on catastrophe theory. When you mentioned it in
your earlier letter, I thought you were referring to a different set of "catastrophists." I refer to things like Alfred Webre and Phillip Liss' *The Age
Of Cataclysm*, the journal *Catastrophist Geology*, and, of course, kindred
voices like those of Velikovsky and friends. (By the way, I wonder if you
are aware of the variety of views now surrounding Velikovsky among his
"supporters"; *Kronos* is simply the extreme position with many other less
supportive but sympathetic wings emerging, particularly in Great Britain
I will be running a general survey of the "schools" in a future *ZS*. I have,
by the way, been corresponding with C. J. Ransom, and he turns out not
to be one of the more extreme V supporters after all.)

Enclosed is a cartoon I think you'll like (if you haven't already seen it)
and a copy of a Paul Curry effect that I think is especially good and which
you may not have yet seen.

Re the issue of logical empiricists, the terminology is by no means consistent. I enclose a section of a survey on same. I recognize the differences
you mentioned but all these "philosophers of science" share more than they
reject. Obviously, the term philosophers of science is already used to include any philosopher's view on science including those who are anti-science.
Whatever our terminology, I think we understand what I meant, i.e., my
strong commitment to science and its method to producing knowledge.

                                        Cordially,
                                        Marcello

P.S. If you haven't yet bothered, check out "Parapsychology" in the *Encyclopedia of Philosophy*. A *terrible* piece.

[Euclid Avenue] 31 May 1978
Dear Marcello

Congratulations on Vol. 1, No. 1! It is packed with valuable material, and certainly a worthwhile periodical to have in the running. You've put everything attractively together, and your essay on the "extraordinary" is a valuable contribution toward clarifying all sorts of basic issues.

The only spot where I think your attempt to play a mediating role became a trifle excessive was in referring to Rogo's book, *Mind Beyond Body*, as on an "intelligent level," and to say that the evidence he presents is not "overwhelming" strikes me as enormous understatement, unless you intended this as satirical. I find Rogo almost as unreliable as Colin Wilson, and almost as gullible—but no matter. You want to avoid the debunker's rhetoric, and I understand.

I hadn't known the details of Curry's new trick, and was delighted to get them. Paul, by the way, is recovering at home from a rather severe heart attack. Enjoyed the cartoon which I hadn't seen either.

Yes, the terminology for schools of philosophy of science is fuzzy. I knew of Feigl's piece, but it was written many years ago before the terminology started to jell. I still can't think of any widely used term for all those philosophers who are strongly committed to science except "philosophers of science." Philosophers who are antiscience seldom if ever call themselves "philosophers of science," and the term has come to stand for pretty much all those people mentioned by Feigl as "allied" to the Vienna Circle movement. Take someone like Whitehead. He is seldom called a "philosopher of science" although he wrote a great deal about science and was strongly committed to it. But he was also a metaphysician, and no one would link him with the "logical empiricist" movement. But as you say—it's a trivial terminological problem, and not worth spending time on. Peirce (whom I tremendously admire) is even harder to classify! Incidentally, my August *Scientific American* column will be about him.

I certainly agree with your estimate of the article on parapsychology in the *Encyclopedia of Philosophy*. (I own the original set, by the way, which I got in return for doing the pieces on logic machines and logic diagrams.) Paul Edwards, the editor, was probably responsible for assigning the ESP piece to a believer. See also Edward's article on Reich, which as I recall is longer than the set's essay on Freud! Edwards sat for years in an orgone book [box] as a firm believer in orgone energy!

All best,
Martin

[Eastern Michigan University]   31 May 1978
Dear Martin,

It seems apparent that our difficulties are indeed linguistic in large part, but not quite in the way I had at first thought, and. I don't think in the way you might mean. I realize that you must be sick of discussing Gauquelin's claims but I think your last letter well worth commenting on. As you indicated, you needn't feel obliged to respond again on this, but I think, I should comment on your last remarks and hope you do not see it as simply my wanting the last word.

You obviously equate "genuine correlation" with one "explainable by a set of causal laws." You really seem to mean by this that, for you, a correlation means a causal relation. And a spurious correlation would be a non-causal situation. *If* you mean by this a direct causal relation, I can only say that you misuse correlation and equate it with causal relation. This is a common error but not one I would expect you to make since you are mathematically sophisticated and surely aware of the distinction. I presume therefore that what you mean is that a genuine correlation must have linkage, direct or indirect (that is, a chain of causation) that would explain it. The example I gave of the escalators being run by the revolving doors is, of course, an example of indirect causal linkage. I think that such indirect chains can be hypothesized to account for Q's claimed correlation—and more on that in a minute—but I think you may miss my point about correlations between things that may not be connected (by causal chains) at all. My example here was of the menstrual cycle and phases of the moon. The universe is full of all kinds of cycles. As you know, there is even a society for studying and intercorrelating such cycles. I presume lots of things follow a pattern cycling over X-days. A correlation between such cycles would be quite high. It would not. indicate any necessary causality. It would certainly be a genuine and not a spurious correlation if computed correctly and measured correctly. It is only not "genuine" in your meaning of there being no causal explanation or linkage. But the use of "genuine" by you in this fashion is, I contend, quite idiosyncratic to you and not what is normally taught in statistics (Ray can correct me here if I am wrong). Certainly within the social sciences a major problem has been trying to keep people from confusing correlations with causes.

So, the first point I must stress is that G's, correlation may be genuine in the sense of being properly computed and measured, without indicating anything about causes whatsoever. We may simply be dealing with cycles that happen to synchronize. Period. And G agrees that no causes may be involved (though he hopes otherwise).

My second point concerns, what I think is your misunderstanding of my proposed hypothetical causal linkage via mesomorphy. I offered this merely as one possible hypothesis of indirect causes; one could conjecture others. But I think you misunderstood my example. This may be my fault. I do not suggest any unconscious awareness of the parents, etc. I do not even suggest any real causal link in the sense of suggesting one between the planetary positions and the births. I suggest a possible causal reason for the difference between the pattern (cycle) of births of athletes and non-athletes, which accounts for G's non-chance correlation. You will recall that G merely claims that athletes are more likely to be born with the planet Mars in certain positions. This happens 4 times each day. The issue concerns the times athletes are born compared to non-athletes (the general population). These times vary each day, but we are speaking of a daily situation and really nothing like what astrologer usually speak of in classical astrology. It is my understanding that one, i.e., Gauquelin, gets a certain correlation between general births and the Mars positions. One gets a stronger correlation between the births of athletes and the Mars positions. It is this difference which is statistically significant. So there is presumably, something different going on with athletes than the rest of us. I simply pointed out that the athlete-mesomorph connection via parents. Also mesomorphs (that is more likely than average to be mesomorphs) might be the hidden factor If the mesomorph parents copulate more frequently, there may be more possibilities for their children to get born at the propitious Mars times than is the case for the non-mesomorph parents. This difference would then show up as accounting for the statistically significant differences between athletes' and non-athletes' births in relation to the Mars positions. This hypothesis may be wrong and indeed you may be able to come up with better ones. But the point we are talking about—all G is talking about—is a simple correlation. He gives no explanation but simply, seeks one. In so far as he would agree that there *may* be no causal link, he would have to agree you in not calling it a genuine correlation in your meaning of "genuine." But the correlation, he would argue, is correctly present and real, even if not what you would call "genuine."

You then make much of the "hundreds" of parapsychological workers versus G's alleged uniqueness, that is, his singularity. Here, I think you are seriously in error. Hundreds of parapsychologists have found correlations that are the same, but the issue with them concerns the interpretation of these correlations. Most of us today would agree that the best parapsychologists use statistics correctly. But there is a big difference between a significant non-chance score and the inference that psi is present. I believe— as I think you probably do—that the correlations are properly arrived at, but we think they are caused by sensory leakage, fraud, lack of controls, etc. To back up the parapsychologist, we need more than the correlation; we need an inference about its cause—which they claim is psi. In G's case, the matter is quite different. First of all, I should note you are wrong about G being the solitary claimant you describe him to be, but more on that in a minute. For the moment, let me assume we are dealing with a single claimant. G claims to have found a correlation by gathering data and correlating it. Anyone can do this with the same data or new data. There is absolutely no problem in replication. It requires no inferences at all. It is like saying "Go look in the next room and see the unicorn." If you look and see no unicorn, that settles that. G's data are birth records. These are public and can be easily checked. The character of his claim is basically very different from that of parapsychologists. We should be very concerned about the inferences as to the causes of the correlations in both cases, but G, unlike the parapsychologists, makes no inferences at all. It is the correlation itself that is presented as anomalous. Now, obviously, if it were merely G claiming to have found this curiosity, we should properly question his methods. Or—as I have so often said—we can choose to ignore it entirely without making any judgment except to say that the probability of his having done it correctly is low. But what are the facts? First re G himself: He claims—and I have little reasons to doubt him— that he serendipitously came upon his correlation while debunking classical astrology. He didn't believe it himself so he replicated with new samples only to find it again. But that the finding continues to puzzle him and that he seeks a solution including one that would "explains away" his finding. Why is that unreasonable? But it goes further than this. The fact is that the Committee Para replicated; his work expecting to find no correlation. They used a big Belgian sample (based on my memory; it could have been a different sample, but the committee Para is a Belgian Committee) claiming that his was a fluke based on a French sample. They found the same damned result he found. They are a highly orthodox group and this presented some real problems for them. They admitted getting the same empirical finding but then insisted that it be interpreted differently. G had some support on that Committee but they published their results with an interpretation of the findings that seemed to debunk G. The problem is they got the same correlation he got. They expected not to get it. Their *post hoc* interpretation could as well have been applied to G's own finding that they earlier believed false. This would have avoided the replication. By doing the replication they admitted the need to do away with his finding, but in this failed. This is all well documented. Now along comes the Zelen test to further control variables in a way that critics predict should eliminate the non-chance correlation. And what do we get? An ambiguous result according to the Committee but a positive result in terms of the original ground rules for the test (which in my view was a mistake that should have been foreseen). So the Committee for SICOP now is replicating—which in itself indicates the first test was inconclusive—under the direction of Dennis Rawlins (who surely is an honorable man)—and we await the results. Using my analogy of the unicorn, it is like G says "Look in the room, I think I saw a unicorn." The Committee Para says "Nonsense, our expert will look in the room and report what he sees. "The expert looks and reports he has seen something very much like what G says he saw. The Committee says an interpretation is needed to explain matters and concludes both G and their expert misinterpreted some stimuli." Along comes your Committee, and sends their expert into the room. He emerges with his description. It matches G's general description. But the Committee says the alleged unicorn was only seen in part of the room and not, as expected, running around all of it, so G must be at least partly wrong. So, the Committee checks into looking into some other room where G thinks they might also find a unicorn (even though G does not claim with certainty that unicorns occupy this new room). But since G thinks unicorns occupy all rooms, he accepts the challenge that the new room (the United States) be looked at. Obviously, the analogy is imperfect, as all analogies are; but it pretty well describes what has happened. G's Mar's effect has thus far been allegedly found in France and Germany by G, and in Belgium by the Committee Para (despite: their interpretation to the contrary). The Zelen test clearly found it only in France as a whole and in Paris especially. Even if Rawlins does not find it in the U.S., I don't see that that will disprove; the other places. G simply would not have a universal effect as hopes he has. We still would need an explanation for the cases where we got it.

Finally, given your special meaning of "genuine correlation" to mean cause (especially direct but possibly merely indirect), such astrological causation would indeed be equal to a claim of telepathy or PK (in fact, it resembles the latter) in its extraordinariness. But that simply is not what is being claimed by G. And even if it were thus claimed, the implications for physics would not be so great as the implications of precognition since the latter would not merely affect our view of space but of time. And I grow more and more convinced that such time inversion might make experimental method impossible in science since the "after" condition might be affecting the "before" state in our basic methodology.

If all the above is seen by you as merely repeating things, so be it and you needn't respond. We certainly have many other things to talk about it.

I trust that you have by now received *Zetetic Scholar* and I hoped you liked it.

Cordially,
Marcello

CC: Ray Hyman

P.S. Do you know Topper Martyn?

[Euclid Avenue]   3 June 1978
Dear Marcello:

If a correlation is left totally unexplained, let's call it a "pure correla-tion." If it is not the result of chance, fraud, unconscious bias, etc., let's call it a "genuine pure correlation."

Science constantly encounters genuine pure correlations, and sometimes decades go by until they are "explained." At the moment, the most mys-terious of all pure genuine correlations in physics is the one involved in the Einstein-Podolsky-Rosen paradox. Two photons (for example), even though many light years apart, remain correlated in a peculiar way. Quan-tum mechanics offers no explanation. It merely formulates a function that correlates them. It leaves the correlation as pure magic, like sticking a pin in a voodoo doll and having someone die ten miles away.

Whenever science finds a genuine pure correlation, the search at once begins for a causal explanations direct or indirect. The assumption is im-plicit that such an explanation exists, unless one wants to accept a universe in which distant objects (e.g., planets and humans) can be correlated with no causal connection of any sort, direct or indirect.

Anyone would assume, therefore, that *if* G's correlation is genuine, there must be an as-yet-unknown causal connection. Now if there are nothing but wild and improbable ways to explain a claimed correlation, the probability rises that the correlation is not genuine. We all know hundreds of such claims in the past. The one I sent you about pi and magic squares *could* be the result of some unknown pattern in the digits of pi. Even though the digits of pi have not been proved patternless, the odds are overwhelming that the correlation I sent is simply a coincidence.

Now back to G. In the absence of any reasonable causal explanation (direct or indirect), the correlation claimed is so extraordinary that the probability is close to 1 that his raw data is inaccurate. It is not "simple" to check this. It means going back to many thousands of birth records, and who is going to bother? I believe the claimed correlation is so wild (i.e., so unsupported by any reasonable theory) that it should not be taken seriously. You seem to think otherwise, or you would not have dignified G by making him a consulting editor.

But here we go, continuing our argument!

Best,
Martin

[Euclid Avenue] 8 June 1978
Dear Marcello

1. Pi=3.14159... [probably referring to a letter by Lobeck in the files about pi and magic squares.]

2. Gernsback's *Science and Invention*, back in the 20's and 30's, had an offer of monetary reward for psychics with Dunninger (as I recall) as the judge. And earlier, *Scientific American* offered a reward with a committee of some sort. I think Rhine was once a *Scientific American* associate editor of some sort! (I have lots of old *Science and Inventions* in the basement, and perhaps could check on details of their offer, if you like. Maybe you mean current awards?)

3. Re: Rogo debunking Geller, etc. My experience has always been that cranks are very good at debunking other cranks. The orgonomists are good at debunking Scientology and vice versa, and nobody is better at debunking one health food fad than a doctor recommending an alternate one. One of the best Armstrong debunking books I own is by a Seventh Day Adventist. Honorton is good at debunking Uri, also; but of course he is convinced his Felicia Parise can move pill bottles. So it goes!

Best,
Martin

[Euclid Avenue] 11 June 1978
Dear Marcello

1. Feynman thinks that the establishment journals *do* print offbeat speculations provided they have a minimum of merit.

2. Even this new journal is trying to screen out "kooks."

3. Question: If Gauquelin submitted to this journal a sober paper giving what he considers the evidence for his correlation, and proposing that it was worthwhile to seek replication, would his paper be accepted or rejected as "kooky"? I think our argument can be focused by this question. Were you the editor of such a journal, you would find G's paper worthy of reprinting, even though you thought the probability of the correlation being a true one was low. I would consider the paper too kooky to print. In the absence of any inductive logic by which offbeat claims can be weighted, each of us would be relying on our intuition. Is this a fair way to put it?

Best,
Martin

[Clipping from NYT, 21 June 1978: *Radical Science Ideas Acquire A New Forum*, by Malcolm W Brown about *Speculations in Science and Technology*.]

[Zetetic Scholar]   23 June 1978
Dear Martin,

I've been neglecting to answer your letter of June 3rd because I wanted to do it fully and have been so busy lately, but I see that time is fleeing, and there is no let-up in sight before I must go to England; so I'm responding now, albeit not as well as I would like. Actually, my delay seems a good thing since I have received a couple of notes from you since June 3rd and this way we will for once; not have our usual problem of ideas crossing simultaneously in the mails.

I think we have come to a very basic difference between us that is not at all semantic. Your letter of June 3 seems to say that a "genuine pure correlation" presupposes the existence of some explanation even if it may take years for us to find one. Now I agree with you that: "Whenever science finds a genuine pure correlation, the search at once begins for a causal explanation, direct or indirect." But I disagree with what you say following: "The assumption is implicit that such an explanation exists, unless one wants to accept a universe in which distant objects—(e.g. planets and humans) can be correlated with *no* causal connection of any sort, direct or indirect." I disagree in that although it is frequently valuable to presume a likely causal link and seek it out, it seems obviously possible that none may in fact exist at all, that you may have a correlation with no causal links, direct or indirect. Now let me be clear. I agree with you that given such a "genuine pure correlation" it is usually sensible to seek a causal explanation. But the world is full of correlations that are spurious not in the sense that that they are invalidly produced but are spurious in that there is no causal linkage that would explain them. The obvious example, which I cited earlier, is the high correlation between the phases of the moon and a particular's woman's menstrual cycle (I earlier referred to a group of women, but the example is better with a single woman). Both have 28 day rhythms so would correlate very nicely, but no causal link probably exists. Given such a correlation, we might be tempted to look for a causal link; and maybe someone creative can come up with one and produce some research. But most of us would leave the correlation at that. Now suppose we got the *average* pattern of women's cycles and found that correlated with moon phases. That might even tempt us further into looking for a causal link, but we would still recognize that no such link might exist and this might simply be the result of synchronous cycles of two independent events.

As I suspect you know, some years ago someone did a study showing a striking correlation between the stork population and baby birth rates. At first, this was treated as just another "spurious" (in the causal sense) correlation. But apparently someone later did a study linking both of these phenomena together through food supply factors (an indirect causal explanation). In this sense, the research generated someone seeking to establish an indirect causal link between menstrual cycle and moon phases *might* be able to come up with an indirect common cause of both rhythms. But it is certainly is quite conceivable that both cycles are simply independent of one another. You seem to balk at the very idea of a genuine pure correlation that can not be causally explained. Yet the world must be full of thousands of cyclical events are synchronous but quite unconnected.

Since I know you are far better versed in math than I am, I find it hard to believe you are making this error and that I understand you correctly. To me, it seems too basic an error. So I have tried to consider what else you might mean. Possibly the key is your use of the word "pure." I would have called your "genuine pure correlation" merely a "validated correlation." But this would translate into merely "genuine correlation," I should think. So possibly your "pure" somehow includes the idea of causality in some (to me) tautological way. I really don't know and you can clarify this at some point.

This seems quite important re our Gauquelin situation. IF you start out with the presumption that a validated correlation implies (that is, necessitates rather than merely suggests as possible) causality, the validation of G's correlation would strike you as alarmingly extraordinary. Such an extraordinary claim would be seen by you as so extraordinary as to make you presume a high likelihood of error someplace along the line (in G's computations, data gathering, sampling or some such). But IF you take my position that the correlation if validated would merely establish it as a genuine correlation *without* implications of cause being necessary at all, the existence of such a correlation would not be such an extraordinary event at all (just as the menstrual cycle's correlation to moon phases is not so extraordinary). Since I view it not as *that* extraordinary, I have less reason to be concerned with the possibility of some sort of error on G's part. Nonetheless, I am skeptical about G's basic work and think the correlation *curious* (rather than spurious) enough to warrant need for replication and further checks on G's work. Even so, I could understand someone, saying that this is so curious that I think it improbable of being a valid correlation so give it a low priority for research effort. But the fact is that G's data have been checked, first by the Para Committee and more recently by Kurtz et al. The former study and the latter revealed a continued curious pattern (though admittedly not so fully as G had hoped to find). Given these replications and given that it remains a curious rather than revolutionary (causal necessitating) correlation, why the strong reaction to it? You are really more "hostile" to the claim than I am supportive of it. This, to me, seems quite unnecessary given our common areas of agreement re such matters. Gauquelin seems to share with you the idea that a curious correlation should have a causal explanation. Because of this, he, feeling confident about the "genuine pureness" of his correlation, goes looking for some sort of mechanism. You seem to think no such search is warranted because you are less confident (actually you are dubious) about the "genuine pureness" of the correlation G claims in the first place. Interestingly enough, however, G—in his conversations with me—seemed to recognize that there *may* be no causal connection (something you don't seem to want to acknowledge), but he has hopes of finding one because he thinks the correlation valid and more than merely what I call *curious*.

In the last two weeks I have been talking to Ray and Persi about our seeming differences and they have been helpful and interested, so I'm sending them copies of this letter.

***

Reactions to *ZS* have been excellent. You might find James Webb's reaction to your review of interest, so I am enclosing a copy of his letter. I had earlier written him mildly apologizing for the review being negative since I had reason to think he might misperceive it somehow (since I am very fond of him and his work and he might have been surprised at the review).

I had an interview with Boyce Resenberger of the *NY Times* re his doing a piece on *ZS*. Don't know how it will come out since he seemed particularly concerned about my views on the Committee for SICOP (after the aftermath of Nicholas Wade's piece in *Science* and the Rawlins reaction, I am a bit shy on opening any wounds). But I hope it goes well for us all and I guess any publicity for *ZS* will be helpful re subscriptions. Also, Ben Bova of *Analog* wrote me a nice note saying he was planning an editorial on *ZS*. That should boost things, too. So, I am optimistic about keeping the thing going. My only problem will be to extract publishable items from people. But things look good at this time, for the future of *ZS*.

Sorry about my failure to recognize PI as pi in the mathematical anomaly you sent me. I should have seen it right away.

You might find my new little general typology of interest. It is: for all *extraordinary* events:

Extraordinary Event	explained	unexplained
empirical/scientific	abnormal	paranormal
non-empirical/metaphysical	supernatural	preternatural

This corresponds quite well to the dictionary meanings and the common usage.

Best,
Marcello

CC: Ray and Persi

[Eastern Michigan University]   23 June 1978
Dear Martin,

I just realized that I had neglected questions you raised in your letter of June 11th.

Your question as to whether or not the radical science journal would accept Gauquelin's paper if soberly presented or reject it as kooky seems based on some false presumptions. I presume that a radical physics journal would reject G's paper because it is not a physics paper. A science is normally defined by its dependent variables (the variables it seeks to explain). In G's work, the planetary positions are the independent variables, which he thinks may explain occupational patterns. Thus, G's work is essentially sociological or psychological (aggregate psychology). The question should therefore be would a radical psychology or sociology journal take his stuff. The answer would have to be *yes* (unless they would reject on grounds that his work was too conventional) because in fact G's papers already appear (on the subject of his astrobiological claims) in numerous *non-radical*, regular journals. I don't think you realize where G's stuff is already published (in addition to his books and the papers put out regularly by his laboratory). Here are some of the regular journals (he also publishes in some of the odd places like the parapsychology journals): *Population; Le Concours Medical; Historia; L'Annee Biologique; International Journal Of Biometeorology; Gyn. Obst. (Paris); Minerva Medica; Bulletin Biologique; Science et Vie; Cycles; The Astronomical Magazine (Moscow); Journal Of Interdisciplinary Cycle Research; Psychologie; British Journal Of Social And Clinical Psychology*; and many papers at regular academic conferences. If you want, I can send you a copy of his bibliography. I really don't think you appreciate how many non-radical science sources do not see G's claims as so extraordinary as you do.

Feynman is correct for the most part and the fact that G's articles have been published by the appropriate regular, establishment journals demonstrates that they have a "minimum of merit."

I took the references to the journals from G's bib he kindly sent me. I don't know all the journals above but took those whose titles seemed the most legitimate ones. I may have goofed on a couple but I know a number of these journals quite well and they are quite established. In fact, G has published several articles in many of the above. He has also published in many odder journals like *Leonardo*, KOSMOBIOLOGIE and others you might think of as fringe places.

Cheers,
Marcello

[No letterhead]   26 June 1978
Dear Marcello:

Congratulations on the marvelous publicity in the NY Time of last Sunday. My guess is that it will increase your circulation by a much higher ratio than it will increase ours. Did you see the latest Gallup poll on the paranormal? I can't believe that the occult revolution has much to do with a serious scientific investigation of such things. The number of such researchers is too small to be important. The overwhelming majority of people in the US are buying the paranormal for reasons that have noting whatever to do with science.

The new counterculture science magazine, as I understand it, is not limited to physics. The NYT described it as devoted to all science, especially physics. But I've not yet seen a copy, so I really don't know how wide a swath they cut.

In response to yours of 23 June, yes of course there are accidental or coincidental "correlations" all over the scene, but are we not caught again in semantics? As you know, there are sophisticated statistical techniques for deciding if a correlation is the result of pure chance, or significant in the sense of suggesting a causal relationship, direct or indirect. Why cloud the issue by bringing up chance correlations, which concern neither of us? My favorite example is the almost identical size of the sky's image of the sun and moon—a truly fantastic bit of "synchronicity", but I know of nobody who considers it anything but accidental. (Well—not exactly—I have a paper by a Protestant fundamentalist who takes this as a proof of the existence of God!)

As I see it, the central point is this. When statistical analysis shows that a correlation is too strong to be the result of pure chance, it then follows that it must have a direct or indirect causal explanation. G's correlation is in this category. If the causal explanations, are wild and improbable, one suspects the raw data. We disagree over how wild the possible causal explanations are. I think they are too wild to take seriously. You think they are sufficiently unwild that one should not call G a "crank," indeed, one can call him a "distinguished" scientist worthy of serving as a consultant to a periodical. It is this subjective difference in our intuition that divides us. All I can do is repeat that I believe my intuitions reflect the science community at large, and yours do not.

Best,
Martin

CC: Ray, Persi

[Eastern Michigan University]   4 July 1978

Dear Martin,

I think we may be getting to the real point of difference between us. *If* I am correct it is not semantic at all. Let me try to explicate: In your letter of June 26th you write: "there are sophisticated statistical techniques for deciding if a correlation is the result of pure chance or significant in the sense of suggesting a causal relation, direct or indirect." You later state: "When statistical analysis shows that a correlation is too strong to be the result of pure chance, it then follows that it must have a direct or indirect causal explanation." I think both of these statements are incorrect. We simply do not have such statistical tools available in the sense that I think you are describing them here. In the broadest sense, first, statistics tell us about likelihoods and not what must be, but my disagreement with you here is far narrower as I will try to explain.

Let us take Gauquelin's case as an example. G found a correlation between athlete's times of birth and the positions of Mars during the days of their births. This simple positive correlation, initially, was surprising on its face. At first, there was no concern, with its relationship to chance distribution in any formal way. The question of statistical chance measure came about as follows: This correlation was compared to another correlation gained by drawing other variables from a similar sample. These twos samples were then compared and it was statistically shown that they were unlikely drawn from the same population. In other words, taking some other occupation in relation to Mars positions did not get you a similar correlation. This simply allows the conclusion that there is something special about the athlete-Mars correlation. It does *not* show that the non chance correlation is the result of any causal relationship. In Gauquelin's case, the Zelen test simply compared athletes in Gauquelin's study to other persons in the same cities born on the same day. This was done *not* to show there was no causality involved but was an attempt to show (and would have shown if the test went against Gauquelin) that the high correlation Gauquelin had found may have been due to some geographic variable that seemed (to most skeptics) more plausible explanation of Gauquelin's correlational finding. At *best*, if would have suggested a geographic causality was operating rather than an occupational one.

I think Persi will back me up when I say that (a) we do not really have the statistical tools to distinguish a correlation from a causality, or at least (b) any such method would be quite complex and has not been applied to Gauquelin's work in any case. From what little I understand of the matter, the use of statistics to establish causality is very much disputed by the experts, anyway, and this is further compounded by the philosophic issues surrounding the reality of causality as a useful construct at all. I would further argue that demonstration of causality can only be achieved through the use of an experimental design in which variables can be manipulated in a way totally impossible with Gauquelin's claim (since you can not really manipulate the major variables, Mars' path and occupational choice, anyway).

I am glad that you accept the reality of purely coincidental correlations. Where we differ is in that you think there are sophisticated techniques for knowing when you have one of those and when they are really causally linked. There are those who would argue, philosophically, that you can never be certain about even the best causal case not being just a coincidence (a kind of ultimate Humean position). I personally don't go that far. I still think the concept of cause is a useful one. But the kind of sophisticated statistics you speak of—and here I must ask for comment by Persi and Ray—(a) don't really exist at this time, and if they do, (b) they are not applicable to a case like Gauquelin's. I know that in sociology Blalock and others have long been concerned with trying to develop correlational techniques that might lead us to causes, but I think merely progress rather than a solution to this question is all they have so far. In short, I think you misunderstand the references to non-chance correlations referred to in the Zelen test when this non-chance element has nothing to do with causal connection being implied.

I hope this clarifies rather than muddies the water between us.

***

Glad you were pleased by the *NY Times* story on *Zetetic Scholar*. Naturally, I was happy about it, and it has brought in some subscriptions (though somewhat less than I had expected). My only regret is that the article referred to the "breakaway group" even though that is really inaccurate since no group has broken away at all, merely me. I tried hard to explain that *ZS* was really a complement to the work of the Committee since we had overlapping people involved, but he obviously preferred to play up the conflicts in his piece and went slightly too far. Overall, however, I think it was accurate.

I go to England and Scotland next week for 3 weeks during which I hope to see many of the people involved with the paranormal. Even hope to meet with Ricky Jay.

Best,
Marcello

CC: Persi, Ray

[Euclid Avenue] 11 July 1978

Dear Marcello:

Your letter of July 4 seems to me to muddy things even more, mainly because you have a way of attributing to me views that only an ignoramus would hold. Whenever I write about statistical evidence for causality I take it for granted that we both know that science operates only with probability estimates, never with certainty. I've said this so often that I feel no need to keep repeating it.

Now there are statistical tests for analyzing data in such a way as to render it probable that some sort of causality is operating. The degree of probability suggested by such tests is vague, and there are sharp disputes among opposing statistical schools. But let's take the simplest of all tests, the chi square, so much used in ESP work. It certainly can distinguish between data that is likely to be the result of pure chance, and data that suggest an underlying causality of some sort. When you write, "I still think the concept of cause is a useful one," I almost fell off my chair. Do you know of *any* philosophy of science, or any scientists, that dispense with the concept? I don't. On the quantum level causality is indeed abandoned at the moment of measurement, when the wave packet is reduced and a quantum system acquires a property being measured. But on the macrolevel (which is what we are concerned with) causality is the very heart of science. I am well aware that "causality" is defined in various ways, but who dispenses with it?

Lest I be again misunderstood, let me say once more that by causality I don't mean just direct relationships such as a ball hitting another one. I mean the "explanation" of the correlation by a natural law or several laws of some sort. The causal relation may be extremely indirect, like your example of escalators and revolving doors. The law involved may be purely statistical, like the laws of entropy or laws that govern card shuffling.

A chi square test may suggest (to a high degree of probability) that the correlations are not pure chance. When this happens, one searches for the explanation, which may be some sort of bias, cheating, or a thousand other things.

So we are back where we started. The data produced by G is clearly too strong a correlation to be explained as pure coincidence. This suggests a causal explanation. In view of the absence of any plausible explanations, and the simpler explanation of bias in the raw data, most scientists are less impressed by G than by a hundred other claimants with stronger evidence crying for replication. I think we both realize that G is in the news only because of the astrology craze; and I find it sad that you are adding to his publicity.

Let me recommend a recent book (1971) that I think you will find philosophically clarifying on the role of causality in statistical explanation. It is *Statistical Explanation and Statistical Relevance*, by Wesley Salmon, Richard Jeffery and someone else [James G. Greeno]. (Pittsburgh Press paperback.)

Best,
Martin

CC: Ray, Persi

[Zetetic Scholar]   1 August 1978
Dear Martin,

Your letter of July 11 really surprised me. First, I regret that you feel that I have imputed ideas to you that "only an ignoramus would hold." Though I think you know I respect both your knowledge and intelligence, I feel that I tried quite hard (and quite respectfully) to (a) quote you directly as often as possible, and (b) demonstrate in my letters how I received my interpretation of your meaning from your words. I frankly still do not see from your last letter exactly where I "went wrong" in my interpretation. In addition, in my few telephone conversations with Persi and Ray about our correspondence, which they have been sharing in large part, both of them said nothing to correct my interpretation. Whatever they think of your views (and I think they both may find flaws in both our cases), they *seem* to have accepted my interpretation as a correct representation of your views. Now, of course, we may all be wrong in our interpretation, but my point is simply that I think they probably would share—at least in large part—the constructions I have placed on your words. If so, it seems to me that the burden must then be on you to make more clear where we (mainly

I) have gone wrong. I do not see this to be the case in your last letter, so don't really know if I can properly respond. Thus I think I had best stop at my end and allow either Persi or Ray (or anyone else you want to let see the exchanges) to clarify the picture for us.

One thing of possible importance, however: We are obviously not speaking clearly to one another re the question of basic causality and its role in science. I think it can be amply demonstrated that a very real debate, exists in the philosophy of science as to the role of causality today. Since I would assume that you are familiar with much of that (Bunge's book, the Hook volumes anthologizing the debates, etc.), I think we must be miscommunicating here. Since we both agree on the need for "explanation by laws" in science, though we might disagree about the role of causality in such laws, this may be irrelevant anyway. In any case, I will try to get hold of the Wesley Salmon book you cite in the hopes that this will make your position clearer to me.

So, the ball is—in my view—either in your court or in a neutral corner awaiting comments from Persi or Ray. Somewhere along the line, I think you and I will have to get together for an evening of discussion if further progress is to be in getting us together.

My trip to England was very useful. I particularly enjoyed the time I spent with Eric Dingwall whom I think is probably the most interesting person alive in terms of life-experience. He seems to have known just about everyone from Conan Doyle to Crowley. I hope someone quickly gets on with writing his biography before he leaves us. His library of some 90,000 works on bizarre matters is alone an incredible thing.

Also spent a pleasant evening with Colin Wilson. His book *The Mysteries* should be out shortly and will sequel his THE OCCULT. In large part, the book deals with Lethbridge. Since he spoke warmly of you, I presume you are in touch with him.

Best,
Marcello

CC: Ray, Persi

[Euclid Avenue]   3 August 1978
Dear Marcello:

There are all sorts of ways of defining, causation, and a thousand difficult problems connected with the concept, but all this is beside our simple point at issue. There is no "my position" that I hold with respect to these highly technical debates. I know of no modern scientist or philosopher (including Bunge, Hook, etc, whom you cite) who deny the role of "causation" (in *some* defined sense) in science. There are statistical techniques by which data can be surveyed, and one can reach probable conclusions about whether the correlations are the result of pure coincidences or whether they follow from laws, which is one way of saying, follow from some sort of "causation." Even, the extreme phenomenologists, who deny that there is any kind, of material substratum beyond experience, recognize regularities within experience and hence "causation." The only point at issue is whether G's claimed correlations are sufficiently strong to be ruled out as pure coincidences. They clearly are too strong for that, and that's why he and you want independent replications made to find out how to "explain" the correlations. There is no need to get into the tangled, difficult questions that divide contemporary philosophers of science over problems involving causation.

By the way, Anthony Standen, a Roman Catholic whom I don't particularly admire, but with whom I exchange friendly letters, has just sent me his latest book, *Forget Your Sun Sign*, published by Legacy Publishing Co. [address withheld]. It is an attack on astrology, and includes one chapter on G. I have not read even this chapter, but perhaps it is a book you will want to obtain and review.

There is a good article on "causation" in the *Encyclopedia of Philosophy*. I certainly agree with you that, as you say in your letter, "a very real debate exists as to the role of causality today." As to its nature and role, yes; but as to the relevance of these debates to the very simple question of G's correlations, and whether they are pure coincidences, or the expression of some sort of causal law, there is no relevance whatever. It's all very simple. Either G has found what you call a "white crow," or he hasn't. His correlations are too strong to be pure chance, so either he is onto something unusual, or his raw data is (as I believe) biased. Let's not unduly complicate this simple issue by getting into philosophical debates over the nature of causation.

> Best,
> Martin

CC: Persi, Ray

[Zetetic Scholar]   8 August 1978
Dear Martin,

The question of causation was, I think, first introduced by you when you claimed that a correlation that was valid implied some sort of causation.

That's why I went into it. I agree that it is irrelevant in light of your apparent substitution in meaning of some sort of law-like relationship for your earlier one of causality. I quite agree that data if true suggest a possible law-like relationship. But I would strongly argue that a law-like relationship can exist without any known or understood mediating mechanism, which is what I presume you meant by causality in your earlier letters. The question I raise (aside from the more major issue of an intervening variable that might account for G's claimed result) is whether or not one can have a law-like relationship between two variables that seem to have no connection between them (no mediating mechanisms) or at least with no concern for possible connections between them. You seem to be insisting that some sort of reductionist argument is necessary to account for a true astro-biological correlation. I suggest that it *may* be the case that we deal here with a non-reducible (at least for the present) "pure" science of astrobiology. We argue along similar lines in other sciences such as Skinnerian psychology or Durkheimian sociology. Such pure science positions argue for a simple input-output model treating what is in between as a black box. Why not a black box solar system as well as a black box cognitive-man or black box society? If G's claims are to be true, the astro-biologists have as much right to argue that mechanisms are irrelevant to their generalizations as Skinner argues that cognitive states and physiology are irrelevant to his Stimulus-Response correlations. The only difference I see is that for most people Skinner is more plausible because his black box (though irrelevant to him) is known to contain some mechanisms thereby making his view seem more plausible. But in G's case, you and others see his "big black box" as empty. But of course G has not claimed the box is empty and has looked for possible mechanisms within it that might explain to, him (and you) why he gets those damned correlations.

Let's not lose sight of our *basic* difference. You think psi is less extraordinary than an "unknown" factor that produces G's correlations. This unknown factor could simply be an intervening variable that is quite ordinary. No such "ordinary" explanation for the existence of psi exists (except the claim of fraud or incompetence which does *not* explain but *denies* the existence of psi). If one accepts psi as valid (not merely a methodological error) and compares it with accepting G's correlations as valid, I simply don't see how you can argue psi is less extraordinary or more plausible than G's correlation. G's correlation says nothing directly about the existence of any new (paranormal) force. It is itself an anomaly but it can potentially be explained quite normally if we can find a third factor that accounts for the correlation. In the case of psi, we are not speaking simply of the "correlation" but the inference from that correlation which the parapsychologists make. The simple correlations that parapsychologists make might be equated with G's correlations and thus be explained by a third variable such as a non-verbal cue. But in the case of G's claims, the difficulty seems to be any third factor since we can not easily conjecture a connecting link such as a cue from a planet to an occupation. I will agree that the correlation G claims is more extraordinary than the correlation the parapsychologists claim for this reason. But our issue has *not* been over which correlation is more extraordinary. Your claim was that the parapsychologist's claim of the existence of psi is more plausible than the claim of Gauquelin that a correlation exists. If Gauquelin claimed sort of strange paranormal force existing between the planets and choice, you would have a parallel case; but he does not. That is the point I think you may be missing that I am trying to make, I hope so. . . .

I really thought the last letter was the last in our discourse on this business until we got together at some future point for direct conversation on it. But I couldn't resist the above last ditch effort at gaining your agreement.

I enclose a copy of my letter turning down the editorship of Psychoenergetic Systems.

Cordially,
Marcello

CC: Persi, Ray

[Euclid Avenue]   14 August 1978
Dear Marcello:

Your last letter opens new vistas of ambiguities. Sure, we can regard the solar system, or the entire universe, as a black box of unknown interior. This is how Koestler explains psi. Jung called it synchronicity. And the claim of parapsychologists about an unknown force is no less plausible than the claims of physicists about the weak force, and the nuclear force, to account for strange correlations. Come to think of it, even gravity and electromagnetism were unknown forces, posited to explain correlations long before there was much evidence that they existed. In short, there is nothing extraordinary about positing unknown forces. Actually, the fashionable explanation of psi at the moment is Walker's QM theory—there was an entire international conference on QM and psi a few years ago, attended by such distinguished physicists as Gerald Feinberg and, Costa de Beauregard of France. In Walker's theory, no force of *any* sort goes from here to there. It is the view favored, by the way, by P and T.

I never altered my meaning of causality. I first used it in the same informal way that scientists use it. It was only when it turned out that you were taking it in some literal sense, such as one ball hitting another, that I became more specific and defined it in the way that it is now universally defined. Of course scientists argue over what "causality" means, and they discuss such fantastic questions as whether it can go backward in time (as Feynman has suggested), and so on, but it's a perfectly respectable word that everybody uses.

Suppose 10 light bulbs simultaneously burn out in my house. This is such a remarkable correlation that I immediately try to think of a causal factor, not of black boxes or unknown forces, and I conclude there was a power failure. G has presented some extraordinary correlations between planetary positions and professions. What leaps to the mind of almost, every scientist, and anyone outside of astrology, buffs and persons who are into, the paranormal, is that the causal factor was a failure to obtain careful records. Were it not for the astrology boom, no one would be taking G's claims seriously. I believe that the probability of biased raw-data is so enormously more probable than finding what you call unknown variables (i.e., indirect causes), or viewing the solar system as a black box, that I don't think G should be taken as seriously as hundreds of other claims of white crows that turn up in the establishment journals every month.

Last week I received a paper from a reputable medical man who finds he can calculate rectal body temperature very accurately by taking the mean between the boiling temperature of water, and the freezing temperature (on an absolute scale) and multiplying by Euler's constant, e. This is probably just a coincidence, but certainly more worth looking into than G's claims.

You could do science a service by printing white crows of these sorts, rather than boost the respectability of a crank who is riding the crest of the astrology boom, even though the believers are too dumb to see that if G is right, all astrology is wrong! I don't think G deserves it, and I look into my crystal ball and make the following prediction. I think you will eventually agree.

Best,
Martin

[Euclid Avenue]   18 August 1978
Dear Ray

Yours of August 8 to hand, just read with great pleasure, and I must say at once that I find nothing in it with which to disagree except, to repeat my previous letter, over intuitive estimates of degree. These are impossible to argue precisely because, as you point out, we have no good rules for distinguishing good from bad science,

I fully grant the marked distinctions between your three categories. I, too, agree that parapsychology is "qualitatively different" from pop occultism. I have no a priori objection against ESP or PK, for example, and nothing in my world view would be shattered if the evidence for such forces became strong enough for me to accept them. My attitude is exactly the same as that of William James, the American philosopher I most admire and whose photograph hangs in my library. I think that James was gullible on many occasions, owing to his lack of knowledge of methods of deception, and his almost total ignorance of mathematics got him into occasional trouble, but he was a Platonist (as am I) in the sense of having a marvelous sense of wonder at the infinite mystery of being, and open to all possibilities.

A word about "qualitative difference." In almost every case (the exceptions being mainly in mathematics), qualitative differences are ways of isolating widely separated portions of continuums. Red and blue are certainly qualitatively different, yet both lie on the same spectrum, and neither has precise boundaries. This brings us back to the point of my last letter. I think you sometimes overestimate differences in degree between parts of spectrums; you think I sometimes underestimate them.

To be specific: I cannot seen *much* difference in degree between the work of P and T and the work of Tart. All three are certainly accepting the rules of science and doing their best to observe them. The question at issue is how competent are they at their tasks. So far as you know, Tart's results are far more sensational than those of P and T—indeed, his results are the best so far obtained by any parapsychologist; they are so high, in fact, that it is one reason why he is looked upon with such suspicion by other parapsychologists. As you know, Tart has taken a leave of absence to work with P and T on their latest remote viewing grant from the government, so all three will be in close cooperation and hard to distinguish from one another. Another example: I do not see *much* difference between my neighbor Gertrude Schmeidler and Thelma Moss. I've heard Gertrude lecture on PK (with fulsome praise of Geller, Ted Serios and Ingo), read her papers on Ingo, and exchanged several letters with her. True, she's a cut above Thelma, but how big is the gap? Obviously there is no way to settle such estimates. I think the gap is "small."

At the heart of my disagreement with Marcello is the same sort of difference in estimating degrees. Marcello has called Gauquelin a "distinguished scientist." In the light of my total experience in reading the works of cranks, and reading about them, and at times meeting them, I find G the very model of a crank. Marcello once wrote that he thought Velikovsky was "probably" a crank. But to what degree? I would have expressed it by saying that V is "almost certainly" a crank. Again, the vague difference in how one estimates degrees.

As you suggest, the problem of demarcation is tied up with the question of finding rules that provide "weights" for inductive hypotheses. Neither Carnap nor Popper (who differed, in my opinion, mainly in the choice of languages, not over anything basic) ever imagined they were anywhere near to finding such rules. In Carnap's terminology, one seeks a way to give a numerical value to c (degree of confirmation) of h (hypothesis) on the basis of e (total relevant evidence). Of course to do this it all has to be expressed in a highly artificial, formalized, logical language. In Carnap's big work on this, *Logical Foundations of Probability*, he got no further than a trivial language with one-place predicates. He thought it might be centuries until a practical "logic of induction" could be developed. Popper denied that "induction" even existed, but then he immediately let it in the back door under another name. At the moment, there is nothing anywhere near a way of evaluating even the simplest hypothesis. Working scientists simply make intuitive estimates of a theory by expressing their belief that it is "probably" true, or "unlikely," or "confirmed to a high degree," etc. At extremes, everyone can agree that the probability the earth is flat is virtually zero, and the probability it is round is virtually 1. But in the gray in between areas science faces the same ambiguities about values as you find in all areas where there are value judgments, ethics, aesthetics, politics, etc. Critics all agree that Shakespeare is a greater poet than Eddie Guest, but on, say, Ezra Pound there are critics who call him the greatest of modern poets and others who agree with Nabokov that he was a "total fake." I go on at length about this because think the situation in parapsychology is similar. Many of the most respected paraphysicists think that quantum mechanics is the key to psi, and treat E. H. Walker with great respect. When I recently asked McConnell what he thought of Walker's views, he responded with one word: bullshit. I find it not at all surprising that you, Marcello and I would have sharp differences in our estimates of the competence of various parapsychologists.

Now about those two cycles being in phase—a menstrual cycle and a python that eats every 14 days. Of course this is a "correlation" of a sort, but I would call it a spurious correlation or a coincidental one, provided (naturally) that there was no indirect way to link then causally by way of, say, the moon's 28 day cycle. Let's take two famous examples from astronomy to make my terminology precise. Venus turns on its axis so as to keep the same face always toward the earth—a surprising fact discovered fairly recently. This correlation is so exact that no astronomer doubts a causal connection. They are in a "resonance lock" and there are many respectable theories to explain how they got that way. Contrast this with what I call a coincidental correlation: the fact that the sun rotates in almost, the same number of days it takes the moon to circle the earth. This spurious correlation has been responsible for many hypotheses that turned out not to be viable. Astronomers now all believe that the rough correlation is a coincidence. It is not exact enough, and there are no plausible hypotheses to explain a causal link, direct or indirect.

I see no connection between such "in phase" coincidences and my debate with Marcello. G's correlations are clearly too strong to be coincidences. If they are not coincidences, one then seeks for plausible explanations via causal chains. Here enters the difference in intuitive estimates of degree. I find all explanations for G's correlations to be so implausible, to such a high degree, that I think it far more probable that his raw data is biased, probably because at first he accidentally hit on some slight initial correlations (due to sheer chance), then hooked himself on a private theory and from then on found what he looked for by unconsciously biasing his interpretations of vaguely defined raw data. I could, of course, be wrong. It is all a matter of estimating degrees of probability.

You ask about gravity waves. Relativity says they *must* exist, so if relativity theory holds up, it is only a question of time until instruments can be made sensitive enough to detect them. Weber's claim turned out to be premature. He did not have "blind" safeguards in interpreting very cloudy statistical data, and most physicists believe that his is the classic case of a scientist finding what he wanted to find. About a dozen replications, some with better equipment than Weber's, has failed to detect gravity waves. But you'll have to ask a physicist for the references. Nobody considers Weber's work pathological in any sense. Nobody calls him a crank, any more than they call Robert Dicke a crank because his modified relativity theory (supported by what he thought was an oblate sun) is now almost totally discredited. To me, what happened to Weber's claim, and to Dicke's claim about the sun, are paradigms of good science. The big lesson here is how quickly both claims were discredited when the disconfirming evidence came in. Contrast this with the impossibility in principle of sceptics ever disconfirming ESP results, because of Catch 22 and various other catches. As I mentioned earlier, Tart replicated his work with a better randomizer, and a way of eliminating a time-delay code, got poor results, but shows no sign whatever of doubting the results given in his U of Chicago book. This is one reason why I estimate a "small" degree of difference between Tart and Targ.

The one place in your letter where our differences in degree came out most strongly was when you mentioned Tart's backing of Ted Serios, and then you asked "Why not?" If one believes in paranormal phenomena at all, you say, "it is not hard to believe" in phenomena such as Ted's. My position is that there are vast differences in degree between the plausibilities of claimed paranormal phenomena. I see an incredibly wide gulf between the plausibility of ESP, and the plausibility of one man's ability to project from his mind to a Polaroid film the exact line-for-line image of a photograph he had seen ten years earlier in National Geographic. And after a highly plausible way of doing his thing was explained by Reynolds and Eisendrath, with the help of Persi, surely you must agree that Tart's willingness to accept Ted's work as genuine differs to an enormous degree from his willingness to accept ESP, PK and precognition.

From my point of view, the wildness of Ted's claim is even greater than the present levitation claims of TM. After all, levitation claims have a long history. They are part of the beliefs of almost all major religions. Both Judaic and Christian scriptures contain examples. Aquinas was sainted because one of the miracles was his levitation in the air while praying. Then there are the levitations of the famous mediums—D. D. Home floating in and out of that window for example. As a matter of simple fact, far more people believe in the possibility of levitation now, all around the world, than those who believe in the Ted phenomena. (What do P and T say about Ted, I wonder? I'd be surprised to hear that they thought he was not a fake.) But my basic point is simple and worth repeating. There are marked differences in degree between the plausibilities of various paranormal claims. Panati once told me on the phone that he took for granted that Ted Serios was a fraud, and I have been told this is now Rhine's view of the matter. So I do not think we should so easily excuse Tart and Gertrude and Scott Rogo, etc., from their continued defense of Ted. It suggests something very basic about Tart that must be taken into consideration when you try to place him somewhere on one of the continuums.

By the way, if you haven't seen McConnell's slashing review of "Future Science," by White and Krippner, write him for a copy. He wrote it for the *Journal of the American Society for Psychical Research*, but they refused to print it, so he is sending it on request. The point of his review is to distinguish between your groups 1 and 3. He argues that the book does a great disservice to genuine parapsychologists (such as himself, of course) by blurring this distinction. My own opinion of McConnell is that the distance between him and Krippner (whom he attacks) is not "enormous." But what does "enormous" mean here? Since we haven't any hard rules, it's just my gut feeling, and little more.

Best,
Martin

CC: Persi, Marcello, Randi

[Euclid Avenue]   25 August 1978
Dear Marcello:

Here's that reference on rectal temperature. I probably remembered it wrong, but I don't have my letter to hand. The point is that it is a curious correlation involving boiling point of water, freezing point, the number e, and rectal temperature. Coincidence? Probably. I mentioned it only because correlations like this turn up literally by the score every month in various "establishment" journals, and strike me as closer to the kind of anomalies by which science makes progress than G's claimed correlations.

I think your proposal is worth doing. Only two or three thoughts occur to me at the moment, and I'm sure you've considered them. One, that you must make absolutely, clear at the beginning that you are concerned *only* with actual hits of the three-digit number itself, and *nothing else*. Otherwise you know what can happen! Parapsychologists will go over the raw data looking for an unusual number of hits close to the number (such as 1 above or 1 under), or an unusual number of correct guesses of two digits, or an unusual clustering of *misses* that are very far from the number (hence indicate that familiar phenomena, so beloved of Rhine, of "psi missing"), and so on. Since there are so many ways to find unusual correlations in a table of 200 guesses and one target, the chance of finding something becomes very probable. So you must stress the point that you are checking on the number itself, and nothing else.

Also, in evaluating the expected number of "hits" the formulas will be a bit different depending on how you supervise the 200 students. If they see each other's guesses, for example, and hence avoid duplications, this alters the formula. This is easily seen by considering a deck of cards. One is drawn, and 52 students each guess a card. If each guesses a different card, the probability of one hit is absolutely certain. Whereas if the guesses are independently made, no one aware of any other guess, there will be many duplicates and the probability of a hit drops below 1.

You might ask Persi if he knows how good the last three figures of the stock exchange report are as a random 3-digit number. I assume they are random, but there are strange statistical anomalies that sometimes hang around such "natural" random numbers—such as the famous one (on which Persi wrote a paper)—that in natural lists of figures (such as populations of towns, etc) the probability that the first digit is 1 is higher than that it is 2, in turn higher than 3, and so on to 9. Very counterintuitive. Probably the phenomenon doesn't apply to final digits, but I'd make sure. The randomness could also be affected if there is a practice of rounding last digits up or down, and so on.

Someone sent me a catalog of September workshops coming up in Sept. at Burklyn, in Vermont. Charles Tart is giving a workshop in cooperation with Marilyn Ferguson, who edits the *Brain/Mind Bulletin.* Other workshop leaders include Christopher Bird, Judy Skutch, etc. Ray recently brought up the point that serious parapsychologists such as Tart are annoyed when they are associated with "pop occultism," yet Tart is surely partly to blame by his constant association with "pop occultism," such as this conference, which is dominated by it. Ditto for Rhine, The back of Mishlove's *Roots of Consciousness* carries endorsements by Rhine and Uri Geller. Etc, etc. I bring it up to strengthen my thesis that the differences in degree between Tart and, (say) Christopher Bird, are not as wide as I believe Ray thinks they are. But as I said in my letter to Ray, these are intuitive judgments that can't be backed by any kind mathematical evaluation.

Best,
Martin

[Euclid Avenue]   30 August 1978
Dear Marcello:

Fulves is still sitting on my sequel to the Uriah Fuller book. Don't know when he plans to publish it.

Jay Marshall, ten years ago, bought from me rights to reprint a long series of articles I did for *Hugard's Magic Monthly,* called Encyclopedia of Impromptu Magic, in which I summarized, in an almost telegraphic style, the basic tricks and stunts that call for no apparatus or advance preparation, and organized alphabetically from apple to zipper. He's finally getting around to issuing it, warts and all, and I have no idea how it will look. The material is old, and I could now expand it to twice the size, but who knows if I'll ever get around to it!

Didn't know about Persi's resignation, and thanks for the copy of his letter. Will Ray be next? I was always amazed that Tart would be willing to serve as one of your consulting editors, particularly with me as another one. As my recent letter to Ray indicated, I cannot see any vast difference between Tart and P and T, and now that all three will be working together with Ingo, I think they will blur together even more.

Yes, I know Mencken's essay on Veblen—one of my favorites. Mencken was particularly amused by Veblen's explanation of why wealthy people kept dogs and cats as pets, rather than cows. (Veblen had his blind spots!)

We are all breathlessly awaiting the first issue of *Omni*, the big new science magazine, to see how it going to play the paranormal.

Best,
Martin

[Eastern Michigan University]   2 September 1978
Dear Martin,

Glad you approve of my basic pre-posting concept. You are absolutely correct about our having to make it clear as to what statistical tests will be acceptable as part of the pre-posting. But it is not just a question, as you point out, of which stock exchange numbers have to be how close, but the actual statistical tests themselves need to be pre-specified to avoid any suggestion of a fishing-expedition post hoc. We must very carefully design this thing and that is mainly where the skeptics will come in.

I am not so sure of the problem of independence that you mention re the students' seeing each other's answers. In the case of cards, with 52 students, your point is partly true. But with a larger number (000-999 and I am now thinking of making it the last 4 digits instead of the last 3), that may be less of a factor. In any case, lets take your 52 card example. If 52 students with perfect memories hear each other and name all 52 cards, there would be a hit, but a single hit would be statistically insignificant the same way a single subject guessing at all 52 cards would be should he consistently say "Queen of Spades" for all 52 of his guesses. If there is any problem here, it might be because a student sees another student write 666 and copies that same number. This would produce 2 hits should 666 actually come up when only one student was really guessing while the other was copying his guess. But even this is probably statistically unimportant just as a subject feeling strongly about the Queen of Spades might say it 3 times in a run of 52 cards and thus happen to hit should the Q of Spades actually come up. (If the cards are really random—as presumably is my stock exchange number—the subject can cancels any special advantage through guessing Queen of Spades 3 times since this is really no different than saying Queen of Spades all 52 times.) In any case, I will consult with Ray and/or Persi on this matter.

I will certainly check with Persi re the actual randomness of the Stock Exchange numbers. I have been told they don't round off the number, but will seek to find out more. We may have to use the last digits of the dollar amount (I presume they don't include the cents and rounding off should be evenly distributed). In any case, this clearly needs looking into. Of course, what I am looking for is simply any agreed upon number generator that we'd all agree is independent. There may be something (some other index) we could better agree upon than the stock exchange volume I have suggested.

One other thing I don't think I mentioned in my memo. I will have a control group making no guesses, simply give me a 3 digit combination. It may well be that certain numbers (e.g., 666) will come up commonly and a chance hit of the market on 666 would then be misleading. In fact, should it be 666 that comes up, and if 666 was culturally more common a guess anyway, it might even be that this culturally common guess would be more culturally common among believers in psi (because of psi's connection with belief in the spiritual and the relation of 666 to biblical salience). So, a control group is an important part of all this. In fact, it might even be better to use a large control (matching) group to guess at "nothing" (i.e., just produce a 3 digit number) as the empirical distribution to compare with the psi-guessing distribution rather than use any purely theoretic statistical distribution at all. I am basically for using empirical rather than statistical distributions anyway since the latter's assumptions may not be met. What do you think?

Finally, re the matter of the 3 categories of Ray's psi researchers. In this area, I am inclined to agree with you in some ways. Tart is a good example. In Ray's defense, Tart's mere association in public with the less responsible researchers really says little about the plausibility of his own work (as opposed to questions of his character and credibility as the narrator of his alleged results). But I think the parapsychological world is really far more multi-dimensional than Ray's categories suggest. I think the issues of honesty versus competence are involved. I would suggest that there are those like von Daniken who I perceive as simply dishonest. This makes me suspicious of anything about which they are the main witness. Here the credibility of the witness is central to everything. But when the noncredible witness presents evidence that is independent of his character (e.g., Todd Zechel's current claim of CIA documents about UFO evidence and investigations), the importance of such evidence may stir us to examine it without much concern about the suspicious character who gives us the documents. Thus, when Targ and Puthoff, suspicious characters, gave us their studies, we were rightly suspicious of their own evidence. Thus I demanded replications by others even more strongly than I urged replications of people like Tart's work. Now that Tart is going to work with Targ and Puthoff, I grow more suspicious of Tart's own work as well. But I still consider the evidence they cite as having an objective character (in Popper's sense) independent of their particular psychologies. That is, if it is true, others can replicate it; if false, others should falsify such conjectures with their experiments. In my view, the objective versus subjective dimensions in paranormal advocacy are really the important ones. I am far less concerned about the integrity of the witnesses and their motivations than I am about the status of their evidence. From all I know so far, I think Tart is honest but not adequately competent (I hesitate to use the word incompetent since that suggests a boob, and he is not stupid) for his tasks. The important thing, however, is that he seems to adhere to my own concern about objective status for his evidence. That is, he wants to play by the scientific rules and is thus, to me, a protoscientist rather than a pseudoscientist. The questions of competency, honesty, etc., all take a back seat to this central common interest in scientific adjudication, for these other issues are also present in greater or lesser degree in all the sciences. I could apply Ray's three categories to sociologists as well as parapsychologists (with people like Vance Packard in the second category, etc.).

Cheers,
Marcello

P.S. Trust you have seen the *Science Forum* article on the Committee, me, the astrology mess at the U. of Toronto, etc.
P.P.S. Thanks for the stuff on the Curious Case of the Rectal Temperature. I may have good use for this oddity.

[Euclid Avenue]   5 October 1978
Dear Marcello:

   Thanks for the article on James, which indeed I had not seen. I admire WJ in much the same degree that I admire Chesterton and Wells, "warts and all." To me, WJ had 3 major weaknesses: he was ignorant of mathematics, he was ignorant of magic (and hence naive in his attitude toward certain mediums), and he launched an attempt to redefine "truth" that turned out to be pragmatically useless (pragmatism is now a dead movement, having been finally finished by Tarski's semantic definition of "truth"). But I love James for his superb writing style, and his sense of the mystery of being which he shared with GKC.
   No, no answer from Puthoff, and none is expected.

Best,
Martin

[Euclid Avenue]   7 October 1978
Dear Marcello:

Joseph May's article is naive, heavily biased, and not worthy of being published by you. His statement "critics have failed to show where his interpretations at any crucial point is a mistaken or an unreasonable rendering of the evidence," just simply isn't true. Sagan's detailed analysis, which May fluffs off with a sentence in a footnote, is filled with details about Velikovsky's mistakes and unreasonable "rendering of the evidence." Not only that, but all astronomers, so far as I know without exception, would agree with all of Sagan's major points. There never was any serious debate over V's views in the community of scientists, only wonderment that his views would achieve such popular success. On not a single point at which Velikovsky's views become outrageous in the light or contemporary astronomical knowledge does May have a word to say. It is not, in my opinion, a responsible contribution to the "Velikovsky debate" (let me repeat: a "debate" that is totally outside the science community), but just another piece of the usual drumbeating for the work of an outstanding crank.

This letter is *not* for publication. I am bored to death with the V flap, and have no desire to participate in the flogging of a dead horse. In *no* sense do the views of V represent what Kuhn has in mind when he talks about anomalies and new paradigms. Like the geology of Price, V's views represent a return to primitive, pre-Darwinian views, developed to justify a literal interpretation of the Old Testament. In Price's case, he wanted to justify Protestant fundamentalism. In V's case, orthodox Judaism. With all the issues around on which there is indeed reasonable debate over anomalies (i.e., the new cosmologies, black and white holes, new views on the origin of the solar system, etc.), why take up space with a scientific nondebate? Publishing May's piece will win you admirers from the V cult (no matter what you publish in the way of rebuttal), but it will diminish respect for you among scientists. And who among responsible astronomers is even going to bother to write any rebuttals? There isn't anything new to say that Sagan hasn't already said (and others). It would be much more timely and interesting if you published, say, an article by Prof. John C. McCampbell, head of the geology department of the University of Southwestern Louisiana, who wrote the foreword to *The Genesis Flood*, by Whitcomband Morris, 1961 (my 1976 edition says its the 12th printing). For every devotee of V, there are 50,000 or more who are taking the new antievolution books seriously. (I assume you know that V also is a creationist; he corresponded at length with Price, with whom he has a great deal in common).

Here again is surfacing our fundamental disagreement over questions of emphasis and degree. You think V is sufficiently noncrackpot to be treated by serious debate in your Journal. I find him about as far out on the continuum of crackpottery as I can imagine for this day. He's just not worth your attention, and May's paper will only prolong pseudodebate that should be allowed quietly to die.

Best,
Martin

CC: Ray

[Zetetic Scholar]   12 October 1978
Dear Martin,

I appreciate your comments on the Joe May article and will, of course not publish them although I do wish you had been willing to write something for that purpose.

I'm afraid I very much disagree with you on the Velikovsky issue. I had been very much an anti-Velikovsky person, I think it would be fair to say, in the early period. My main objection to things had been the overreaction by the Establishment (a la de Grazia's book's description). I read an early version of the Sagan paper and was, I must say, then impressed by it (even though I even then noted that he disagreed with some critics of Velikovsky including yourself about some of the issues raised, e.g., the problem of the stalagmites, etc., which Sagan explicitly said you were wrong about). But I then went on to read the *Kronos* rebuttals to Sagan and the others, and I am by no means convinced that this is a dead issue. Quite the contrary in the sense that I am now reasonably convinced that Sagan simply did not do his homework. In fact, I now have a copy of the tapes of the AAAS symposium and must agree with the pro-Velikovsky critics that Sagan changed a great deal in his paper between then and its publication. Changes which further show his initial sloppiness.

I do not claim that Velikovsky is right. What I do claim is that we are dealing with a legitimate problem and one on which the Sagan and other critiques have been reasonably well answered. Two facts stand out to me: (1) Rebuttals exist which are at least manifestly reasonable and check out well where I have checked the evidence (for example where it is claimed Sagan misrepresents Velikovsky's position—I checked the text and he did). These rebuttals have not been answered by Sagan or anyone else and until that is done, the lid is not on the case. (2) There has been a recent upsurge of interest among respectable scientists in the claims of Velikovsky. The *SIS Review* is a better example than *Kronos*.

I will certainly agree that there is fanaticism among Velikovsky's adherents in many cases. But there is a spectrum of views on this side, too, and Joe May is not among the most extreme by any means. You refer to May as naive, and cite his statement "critics have failed to show where his interpretations at any crucial point is a mistaken or unreasonable rendering of the evidence." After reading Sagan and the response to Sagan in *Kronos*, I can see why May would say that. I like others (I suspect you and Abell are among these) was initially seduced by Sagan's piece. But I must say the *Kronos* rebuttal raised serious doubts. It does not settle the matter, but it certainly does display the weaknesses in Sagan's arguments.

As I see it the central question remains as to whether the matter is open or closed. I certainly see it as still open. Velikovsky's conjectures have not been falsified to a reasonable degree yet, and I am by no means alone in thinking so. And I really would be surprised if you would still think so after reading the rebuttals in *Kronos, SIS Review,* and *Penseé*. From your letters on this, I wonder if you have read these.

You are very likely *wrong* about Velikovsky's trying to justify orthodox Judaism. What is your basis for that? As May points out, the same could be said for V's support for orthodox Hinduism. Even if your motivation imputation is correct, I am not sure it is relevant to the arguments being offered. Certainly, this is not the motivation behind the scholarly supporters of Velikovsky. I did not know that Velikovsky was a creationist, but this could be said of Karl Popper, too, so I don't know what it really means relative to the points at hand.

In any case, the paper will be presented in *ZS* as a controversial but reasonably stated paper. I think you will agree that it is that. The purpose of *ZS* is to promote debates and dialogues on unsettled matters. You seem to want to treat Velikovsky as a closed scientific matter but I think it is easily demonstrable that it is not. These are, after all, matters of degree. We obviously do not agree about your characterization of this as "scientific nondebate." But I hasten to remind you that there are those who would argue the same way about some things like parapsychology where you might be on my side of the fence.

You mention the possibility of something preferable by John C. Mc-Campbell. Fine. But he did not send me an article while May did. The fact is that I am in no position to yet commission articles (especially without payment and a limited readership). I will try to publish the best stuff I can, but so far I have had to produce most of the material myself and have really had little even from the consulting editors.

I will make a special mention reminding readers of *ZS* that the consulting editors are not responsible for the pieces in *ZS*.

I hear by the grapevine that you may have been invited to come to the

U. of Michigan for a talk (I think by the math dept.). If so and if you come, I hope to see you then. The new issue of *ZS* is about ready. I think you will like much of it. Hope so anyway.

Cordially,
Marcello

CC: Ray Hyman

[3 page article on correlations between astronomical and subatomic constants, by Albright]

[Euclid Avenue]   15 October 1978
Dear Marcello:

    I read your 12 October letter with growing amazement. That you can regard the Velikovsky affair as "open," that Sagan did not do his homework and has been "reasonably well answered," that I am wrong in describing V's motives as justifying his orthodox, fundamentalist Judaism (haven't you read any biographical articles on V?), that May's piece is a "reasonably stated defense," and that V has not been "falsified to a reasonable degree," is beyond my credulity. (Incidentally, when I spoke of V as a "creationist," naturally I meant an opponent of evolution, not a "creationist" in the sense one could call Popper one.)

    When I say that the V case is "closed," I mean it was not even "open" in the scientific community. By scientific community I mean 99.99 percent of astronomers and physicists. That you feel otherwise I can only attribute to the fact that you are reading the V literature as a sociologist, not as one informed about modern physics and astronomy.

    Your letter points up such a vast gulf between our respective ways of viewing the continuum that runs from responsible scientific controversy to crankery that I now see that there is no way to bridge the gap. Please remove my name from your list of consulting editors. May's piece is unworthy of you. If I can't convince you of that, then I am in no position to be of any use to you in any capacity. Moreover, I do not wish my name associated with a publication whose editor looks upon V's views as a genuine challenge to establishment science in the sense intended by Kuhn.

    Since I will be sending our exchange of letters to several friends, let me close by summarizing the essence of V's claim. He believes that within the past ten thousand years (actually less), a giant comet erupted from Jupiter, passed close to the earth (causing violent catastrophes which provide rational explanations of such OT tales as the sun standing still, manna from heaven, the parting of the Red Sea, and so on), then finally settled down as the planet Venus. I know of no astronomer anywhere in the world who thinks this is anything but preposterous in the light of several thousand facts.

Marcello—what is happening to you?

In total bewilderment,
Martin

CC: Ray Hyman, Persi Diaconis, Paul Kurtz, Carl Sagan

[Zetetic Scholar]   19 October 1978
Dear Martin,

I was flabbergasted at your letter of October 15th, but I will of course immediately remove you from my list of consulting editors as you requested.

Of course 99.99 percent of the scientific community views the Velikovskian claims as a closed matter contradicted by what they believe to be "several thousand facts." But that does not make something a closed scientific issue. The opposition (the proponents of Velikovsky's views) contend that this simply is not the case: that most astronomers are not aware of Velikovsky's actual position, have depended on people that have misrepresented Velikovsky's position) and that there are no such "thousands" of facts that run contrary to Velikovsky's views. Now it should be obvious that this is a matter that can be rationally discussed and adjudicated. It is also obvious that scientists (those who publish in *SIS Review* and in *Kronos* are certainly among them but are not all of them) share the minority opinion of Velikovsky's proponents/defenders. In addition, there are many of us undecided on the matter who believe (a) Velikovsky was confronted by more dogma than rationality by you and others in the early rounds (this is not only documented in de Grazia's book but has been publicly stated by many people including a mild version of such a statement by Sagan!), and (b) the more recent attempt to "deal" with Velikovsky has simply not settled the matter. I don't see how you can honestly read the criticisms on Sagan's paper, for example, and think that he has shut the case. I am not saying that there is no case against Velikovsky; I am saying that the case is by no means shut against him. *Have you read the criticisms of the AAAS volume?* I have and find these criticisms in need of an answer. The criticisms are *not* just a bunch of red herrings.

IN ANY CASE (I seem to picked up capitalizing things from Phil Klass), let us assume you are completely correct and the pro-Velikovsky claimants are, completely wrong and stupid. Why not simply say that in the pages of *Zetetic Scholar* and point out the idiocies. If you simply think the answers are already in the literature, and you don't feel it worth your taking the time restating them, then just say that and refer people to the literature you approve of. Obviously, though, you simply don't want to legitimate the problem as being a serious problem in the first place. *But that is something the arguments will reveal and can not be judged a priori.* Christ, Martin, what is there to fear from such discussion in *Zetetic Scholar*? Won't the facts emerge in the wash? And if you are proved right, won't you be able to simply say "I told you so." What is the cost of my taking up this matter to anyone?

Of course, you have a perfect right to be associated or dissociated with anything you want. That is your privilege. And I respect that. But do you think the other consulting editors all think every problem that will be taken up in *ZS* is an equally important or serious problem? As you know, I get flack from both sides on these matters. It seems to me that your attitude indicates not merely that you don't want to engage in a controversy on Velikovsky but that you don't want your name associated with the idea that anyone else thinks it worth discussing. The whole point of *ZS* is to find responsible scholars tolerant enough about extraordinary claims that they want them publicly and rationally aired. You seem to be less interested in seeing such controversies get aired than in blocking what you see as illegitimate discussions—or at least withholding your own support for such discussions. Obviously, you see Velikovsky's views as constituting a claim like Freud's example of the man thinking the earth's core is strawberry jam. You think the ideas are so outlandish in the first place that they don't warrant serious discussion. But the same view has been held by many scientists about many issues that others disagreed about. I guess what I am concerned about is my severe disappointment that you are so willing to dogmatically judge what can be seriously discussed by others. I thought you would *in principle* support free and open discussion of any outlandish ideas. But your request to be dissociated from *ZS* seems to indicate that you would feel stigmatized in having your name associated with some discussions of some ideas you think too trivial or too stupid to bear discussion by serious people. I could understand your point if *all* the serious scientists felt Velikovsky's ideas were such "strawberry jam" that they would not waste time on them. But the fact of the AAAS symposium alone evidences that others do not think so. (You seem to view the AAAS Symposium as either a mistake, too, or simply an attempt to preach to the general public to inoculate them against the heresies suggested by Velikovsky.) As I see it, *ZS* upholds a kind of civil liberties for scientific or would-be scientific ideas. I would have hoped you would support the principles involved even if you did not always like the subjects being contested. The irony of all this, to me, is that you seem not to have disapproved the Velikovsky issue of *The Humanist*. This may be because that issue was "stacked against" Velikovsky. But if you and the other critics take advantage of the space offered you in *ZS*, Velikovsky's views can be equally damaged.

Well, what more can I say. I hope you might reconsider, but I presume your mind is made up on this matter. You ask "what is happening to me?" but I don't think my views expressed on this matter are remotely inconsistent with things I have stated to you in the past and seem to surprise no one but you thus far. I will very much miss your support of the journal, but despite our differences here, I am not so foolish as to underestimate your knowledge and talents elsewhere; so I trust we can remain friends in spite of such a thing as this business. I still think we agree on a great deal.

I suspect that before it is all over, most of the other consulting editors will also resign because I am willing to take seriously either a claim or its opponents when they see such as irresponsible. As you know, Tart has already threatened resignation for such reasons.

In any case, thanks for your support of *Zetetic Scholar* for at least its first issue. Should you eventually decide to rejoin my efforts with it, you will always be more than welcome. I respect you just as I respect our differences. As L.B.J. said of Edgar Hoover, I'd rather have you pissing from inside the tent to outside than the other way around. But if you would prefer to criticize the writings in *ZS* from outside its circle, I guess that's the way it will have to be; but I think we will all be the losers for it. Since you sent copies, of your letter to the others below, I am doing the same with this one.

Sincerely,
Marcello Truzzi
CC: Ray Hyman, Persi Diaconis, Paul Kurtz, and Carl Sagan

P.S. Luckily, your resignation got to me before I put the new issue of *ZS* to press this week end. So you will not be listed in the second issue. I will mention your resignation in my general editorial and will send you a copy of the issue when it emerges from the printer.

P.P.S. You seem to think that May *completely* supports Velikovsky's ideas. He does not.

[Euclid Avenue]   23 October 1978
Dear Marcello:

As usual, our disagreements are over issues of degrees of probability in evaluating unorthodox claims. Yes, I do regard V's claims as very close in degree to Freud's example of the claim that the earth's center is made of jello. To use Huxley's metaphor, which I think much superior to "white crow" (after all, a white crow is no more astonishing than a black swan or a blue apple), V's claims are much closer to the ancient claims of having seen a centaur. By the way, let me recommend your looking up Huxley's book, *Hume*, for his chapter on Hume's famous essay on miracles. Huxley has some marvelous passages on the continuum of credibility that runs from credible unorthodoxy to incredible unorthodoxy.

You seem to think that the debate between the astronomers and V can be "settled" by rational debate. I find this naive. Sagan told me he once decided, in conversation with V, to stick to just one tiny topic, the success of astronomy in accurately predicting the paths of planet without recourse to disturbances by magnetic fields. How did V account for this? V paused, complimented Sagan for his perspicacity, and said, "I would like to make a prediction." Sagan awaited the great revelation. I predict, said V, that someday you will decide that I am right. At that point, said Sagan, he realized that rational dialog with V was hopeless. The V case will no more be "settled" than Voliva and his followers were ever convinced by astronomers that the earth is round. It will be settled only by dying out, and even then only to the degree that V takes his place with Price and Donnelley and other crank "catastrophists." There is no way Sagan can ever be "proved" right because there is no way to "prove" anything in science.

I disapproved of May's article because it (1) raised not a single new point that has not been made over and over again by Grazia and others, (2) raised not a single substantive scientific issue that divides V from astronomers and geologists and historians. Because it is such a poor article, old hat and dull, I thought it unworthy of your journal—a poor editorial choice. With so many new and exciting unorthodoxies, centaurs if you will, to write about—why keep going the V debate—or rather nondebate since no astronomer takes V seriously, and indeed, doesn't want to waste time even reading one of V's books? When I called the V case "closed," I clearly meant it in the sense of a degree very close to zero, since 1 and 0 are the unreachable ends of the continuum of credibility. But why go on repeating myself?

                                        Best,
                                        Martin

CC: Sagan

[Euclid Avenue] 24 October 1978
Dear Marcello:

Yours of October 22 just arrived. Your statement of my reasons for resigning are admirably expressed, and I have no criticism to make of them.

I don't give lectures anywhere, as my form letter (enclosed) indicates. I *did* visit Ann Arbor last year, to gather material for my column on Berrocal, and to visit Frank Harary (a graph theorist), but it had to be such a short visit that I did not try to contact you. Charlotte came with me, and my son James (who works in Bowling Green) joined us. If I fly or drive there again for any reason, I will certainly try to stop off at Ypsilanti.

I've been going through my files, preparatory to writing a chapter on "Science: or Why I am not a Paranormalist," for a book of essays tentatively titled *The Whys of a Philosophical Scrivener*. In doing so, I came across two items that may be of interest. One is an extra copy of an essay by Norman Gridgeman, an old friend in Ottawa, that perhaps you have not seen. I send it only because it contains some information about a Canadian test of precognition, somewhat similar to the one you are planning. Gridgeman is a statistician who works for the Canadian government, and occasionally writes about prescience. He contributes often to the *New Scientist*. I found I had a copy of an offprint of the same article, so this can be kept. The other is a copy of an amusing entry from *Scientific American* that points up how easy it is to squeeze correlations out of a mass of statistical data.

Give my best to Ricky. I am one of his greatest admirers, even though we don't see eye to eye about the value of exposing psychic charlatans such as Uri Geller and Suzie Cottrell (Suzie's work with autistic children, in Wichita, by forcing alphabet cards that spell CAT and DOG, is shameless). Uri has resurfaced in NYC, by the way, and is trying to persuade *Omni* (the new science magazine published by Guccione, of *Penthouse*) to do a major article on him. As usual, he doesn't care about the angle, so long as they spell his name right. I also enclose a letter from Uri's former manager, Yasha Katz (you know about his interview on an Italian documentary on the occult). Even though he helped Uri cheat on many occasions, he still thinks Uri's powers are genuine *at times!*

Pip, pip,
Martin

[Attached was this notorious form letter Gardner used but he later regretted its tone:

Martin Gardner regrets that it is impossible for him to:

1. Evaluate angle trisections, circle squarings, proofs of Fermat's last theorem, ...

2. Give advice on, or supply references for, high school science or math projects.

3. Inscribe books for strangers.

4. Give lectures, or appear on radio or TV shows.

5. Attend cocktail parties.

6. Make trips to Manhattan except under extreme provocation.

7. Donate books to libraries.

8. Provide answers to old puzzles.

9. Prepare material on speculation for toy companies or advertising agencies.

10. Put the reader in touch with Dr. Matrix.]

[Euclid Avenue]   10 November 1978
Dear Marcello:

You may know all about the Tychonian Society, but if not, you ought to get on their free subscription list. I heard of it only because the editor sent me their August issue which contains a lengthy review of my *Relativity Explosion*. A very favorable review, by the way, even though the editor thinks Einstein is all wrong. Indeed, the writing in the magazine throughout is unusually literate and well informed. Everything is scholarly referenced, with excellent bibliographies. Do you know anything about van der Kamp? Is it possible that the society is a put-on? I assume not.

Big coincidence department: the final revelations about Soal and the final revelations about Cyril Burt, both breaking at about the same time. Boyce Rensberger had a good story on the Burt case in *NY Times* a few days ago. *New Scientist* has a funny piece on Taylor's letter in *Nature*.

Best,
Martin

[*The Bulletin of The Tychonian Society*]

[Euclid Avenue]   26 November 1978
Dear Marcello:

Vol. 1, No. 2 just came. Most impressive. Like No. 1, swarming with items of great interest and valuable reference lists.

Omni seems teetering on a brink, with Guccione pushing for pro-paranormal, and the editors pulling the other way. No. 2 (Nov) featured an article on psi by Lyall Watson, of all people, but No. 3 (Dec) has a satirical piece on psi and little to complain about. There are two ads for Scientology, which *Scientific American* would not accept. Art Ford, who represents himself as "advisor to *Omni*" was on the Candy Jones show last night pitching his latest book. The title is something like *The Great Acceleration* (I was half asleep when I heard part of the show), and is all about UFOs being life forms, kindly disposed toward us, and compounded of what Ford calls "microneutrinos" which are 100 times smaller than neutrinos. Candy said the book could be ordered through Gotham Book Mart, NYC, so it may be a privately published thing. As I recall, Ford says it is about 400 pages, and largely "undocumented" because, as he said, the documentation would have required another 400 pages. Ford telephoned me before the first issue of *Omni* came out asking if I would be on the staff, and setting up a date to come out and see me. Evidently he later found out more about me, because the date was canceled and I never heard from him again. I have no idea who he is. I think he wrote an earlier book arguing that Kennedy was killed on orders from Nazi businessmen. It will be interesting to see if Oberg reviews Ford's UFO book unfavorably, since Ford and Guccione appear, to be (according to Ford) personal pals.

I was glad to see your discussion of "anomaly" as a fuzzy word. It was Wittgenstein who used the word "game" (as you may know) as an example of a word that has meanings loosely connected like a family-tree graph, so that one meaning may overlap with another close to it on the graph, but widely separated meanings may have, nothing in common at all. For example, there is very little in common with "anomaly" in the Fortean sense, and "anomaly" in the sense' of a scientific result so astonishing that it leads to what Kuhn calls a new paradigm. The classic recent instance, as I've so often stated, is the Michelson-Morley experiment, which was not adequately explained until the theory of relativity was developed. Miller's repeated failures to confirm it then became an "anomaly" in a much weaker sense—i.e., it didn't lead to anything, and was overwhelmed by the thousands of confirmations of the MM experiment. It's unfortunate that we don't have a better vocabulary for distinguishing such things, what struck me was the sharp contrast between the way the word is generally used by philosophers interested in the sociology of knowledge, or the sociologists, and the kind of anomalies stressed in your article. I can't imagine Popper, for instance, taking seriously *anything* in the four books you list as written by "four major anomalists." Nothing reported by Fort has ever led to any scientific change, even on a slight level—both Fort and Thayer regarded Fort's collections as gigantic jokes. Put another way, I think there is only the most tenuous connection between the names cited on page 69 (and the kinds of discussion now going on about the sociology of knowledge) and the trivial sort of material, largely anecdotal, to be found in Fort, Gould, etc. I have no objection to letting "anomaly" stand for such things as sea serpent sightings or rains of frogs, but then we need a different word to stand for such revolutionary surprises as the MM experiment, or the paradoxes that led to quantum theory.

A unicorn, for instance, is an anomaly in the sense of oddity, but has almost nothing in common with the MM test. I mean a real unicorn, if one were found, and not the myth. There is no reason why evolution couldn't, produce a horse with a horn—it produced the narwhal easily enough—so if, say, a fossil of a genuine unicorn turned up, it would be a surprise, but nothing that would upset the theory of evolution, or anything in orthodox geology and biology. I always assumed that when you talked about anomalies you meant significant anomalies that led or could lead to major scientific upheavals. The Geller effect, for example, if validated, would be an anomaly in the latter sense, as John Taylor and others quite properly realized. Koestler is absolutely right in regarding the effects of psi, if true, as heralding a new Copernican Revolution. But Fort? Thayer once described him as laughing uproariously every time a reader took him seriously.

When you say on the last page that a single unicorn exhibited publicly would prove its existence, I assume you are writing shorthand for a single unicorn, carefully examined by reputable zoologists, and pronounced genuine. It's an important distinction, as you know, that runs through all the current crank literature on UFOs etc. Barnum publicly exhibited the Cardiff Giant, which not only was a fake, but a fake of a fake. Of course I know you agree on all this, but my basic point is that I sense in the *ZS* a certain amount of confusion between claims for trivial anomalies (e.g., a rain of frogs, a white crow, a unicorn) and claims of anomalies that are potentially revolutionary if they can be established. It is the latter sort of claims that are behind the "occult revolution" that is dominating the minds of so many young people. I think it would be good to keep in mind that extremes of continuums require different words, otherwise useful distinctions get blurred and rational debate is made harder.

Best,
Martin

CC: Ray

[Zetetic Scholar]   30 November 1978
Dear Martin,

First of all, let me say I am delighted that you enjoyed the second issue of *ZS* and apparently find nothing of which you harshly disapprove in it. (I still hope to lure you back to the editorial board someday.)

Re the issue of anomalies and your distinction between serious (e.g.. what you think of as Kuhnian) anomalies and lesser (e.g., Fortean), anomalies, I find your comments both surprising and agreeable. You seem in your letter to be saying much of what I have been saying but I don't think you realize the full implications of your comments. Thus, we fully agree about unicorns (if validated properly) being relatively undisturbing anomalies and both seem to agree that some anomalies are theoretically more significant than others. But where we seem to disagree is that which anomalies are severe and which are not may be more controversial than you seem to want to grant. For example, you and I would agree that the reality of psi would be very theoretically important indeed. But our mutual friend Ray Hyman would disagree, for he thinks that even if psi were true, it might be quite scientifically unimportant (since Ray thinks psi may be so unreliable and irregular that it would be much like the universe occasionally hiccupping, a curiosity but scientifically unimportant, for those seeking a regularized predictive view of the universe). Also, you and I disagree about the theoretical importance of Gauquelin's claims. You think they are so preposterous and contrary to the way we view the world that it is comparable to someone saying the center of the earth is made of jelly. Whereas I think that Gauquelin's claim, if true, is really more like the unicorn case—it might be explainable in ways that are hardly paranormal at all (e.g., a third common factor causing the correlation Gauquelin claims exist). My point is that the degree of extraordinariness of a claim needs to be viewed as something relative to a particular body of existing theory, which it may contradict. That is, the degree to which something is anomalous is some quite relative.

Now, I appreciate the distinction you wish to draw between Kuhn-type of revolution-bringing anomalies that destroy paradigms and silly little ones of the Fortean kind. But I think you oversimplify greatly the extent of this demarcation problem. I say this even though you and I would probably agree about many if not most individual cases of anomalies. Also I think you also may be misreading Kuhn and his actually rather complex and ambiguous definition of what constitutes an anomaly. I don't know if you are already familiar with *Criticism And The Growth Of Knowledge* edited by Imre Lakatos and Alan Musgrave, which is a symposium on Kuhn's views (with Kuhn's reply). I am very impressed by the criticisms of Kuhn in this volume, particularly by Lakatos and by Feyerabend. Lakatos emphasizes and beautifully documents that the kind of critical test based on what you seem to think of as a serious anomaly is largely an historical fiction. In fact, his discussion of the realities of the Michelson-Morley experiment, an example you mention, is largely a post hoc reconstruction of it as a critical test. Again, the point is the relativity of an anomaly to a body of theory, in Lakatos' case what he calls a research program. The point Lakatos makes repeatedly, and to me convincingly, is the reality of a veritable sea of anomalies always surrounding research programmes and our frequent post hoc reconstructions of many of these as representing critical (important) test cases that falsified (or corroborated) a program. It seems to me that the very fact that you, Ray and I can disagree about specific anomalies and their importance in itself demonstrates a degree of relativity in these matters that you seem not to fully admit. And just as the three of us would, agree about the relative unimportance of many of the anomalies cited by Forteans, I think we could find reasonable and responsible scientists that might—on rational grounds—disagree with us.

The thing I most hope to do with *Zetetic Scholar* is work out some of these demarcation problems. But I hope we can get more agreement among us that there really are such problems. This is why you and I seem to so often disagree, I think. In my view, until this letter you just sent me, I have felt that you wanted to downgrade the existence of the problem itself by relying too much on your own intuition and historical experience to make judgments about what anomalies were to be taken seriously and which ones are simply crank claims. Even though I am inclined to agree with many of your opinions in this area, I see these as prejudices we share rather than purely rational cases we can make. Thus, since I am truly trying to make *ZS* a scientific journal that ignores my own prejudices where I can do that, I feel it essential that the journal take the posture that such matters are far more problematic than you and I may be inclined to think their solutions are. It is the old problem of deciding between a Type I and Type II error. I think that we should—in *ZS*—be particularly careful about not making a Type II error and miss something important that may be there but obscured by the forest of "normality." That is why I think it essential that someone like Velikovsky be discussed in *ZS* pages; not because I agree with Velikovsky nut because I truly think reasonable men can differ about the value of some if not all his ideas. Most journals run in fear of making a Type I error (thinking something special is happening when nothing really is), and that is the way it should be; but the Type II error also exists, and I am in many ways more interested in *ZS* avoiding it. Since mere discussion of these issues (like Velikovsky) does not, in my mind, constitute an endorsement, and especially since *ZS* is hardly an influential journal as far as main-line science is concerned, I see little harm in concentrating on the Type II rather than the Type I error in *ZS*. And I do see the need for some journal other than the pro-paranormal ones being concerned with both kinds of error.

You consistently surprise me with the wide scope of your reading, so you may already be well familiar with Lakatos, and you may even reject his ideas. But if you are not familiar with him—and I think you would also enjoy his stuff on mathematics—let me strongly urge you to read his "Falsification and the Methodology of Scientific Research Programmes" in *Criticism And The Growth of Knowledge* (Cambridge U. Press, 1970). Lakatos is the leading (recently deceased) extender of Popper and the principal critic of Feyerabend's anarchistic science viewpoint. Though I find much I agree with in Feyerabend, I still prefer the rationalist view of science that Lakatos-Popper represent. '

If you have not yet looked over Feyerabend's new *Science in a Free Society*, I hope you will do so. In fact, I'd love to see you review it somewhere. I think the NY REVIEW OF BOOKS would be natural place for it given the very controversial ideas in the book and Feyerabend's heavyweight reputation as the bad boy Puck of modern philosophy of science. I am very interested in your opinion of this book. He may be able to convince you where I have not.

Best,
Marcello

CC: Ray Hyman

PS. Interesting article in the December 1978 issue of *Human Nature* magazine: John Beloff, "Why Parapsychology Is Still on Trial," pp. 68-74. Martin, I hope you will send me a copy of your reply to Puthoff's recent letter responding to your earlier one.

[Euclid Avenue]   13 December 1978
Dear Marcello:

I fully agree, of course, that "anomaly" like almost, any other noun, refers to a continuum, so we are back to our old problem of how rational men give labels to parts of continuums. I've just been working on an essay for a book I hope to complete someday, titled, (the chapter, not the book): "Beauty: Or Why I am not a Relativist, Part I." It is a defense of Dewey's position that, given a common human nature, it is possible to set up a rough (very rough) set of standards that transcend individual taste. To put it crudely: it is not nonsensical to assert that Shakespeare was a greater

poet than Eddie Guest, or that Mozart's music is "better" than the sounds
of a punk rock band. The problem here is very similar to that of stating
criteria for "truth" (Tarski sense) of a theory. We know very little about
such criteria, but at extreme ends or the spectrum, we can say that a
reasonable man will find the round earth theory "truer" than the flat earth
one. Similarly the absence of criteria for other values (beauty, goodness)
does not make unreasonable (given certain 'assumptions almost everybody
agrees on) the placing of one work of art above another. In the case of
anomalies, we are in the same relativistic bind, and the best we can do is
use our own informed judgment, with due respect for our peers.

It is here that you and I disagree on matters of emphasis. If Ray Hyman actually believes that if psi is real, it is unimportant, I would strongly
disagree. Even if true, but unreliable and irregular, it requires a new kind
of force in science (or "interaction" as physicists prefer to say) that surely
would lead to the opening of incredible doors of understanding of the universe. I believe, and I think most scientists would agree (here is my respect
for my peers) that, say, the "Geller effect," if true would be a revolutionary anomaly as compared with the sort of items that Fort clipped from
newspapers and copied from old periodicals. It is, of course, all a matter
of how one uses words. I would prefer to use the word "curiosity" for the
trivial anomalies and confine the word to the more significant departures.
Of course scientists and philosophers differ over significance of anomalies
fairly close on the continuums, but my fundamental point, which I must
keep referring to, it is that when the difference is great enough, a reasonable
man can make distinctions with a high degree of confidence (probability,
dogmatism, etc.) I know of no astronomer, anywhere in the world, for example, who thinks it possible that Venus is a former comet from Jupiter
that has been in its position only in the past few thousand years. This
is V's main thesis, on which all else hinges, and to defend it requires the
abandonment of so many well established facts and laws as to place his
central thesis on a par with hollow earth theories. The place for an article
defending V, is in *Fate* magazine, not your magazine; and indeed, such an
article *is* in the current issue of *Fate*.

Yes, I'm very familiar with both Lakatos and Feyerabend. I clipped L's
series on the evolution of Euler's Theorem about polyhedrons when it ran
as a four-part series in the *British J. of the Philos. of Science*, to which I've
subscribed for 20 years. I was not much impressed by it, mainly because he
made it appear as if the discoveries of new kinds of polyhedrons (e.g., a cube
with a square hole through it) were somehow counterexamples to earlier
theorems, when in fact they were merely generalizations of the sort that
go on constantly in mathematics. Euler was fully aware that his theorem
applied only to polyhedrons topologically equivalent to a sphere. Once he

hit on the fundamental formula, mathematicians began to generalize it to polyhedrons of other types. Lakatos was a stimulating writer, but not, I think, in the same league with Popper and Carnap, etc.

Feyerabend's point of view is regarded by almost all philosophers of science as extremely eccentric, and I personally find it of no interest. The notion that in science "anything goes" is such a wild exaggeration that it leads to more noise than enlightenment. Did you know that F once said he bases his view on Kierkegaard's *Concluding Unscientific Postscript?* For a good negative review of his *Against Method*, see Mind, April, 1977, pp. 294-295. It is by Prof. Harre, of Oxford. Politically F is an anarchist (it follows!), his writings are filled with abusive comments about just about everyone except himself, and it is certainly true that he has very few admirers among contemporary philosophers of science, who regard him as slightly unbalanced. You recommend that I check on his *Science in a Free Society*. I can see why. Let me cite, for Ray's information, some items from Chapter 6, which is on astrology. A footnote lambastes the opponents of Velikovsky, and defends de Grazia's book, which I regard with considerable contempt. Footnote 14 cites Lyall Watson's *Supernature*, and, jumps on the establishment for not seriously considering Watson's "scientific studies" (e.g., Pyramids that sharpen razor blades, Ted Serios, plants that respond to human emotions, etc.) Note 16 is a favorable reference to S. W. Tromp (see my *Fads and Fallacies* for his truly crazy book on dowsing, in which he firmly believes. You know, of course, of Gauquelin's close relationship with Tromp.) In an overall way, he defends astrology, with the following caveat: "The remarks should not be interpreted as an attempt to defend astrology as it is now practiced by the great majority of astrologists." I am relieved to read that you prefer Popper to Feyerabend, because I know of nothing in my general attitude toward the questions about which we argue that I do not think Popper would fully accept.

Best,

Martin

CC: Ray

[Zetetic Scholar] 19 December 1978
Dear Martin,

This is just a quick reply to your letter of the 13th.

Your chapter on consensus criteria re aesthetics sounds most interesting. I generally agree with what you say. I have long felt (since I am largely a pragmatist) that non-empirical truths rest on a criterion of consensus that emerges between discussants in a dialectic (Plato's sense) pattern which may ultimately be related to common biological and socialization denominators that direct the, consensus. This "truth" is relative to human beings, but it may be treated as ultimate since we have little to converse and argue about with Martians (at least so far). Since much of what is science is also linked to the non-empirical realm, such consensus factors seem to play a part in what we usually think of as objective science as well. And, of course, it is this twilight area of how important such actually subjective (but consensually agreed upon subjective areas, so they are in practical fact inter-subjective) factors are for science that frequently divides not just you and me, but most of the philosophers of science that we have both brought up from time to time.

I agree fully with you that whether or not we label an anomaly serious or a mere curiosity is partially semantic. But the implications are often quite great, and I think our differences go beyond that. For example, if your premises about Velikovsky's theory requiring the jettisoning of well established facts and laws were correct, I would agree with your conclusion. But it is that, premise which is being argued by the scientists who defend Velikovsky and by Velikovsky himself. The best argument I have seen for your premise is that by Sagan. But (a) Sagan disagrees with much that you have said (e.g., he rejects your suggestion of the problem with the stalagmites, etc. and he seems to agree with de Grazia about the earlier episodes of criticism of V), and (b) Sagan has been in substantial part replied to by the *Kronos* and *SIS Review* defenders of V as well as by V himself, and these replies—I think—present reasonable objections which do not completely refute Sagan but certainly open the door for much needed rebuttal since they do enough damage to Sagan's case to force open the door on further dialogue when you and others may think the door now deserves to be closed. I must emphasize once again to you that my concern is not so much with V's vindication (I happen to think V is largely wrong); my concern is with scientific fair play and the problem of avoiding the blocking of inquiry. When you see the May article and the replies, I hope you will think better of what I am doing. There will be some very strong negative things said about V. If all who have promised to respond do so, there will even be more (e.g., I have not yet received the promised replies of several people like L. Sprague de Camp and John Boardman).

Two corrections to statements in your letters. I have made inquiries about Velikovsky's alleged support for Creationism. His supporters tell me that is absolutely not true. I must therefore seek further evidence from you on your claim on that point. Secondly, you describe Feyerabend incorrectly as a political anarchist. That is not true and is explicitly denied in *Science in a Free Society*. I also correspond with Feyerabend and my exchanges with him would seem to confirm this further.

I will read Harré's review of Feyerabend, but note that he replies to Harré in *Science in a Free Society*. On one thing I do agree with you. Feyerabend does not cite good examples of paranormal claims. His citation of Watson is unfortunate but better studies could have been cited which reinforce his central point. Given your comments, I would urge you carefully examine the whole book since my impression (which may be inaccurate but is my inference from your letter) is that you may only have carefully looked at the astrology chapter (which I find fun but not of central import).

You will probably enjoy Joseph Agassi's attack on the Dean volume on astrology in my next issue. (I will have a review symposium on that book). On the other hand, Agassi is on my side—as far as I can tell—re the Velikovsky matter. So, once again, we see how complicated this maze is since so many of us "reasonable fellows" can disagree among one another about what is pseudo and what is protoscience.

I guess the differences between you and me and Ray and various others are for me the best proof that the matters are not as settled as you want to make them. In a sense, Martin, you seem to have arrived at some well reasoned opinions. I don't agree with some of them, but I respect your honesty and intelligence in arriving at them. But I honestly can not say that I have yet achieved such firm opinions as you seem to have. In a very real way *ZS* represent my hope that I will get the best inputs I can to help me (and other readers) to reach more solid opinions. The end result may be that I will agree with you more than I do now. I would really not be surprised since you obviously are way ahead of me on reading on many of these matters. But I guess I will have to learn whatever I learn the hard way through watching some debates re-staged for my benefit that you may already have been through. I can only say that I am very thankful that interaction with you—even when we disagree—remains part of my continuing education.

I am afraid I think much more highly of both Lakatos and Feyerabend than you do. The fact that most philosophers of science put down Feyerabend to me is relevant only in so far as they present valid arguments against him. F is not without faults and, as I have said, I disagree with much of his perspective. But I think he clearly is a "heavyweight" and deserves a better rating than you seem willing to give him. Re Lakatos, I generally find him improvement on Popper and I think that view is a growing one.

By the way, having just finished the reply of T. Kuhn to his critics including Popper, I wonder what your position is there. I gather you admire both Kuhn and Popper, but my position seems far closer to Kuhn's (in its revised form) than your own.

Merry Xmas and a Happy New Year,
Marcello
CC: Ray

P.S. I keep forgetting to mention: I teach at EMU in Ypsilanti but live in Ann Arbor (9 miles away). So I apparently missed you when you were in Ann Arbor earlier. Great pity. Remember for next time.

[Euclid Avenue]   26 December 1978
Dear Marcello,

Some thoughts on your letter of Dec. 19. I think Sagan is wrong in his remarks about the Carlsbad formations, but it is hard to argue because it is a question of how diastrophic is V's diastrophism? There are spots in Carlsbad where 20-foot-long stalagmites join their 20-foot long stalactites by a thread so slender that visitors are not allowed to touch. I believe that the majority of geologists would agree that the kind of upheavals the earth would experience if V's theory of the earth's contact with Venus is assumed would be sufficient to break these threads. It is worth noting that V himself agrees with this. Several years ago I had a long correspondence with one of V's disciples on this question, and, if he is to be trusted, he told me that V was in full agreement on this point, but maintained that the formations in Carlsbad were only a thousand or so years old, rather than the long ages assigned by geologists. He sent me a photo from a pro-V journal that showed a long stalactite that had formed in a few weeks as a result of chemical leaking into the ground from a plant. My reply was that every winter I see water stalactites that form overnight on the eaves of my house. I urged him to consult some references on limestone caverns in which reasons for the ages of the formations are given, but (typically) he never looked up the references.

When you say Feyerabend is not a political anarchist, perhaps once more we are debating the meaning of a word. There are anarchists of all varieties, and some like to call themselves anarchists and some don't. Is Prof. Nozick an anarchist? Paul Goodman and Aldous Huxley called themselves anarchists, but held views almost indistinguishable from those who now call themselves' 'libertarians." Milton Friedman's son has just written a book on anarchism. Is he an anarchist? That Harre viewed F as a political anarchist is evident from the enclosed portion of his review. I also recommend Ernest Gellner's long review of the same book in the *British J for Ph. of Science*, Vol. 26, 1975, pp 331-342. I will enclose a copy of the last page because I believe it expresses the opinion of most philosophers of Science, including Nagel who has voiced similar sentiments.

Perhaps we are also arguing over a word when we ask if V is a "creationist." Here again there are a variety of views that can be labeled one way or the other depending on the emphasis. My information is based on a correspondence between V and Price, reported to me many years ago by Price. The issue is fuzzed by the fact that V interprets all Old Testament miracles as the result of natural forces, most of which he invents for that purpose. Thus the Red Sea parted at just the right instant to save the Israelites. For V, Moses was a genuine prophet of God, yet the miracle he invoked had a natural cause—i.e., Venus. I have no doubt that V believes that life on earth arose from what he considers natural causes, but viewed from outside his system, it would be described as creationist. Do you know, for example, what time table V uses for the events of Genesis, or what interpretation he puts on the story of Adam and Eve? Until V writes explicitly on such matters (if he ever does), I think we had best not trust to reports by his disciples.

You ask my opinions of Kuhn and Popper. I am not much of an admirer of Kuhn, and where Popper and Carnap clashed I am on Carnap's side. I think the differences between Carnap and Popper were largely differences in language, but with Carnap choosing a language more in line with common usage and therefore one that causes less confusion. I can't recall if I ever sent you my review of the Schilpp 2-volume work on Popper—if I did, I don't want to send it again—but in it I tried to make this point clear with respect to Popper's denial of "induction" and Carnap's defense of the term. Although I edited Carnap's *Philosophical Foundations of Physics* (now available as a paperback called *Introduction to the Philosophy of Science*), the philosopher with whom I am most in agreement on such matters is Russell, especially his *Human Knowledge, It's Scope and Limits*, which I think is currently underestimated. I am also a great admirer of Reichenbach, and his student Wesley Salmon. To me, reading Feyerabend is something like visiting, a 10-in-1 show and being amused by the freaks. I cannot take him very seriously.

Prediction: No matter what the final results are of the current Venus probes, V's supporters will be trumpeting them as marvelous new vindications of V's theory of Venus as having sprung out of Jupiter. By the way, has V ever explained how Venus managed to get itself into such a nearly circular orbit? The odds against this, for a comet, are pretty astronomical.

All best,
Martin

CC: Ray

Dear Martin,

    Just some reactions to your letter of Dec. 26th.
    Re the differences between you and Sagan re Velikovsky, you may well
be right in your criticism. My only point concerns the fact that the critics
of V are by no means in accord yet these differences don't seem to matter
among critics who seem to avoid criticism of fellow critics on such points.
And, of course, I don't suggest this makes V right. I only suggest that
what we have been seeing is not a real dialogue at all but an attempt to
win debating points and, to me, that's just not proper science. In V's case,
however, I think we have a special problem in trying to deal with him via
his disciples, many of whom also disagree among themselves, and—more
important—many of whom disagree about what V is claiming. Thus, your
informant re V's creationism sharply disagrees with my informants. This
may, as you say, be another semantic problem, but it seems unlikely. As I
understand it, creationism is supposed to mean some sort of fundamentalist
view of the universe's creation (or at least man's creation) as described in
Genesis. Now though V supports some of the miracles in the early books of
the bible as having actually (though supposedly naturalistically—in terms
of his theories) taken place. I know of no direct evidence that he has ever
espoused anything like Adam and Eve and the Garden of Eden mythology
as having actual historical reality. So, if I am correct in this, I don't see how
you can stretch his pro-biblical history views into any kind of creationist
position as that term is usually meant. Your informant Price may know
something I and my informants do not; but my informants regularly deal
with Velikovsky in correspondence and it seems unlikely they could be so
far off.
    Re Gellner's article, I don't see how you ignore Feyerabend's lengthy
reply to Gellner in *Science in a Free Society* yet you seem to. Also I call
your attention to the note on Rom Harré in *Science in a Free Society* on
pp. 131-32, note 4. See also the note on page 150 re Harré. Given Nagel's
views, I can understand his similar reaction to Feyerabend, but from what
I can see, you badly misjudge the contemporary importance of Feyerabend
and Lakatos. I think the same is true for Kuhn (whom I really denigrated
until I read his reply to his critics in the Lakatos and Musgrave volume
examining his views against those of Popper and his allies.)
    I had not realized that you had reviewed the big Popper collection. I
would love to see a copy of that review (if it is too long for you to copy for
me, let me have the citation so I can get if from the library). I noted your
quotation on the Popper autobiography issued separately; I presume it was

from the review you mention.

Your comments re Popper, Carnap and Reichenbach are most interesting to me. Interesting that you should mention Russell in the context you did for he is the closest thing to my own idol. By the way, you might want to try reading Randall Collins (a sociologist himself, by the way) new novel *The Case Of The Philosopher's Ring* in which Russell teams up with Sherlock Holmes against Aleister Crowley who has stolen the mind of Ludwig Wittgenstein. A silly novel in some ways but with good episodes despite its (I think) falling apart after the first half. A fun premise and lots of cute dialogue between Holmes and the other Cambridge philosophers, too. I think you'd like the quasi-mathematical parts, too.

Martin, can you tell me anything about Roslyn Targ? She is a literary agent and related to Russell Targ. That's really all I know. She has asked to represent me as an author. From what I can gather, she might be excellent but I'd love your opinion before I do anything rash. I have finally gotten started on my book on the paranormal. I enclose a copy of my outline as it now exists. I suspect it will be largely self-explanatory since you know my views in detail anyway. As you can see, it will incorporate much we have discussed. As always, I would welcome your comments, if any.

Best,
Marcello Truzzi

CC: Ray Hyman

[Postcard] 17 January 1979
Dear Marcello:

An old friend, Herbert Alexander (retired President of Pocket Books), tells me that Roslyn Targ is the wife of Wm. Targ, former editor-in-chief of Putnam's and father of Russell Targ. Roslyn is Russell's *step*-mother (Wm Targ had a previous marriage). She is a highly respected agent. Herb knew nothing about her views on the paranormal—but of course Wm Targ was responsible for Putnam's' flood of books on the topic (Ed Mitchell's big book, etc, etc,. etc.)

Best,
Martin

P.S. Thanks for the Azar paper that just arrived. I found it filled with naive misunderstandings of many aspects of current (and legitimate) controversy.

His point about the "circularity" of fossils → evolution → fossils → ⋯ is
the core of G. M. Price's attack (see *Fads and Fallacies*).

[Eastern Michigan University]   23 January 1979
Dear Martin,

A while back you wrote me something about Gauquelin being involved
with S. W. Tromp and dowsing (I presume you got the stuff Tromp sent me
to pass on to you which I mailed out to you last week). When your letter
saying this came to me I was curious so asked him if he had been involved
with Tromp and dowsing. I just got a letter from Michel and I quote his
comment (in imperfect but clear enough English):

"Maybe I shall try to contact Martin Gardner. [I had earlier encouraged
him to do so.] But I cannot exactly imagine the best way for that. Anyhow
he is wrong about the fact that I was associated with Dr. Solco W. Tromp
in dowsing beliefs and experiments. I just wrote to Solco suggesting he will
keep you directly informed on this interesting matter."

I presume this may be why Tromp sent me the articles to forward on
to you. I have never communicated with Tromp myself and he enclosed no
letter of explanation when he sent the stuff for you with the card I sent
you.

Thanks for sending the review of the Popper symposium book.

I am very anxious to read your chapter for the Abell book from what
Ray tells me about it. Any chance of my getting a copy from you?

I hear from a couple of parapsychologists that they believe you and
Randi met with Wheeler prior to the AAAS and steered him in the direction
he went. Seems unlikely given the tone of your letter to me re Wheeler's
anti-psi blast (you seemed to just have heard of it yourself from a friend).
Any foundation to the rumor?

I am currently working on a typological piece on ufologists in which I
make some unusual evaluations of Phil Klass. As you know, Phil's first
book on UFOs claims that many of the reports were of actual sightings but
that these were in reality extraordinary plasmas of a kind he postulated
but which are generally unacceptable conjectures among plasma specialists. Phil has since written his second book which mainly claims error and
frauds but mentions no such plasmas even though he has not repudiated
his first book and informs me he sticks with the first book's arguments.
This means that Phil is actually claiming an extraordinary explanation
for critical UFO sightings. Hynek and he agree upon the sightings but
do not accept Klass's plasma explanation. But both agree about a set of
cases which do not have ordinary explanations. Since Hynek offers no solutions for these cases but merely asserts they represent a puzzle, while Klass

asserts they are extraordinary plasmas which act in ways not acceptable as explanation by plasma specialists, this means that Hynek (for this set of cases) is actually more conservative a scientist than Klass. What has been overlooked by so many people is that Klass claims to identify a number of critical UFO reports as actual sightings while using a highly controversial explanation to identify them as plasmas. This is a major difference between Klass and Menzel, Oberg and others. I am told—and am just beginning to look into this—that Klass's first book was received rather badly by the scientific community who also disliked Phil's plasma theories. I would be most interested in your comments on any of this.

A major point of my typological piece will concern the fact that most controversy over UFOs is really not over whether there exist unidentified flying objects, but on the identifications being offered. Thus, Phil is really a believer that he has identified the objects he has studied and is really arguing with those who claim some extraordinary identification of them (as extraterrestrial, etc.). The truly rational reply to the commonly asked question "Do you believe in UFO?" should be "yes" since it is obvious that many have not been clearly identified even though we may have pretty good guesses about many of them. The skeptics reply of "no" should be to the question "Do you think UFOs have been identified as extraterrestrial or metaphysical in origin?" The whole business is a semantic mess. In all of this, we have lost sight of the fact that some of the identifiers of UFOs, in their rush to avoid identifying them as extraterrestrial or paranormal, have been terribly sloppy. Phil is a good case. Because Phil has the common "enemy" of the extraterrestrial proponents with Menzel and Sagan, he is perceived as an anti-UFO critic when he really favors a plasma identification theory that is itself highly controversial and dubious. And when Oberg suggests that President Carter and his son saw the planet Venus, they can ignore the descriptions they gave ("as big as the moon1", etc.) with everyone treating the purported debunking as successful. (If you avoid the descriptions given by the witnesses, as in this case, any explanation is equally good; why not say they mistook an eagle since that is no less a fit than Venus seems to be.) Any comment?

Best,
Marcello Truzzi

CC: Ray Hyman

P. S. Thanks for the information re Roslyn Targ. Just heard from her that she has decided not to handle me or my books. Surprising letter since she even asked me to reimburse her postage in sending the stuff back to me (75 cents) after she was the one who solicited me to write her in the first place.

[Postcard]   28 January 1979
Dear Marcello:

   Wheeler is in the physics department, Univ of Texas, Austin. His blast
at what he calls SCESP (so-called ESP) is an appendix to his paper on QM
and consciousness.
   Best,
   Martin

[Euclid Avenue]   26 January 1979
Dear Marcello:

   I don't recall exactly what I said about G and Tromp, but I didn't mean
to suggest that G was associated with Tromp's dowsing experiments, only
that he was associated with Tromp. I am told they are good friends. G
has published papers in the Dutch *Journal of Interdisciplinary of Cycle
Research* (or some such title) of which Tromp is editor, and I am fairly
certain I've seen favorable references to Tromp in G's writings. (I don't
own any of G's books; and have no desire to keep copies but you might
check the index, if there is one, of *Cosmic Clocks* to see if Tromp is cited
there.) I know nothing about G's views on dowsing. You might ask G
sometime what they are, just for fun! I do know that Tromp is a firm
believer in astrology (see his big book on dowsing), so perhaps this brings
him and G together for at least friendly debate. If you can get hold of
Tromp's "classic," I think even you would agree, looking it over, that if
Tromp is not a crank, nobody is. To sum: the quote you got from G says
no more than that he is not associated with Tromp's dowsing work. But
does he consider Tromp a genuine scientist (a matter of degree, of course)?
   Neither Randi nor I have ever met Wheeler. I've exchanged a few letters
with him on cosmological matters, and he sends me offprints, but I always
regarded him as likely to *favor* parapsychology, since he is so often cited by
P and T and others for his views on the role of the observer in QM. I first
heard of Wheeler's blast when I saw the report in *NY Times*, and Randi
called me the next day about it—he, too, finding it a welcome surprise. (It's
possible that Wheeler was in touch with some friend who has communicated
with me and/or Randi on such matters.)
   Between you and me, I was not favorably impressed by Phil's first book
on UFOs, and I think he too is probably unhappy with it now. As for
Carter's UFO sighting, I know none of the details and have no idea whether
the Venus explanation is plausible or not—and couldn't care less! I *can*
report that Guccione (of *Omni*) wants Spielberg to make a big documentary

for TV, on flying saucers as extraterrestrial spacecraft, etc., so it will be interesting (if this comes off) to see how long Oberg can hang onto his job at *Omni*. Am having dinner tonight with *Omni* editor, Frank Kendig, so maybe I can find out more details.

Prefer not to circulate any more copies of my paper for Abell for fear one might get into the hands of Walker. (Also, it is *very* long and tedious to copy.) He has been writing me letters of almost unbelievable insults and near paranoid hostility. Now Abell is getting similar ones, demanding a right to reply in the Abell book, etc., etc. The whole psi scene is beginning to bore me to death, and I've decided to cut down as much as possible in giving time to it. Especially since I'm not paid anything for it.

Are you planning any coverage of the Soal scandal? I expected to see reports on it in the media, but all has been silence. Too negative a story, I assume. Surely it must have received *some* publicity in England?

Chances now better than 50/50 that *Time* will launch a new science monthly next year. Leon J. is taking time off to prepare issue No. 0, the dummy, which should be finished about March. Publishers tell me that books on the paranormal are no longer selling. Apparently those who read books are getting bored with the trend, while it moves out into the hinterlands and provides high TV ratings. Did you see the NBC documentary with Raymond Burr as narrator? I couldn't imagine a show worse than the Burt Lancaster one, but this managed it. And now Orson Welles is narrating a movie based on *The Late Great Planet Earth*, the best seller about how the world is about to come to an end (via Christ's second coming—TV ads in NYC are showing computer analyzing names of world leaders to see if they add to 666, the number of the anti-Christ!). Of all people—Orson!

Best,
Martin

Dear Martin,

Just a quick note. Re Tromp, Gauquelin does quote him on the notion of borderline sciences in *Cosmic Clocks* but not in favor of dowsing. I am reasonably sure that Gauquelin is a friend of Tromp's but I guess what bothers me what I sense as a strong hint of guilt by association on your part. After all, I consider myself a friend of both you and Gauquelin but I disagree with both of you about many things. I suspect that Gauquelin and Tromp also share honest differences and, like Gauquelin, Tromp has shown no signs of being a charlatan or fraud. At worst they are simply both misguided or wrong in their conclusions; but I have no reason to doubt their honesty, and I hope you don't either.

Glad to hear your confirmation of my interpretation that Wheeler's AAAS paper was an independent phenomenon. I will pass that along to those who think they find a conspiracy.

My own correspondence with Phil demonstrates that he completely sticks with his first book and his plasma explanations. I will look forward to hearing more from Phil on this in due course.

Sorry about not being able to read your paper for Abell at this time since I am sure I would learn much from it. Naturally, I will wait for it and look forward to reading it when it is available. I had heard that Walker was giving you a terrible time. He sounds like a bastard from what I have heard. I have no communication with him and don't intend to start any.

Re the Soal scandal, you are incorrect about the lack of notice given it if you mean the psi media (mass media is another matter). Since the psi proponents have pretty uniformly accepted the discrediting of Soal for some time, it would be beating a dead horse somewhat to make too much of it now.

The Raymond Burr narrated film was actually a 1976 film that was re-run. It was mainly a movie-house film and not a regular TV documentary. I agree with you that it was absolutely dreadful (I made a videotape of it). Though disappointed in Orson Welles doing the narration of *Late Great Planet Earth*, I am not really surprised given his other public utterances on psi, etc. He has been "for sale" for a long time; I just hope it helps him produce some good movies of his own since I gather that is how he works things (takes any jobs to get financing for his own films since the studios will not finance them).

I have hopes of having a panel for the upcoming (Sept.) American Psychological Association meetings (in NYC) dealing with experimentation on the paranormal. If they approve it, it will include Ray, Jim Alcock, G. Schmeidler, T. X. Barber, John Palmer and Bob Morris. Will let you know the details if the thing gets accepted.

> Best,
> Marcello Truzzi

[Book review by Gellner]

[Euclid Avenue]   17 April 1979
Dear Marcello:

Many thanks for the offprint of Westrum's piece, which I look forward to reading this evening. (Have now done so. Excellent.) About 30 years ago I once planned to write a similar article, but I never got around to it, though I included a paragraph on the topic in the opening chapter of my old *Fads and Fallacies* (p. 9) where I call it the "most notorious instance known to me of "scientific stubbornness." A different but related topic that Ron W. might consider tackling is why Galileo refused to accept Kepler's theory that the moon caused tides (for which Kepler had excellent correlations!), as well as Kepler's elliptical planetary orbits (again, Kepler had good reasons for the theory).

Glad to hear your "Graph of the Zodiac" is getting published. Persi tells me (phone) how much he enjoyed his visit with you. Next. issue of *NY Review of Books* (dated, I think May 17) will have the Wheeler appendices (and his letter to the AAAS) for which I wrote an, introductory short piece to give the quantum mechanics background. I think Sarfatti is going a bit mad. (Did you see the interview with him in *Oui* a coupla months back?) He has gone underground, fearing assassination by the est crowd! Don't know if you're on his mailing list and have been receiving his wild attacks on Erhard as a native fascist. Erhard keeps rolling, his latest distinguished convert being Bucky Fuller, who is appearing on the platform with him to give speeches.

The number trick you mention is an oldie, going back centuries. I once had it in my column with the following presentation. One person writes XYZXYZ on a slip (he chooses XYZ but tells no one the number). Hand it to someone with the request he divide by 7. You predict no remainder. Repeat with two other people, using 11 and 13, each time making the same "no remainder" prediction. Last person hands his slip back to first person, who finds on it his original number XYZ. Yes, there is a ridiculous simple explanation of why it works. 7, 11 and 13 are the prime factors of 1001. Multiply any XYZ by 1001 and it's obvious the product must be XYZXYZ. Successive divisions by 7, 11 and 13 naturally reverse the multiplication process, leaving the original XYZ.

You have surely heard about Allen Spraggett's arrest in Toronto, on charges of molesting a young boy. I had the pleasure of telephoning Canon Rauscher to give him this news, knowing that he and Spraggett were at work on another book collaboration. Rauscher reacted in typical minis-terial fashion by saying, "This is going to reflect on *me*." I had many years ago belabored Rauscher about the sly innuendos about homosexual-ity in the book he and Spraggett did on Ford (especially low was a footnote implying that Kreskin is gay), and Rauscher had assured me that this was all Spraggett's doing. What a con man Spraggett is! The arrest has killed his TV show (very popular in Canada) on the paranormal. Last month Ira Einhorn, a young counterculture type who was into parapsychological warfare, was arrested in Pennsylvania. Seems his girl friend had been missing and police got a search warrant. Found her body in a trunk where it had been for several weeks. Haven't seen any news clippings, but if you're interested, Martin Ebon has the info.

You were right in telling Gauquelin I probably wouldn't want to see him. He's written me twice about getting together, but these are among the some 50 letters a week that I don't answer. Jeane Dixon's latest (Houghton Mifflin) is a book on horoscopes for dogs. As you may have heard, Fawcett set an all-time record of price paid for paperback rights: 2.25 million bucks for Linda Goodman's *Love Signs*. (Bantam bid up to 1.7 million.) And Orson Welles is the narrator of the highly profitable film, *The Late Great Planet Earth*. (This a really *big* trend in the U.S.; it is zooming up while occultism is going down (except for movies and TV.)

Scribner's has reissued Mulholland's *Beware Familiar Spirits*, for which I did an introduction. Would send you a copy, but Scribner's gave me only two (it's paperback, by the way), and I gave my spare yesterday to Randi who was out for photocopies of this and that for his forthcoming Crowell book *Flim Flam*.

Cheers,
Martin

[Postcard]   13 June 1979
Dear Marcello:

Yes, you're quite right. I view Velikovsky as only a small cut above flat earthers and creationists (see my Sagan review in *NY Review of Books*, June 14), so I see no reason ever to take him seriously.

Gertrude was dead serious in her *Games* piece (her original submission was even worse).

Have you seen Tart's report of replication *failure* in the *Journal of the American Society for Psychical Research*? (Even funnier than Gertrude's piece.) I've *finally* realized what your position re off-beat science is. You're an old-fashioned Fortean!

Best,
Martin

[Zetetic Scholar]   16 June 1979
Dear Martin,

Re Velikovsky, I am glad that I seem to understand your position correctly even if we continue to disagree about the merits of the matter. We obviously are using different criteria for assessing various paranormal claimants as more or less reasonable/plausible. Our earlier exchanges on Gauquelin in relation to parapsychology represent a case in point. As you know, I have been trying to explicate my own criteria for rank ordering various claims as more or less extraordinary. I can not find similar criteria for determining degrees of extraordinariness in your own writings (your early piece in the *Antioch Review* seems to come closest, I guess). I hope you someday explicate yours (or if you have already done so and I have missed out, let me know what to read of yours that does so). The thing that continues to surprise me, however, is that you seem to have responded to my pieces that sought to explicate my own criteria with what I have taken to be agreement (possibly I have misinterpreted your silence as consent in some cases). Since I have tried to base my judgments on the criteria I have set forth, I don't really understand why our specific judgments seem so far apart.

Re the Tart paper, I fully agree with your remarks (I presume you refer to the April 1979 *Journal of the American Society for Psychical Research* article). You'd think the damned thing was successful from the way he writes it up. Re the Schmeidler piece, if she meant it fully seriously, I can only say she is becoming rather sad.

Re your remark about my being a Fortean, you are largely correct, but I distinguish Forteans from myself in that they seem to be interested in anomalies whether or not they are scientifically validated. I am not. I do feel much sympathy with the Fortean spirit. I once thought that you did also, but I think this may be less true of you today than 20 years ago. As Westrum and I outlined in our paper on anomalies, Forteana is usually merely alleged rather than validated anomalies. To Forteans, anomalies seem to be ends in themselves; I see them as purely instrumental in helping to expand and develop new science.

As you may know, I am currently conducting a poll of the members of the Psychic Entertainers Association on their attitudes towards psi, exposers, frauds, etc. So far the results are interesting and confirming of my expectations, which include the strange (to me) fact that those who practice phoney psi believe in the reality of psi more than their fellow magicians who do not specialize in mentalism. I hope to do a parallel study on non-mentalist magicians.

Did I mention to you that the next issue of *Zetetic Scholar* will include a poll of college professors (national sample) and their attitudes towards psi. Startling results include (a) about 66% consider psi an established scientific phenomenon; (b) psychologists say "impossible" to psi by 53% (thus a widening gap between them and other academics); (c) about 38% of the academics believe psi in plants and animals is established as scientific fact; and (d) those who reject psi seem to do so on "a priori grounds" while those who believe seem to have read more of the literature. Since psi is not scientifically established, and since a prior grounds are not adequate to reject psi scientifically, I feel that this simply means that we have irrationality on the matter by both extremes. (Belief in psi as established goes up to 77% among professors in humanities and education!) Since the Gallup polls earlier showed that about 51% of the public believes in psi and that this goes up to about 56% among the college educated, it looks like the believing professors may be the source of psi belief for the college educated. Who said education decreases gullibility?

What did you think of what I perceive as the almost pure ad hominem attack by Klass on Hynek in the current *Skeptical Inquirer?* Though you may have liked it, I hope you recognize it for what it is, the attempt to impute motives to Hynek that suggest that money and fame rather than conviction, caused Hynek's shift on UFOs. Even if it were true, it is certainly irrelevant to the evidence being put forth by Hynek. I don't see how this sort of piece will enhance the credibility of the Committee among those who remain agnostic about these matters and are interested in the arguments and not the personalities involved. I also happen to think it is an inaccurate article based on the historical facts about Hynek; psychohistory is bad enough without pseudo-psychohistory.

                              Best,
                              Marcello

[Zetetic Scholar]   23 June 1979
Dear Martin

Leroy Ellenberger is an interesting correspondent of mine who is particularly interested in Velikovsky (he is an admirer of V but not a hard-liner or true disciple). I recently spoke with him about V's relationship to Einstein and V's book (unpublished) on Einstein. I am particularly interested in Einstein's views on deviant science and you obviously have interest in both Einstein and his views on parapsychology, so I thought you'd find information on Einstein's views on V of some interest, too. Ellenberger has read

V's book, Einstein's letters to him, etc.; so that part is apparently correct. (I have no reason to doubt Ellenberger who has been very reliable on such matters in the past. He frequently sends me documents on this and that of a Fortean nature.) Ellenberger is now trying to get to see the Einstein annotated volume of *WORLDS IN COLLISION*. Though the "party line" is that Einstein was just a very kind man who humored many cranks, it appears that his views of Velikovsky were more complex and that he took V more seriously than you do. You may already know what's in Ellenberger's letter to me; but if you don't, I think you would find it of interest.

Cordially,
Marcello

P.S. Of course, none of this is to say V is right; it merely indicates that Einstein would almost certainly would have considered it appropriate for a journal like ZS to seriously discuss V's works while you seem less tolerant.

[Postcard]   27 June 1979
Dear Marcello:

New book you may want to review: *The Prophecies and Enigma of Nostradamus*, by my old friend Everett Bleiler (former editor at Dover for some 30 years). Bleiler published it himself, using the anagram pseudonym "Liberté E. LeVert," and the imprint of Fireball Books [address withheld]. Edition limited to 750 copies, at $15 each. It's a marvel of scholarship.

Best,
Martin

[Euclid Avenue]   27 June 1979
Dear Marcello:

Yes, Einstein was extremely tolerant of cranks, and wrote letters to a great many varieties, always patiently explaining their errors. I used to write patient letters to angle trisectors, explaining their mistakes, until I found that it accomplished nothing—or rather, worse than nothing because it often prompted phone calls, and in one case, the crank showed up at the front door just as we were sitting down to dinner, and was miffed when I sent him away!

I believe that Bernard Cohen's *Scientific American* article was *not* biased, an opinion confirmed by several friends of mine who knew Einstein well. Only V and his "admirers" think it was biased. But I'm too bored with all V matters to want to argue about it, especially after having waded through drivel that his "admirers" sent to the *NY Review of Books* to protest my review of Sagan's book. Obviously when a crank covers hundreds of topics, he can't be wrong on all counts. You judge a crank by his central dogmas, not by peripheral matters like the temperature of Venus or lightning bolts on Jupiter. Please, Marcello—let's drop the V matter. I believe it is a waste of your valuable time as well as mine.

Best,
Martin

[Euclid Avenue]   8 July 1979
Dear Marcello:

Yes, magicians who get into mental magic do indeed, in my limited experience, have a stronger tendency to believe in ESP than other magicians. Bruce Elliott and I used to talk and wonder about it. Doc Tarbell did a fine eyeless-vision: act, but he liked to talk about how at times when he didn't actually glimpse something handed to him, he would correctly guess its color, etc. One night at a SAM show in Chicago, as a joke, someone turned out the lights, in the middle of his eyeless vision act! Dr. Jaks believed in ESP as well as in graphology and other things. I'm not. sure about Dunninger. You might ask his old friend Walter Gibson [address withheld]—Walter is a firm believer in ESP, and his wife Litzka is a palm reader. I've never met Kreskin, and don't know if his professed belief in ESP is a put-on for his PR image or not.

A point you might want to make is that parapsychologists and true believer's in ESP are more upset by the mentalists like Kreskin than are people like Randi and myself. Like all things, there is a continuum from a performer like Geller to stage performer like Blackstone, Jr., who includes mental effects in his show and in between are all stages of emphasis on the genuineness of ESP. I have no objection, to a mentalist omitting the "disclaimer," and draw the line between good and bad when the mentalist starts getting himself authenticated by parapsychology labs, and becomes "more" of a professional "psychic" than an entertainer. Billy D. is a good case in point. He doesn't do any kind of act—he just (or rather used to) tries to get himself certified as psychic by doing a dull series of card tricks an based on "estimations" and what Persi calls "multiple end-points" ("outs"). For a concise, quotable expression of my views, see my introduction to the new Scribner edition of Mulholland's *Beware Familiar Spirits*.

As you probably know, Bascom Jones and his periodical *Magick* have long advocated the dropping of all disclaimers by mentalists—you might ask him to send you some of his editorial comments on this if you don't see his sheet. Some of the strongest attacks on mentalists, Kreskin in particular, have been made by—Canon Rauscher, who collaborated with Spraggett on the Ford book. There is a long section in the Ford volume that lambastes Kreskin, very unfairly in my opinion (one footnote suggests that Kreskin is gay—Rauscher says Spraggett is responsible for it!) I once asked Bill Rauscher how come he goes out of his way to give away Kreskin's main secrets (the clip board, the nail writer, etc...) but has never written about Uri's methods. Rauscher's response at the time was that he hadn't decided yet whether Geller was a fraud or not. Don't know what he thinks now, but he still hasn't written a line about Uri, although he has done several articles that portray, Kreskin as doing untold harm to parapsychological research! (Ask him for offprints—some are in magic magus, I think.)

Rauscher, as you must know, has a magic background and does a full stage show. His private phone is [phone number withheld] (only he answers it). Don't have his address handy. Somewhere in NJ.

I suppose Thurston is the outstanding example of a famous stage magician who believed in the paranormal. Today, of course, it is Don Henning. Hereward Carrington is the classic example of a magician who did his best to expose the methods of fake mediums, but at the same time believed in genuine ones. Then there is that marvelous book by D. D. Home in which *he* wrote about the methods of all the phony mediums of his day, carefully excluding, of course, his own techniques!

Have you ever seen one of the Nelson Enterprises old catalogs? Robert Nelson was well respected in the magic community, yet his firm sold apparatus to fake mediums all over the world (I'm told he had purchasers in India!), and to professional mentalists. I've never heard of Milnov. Usually these thousand dollar manuscripts turn out to be "pure gold" only in the sense of pure gold for the seller. But maybe he has something new. If you find out what it is, please let me know and I'll do the same.

An unbelievable 4-page letter to *NY Review of Books* from Ellenberger. If you think I use ad hominem arguments you should see how he blasts both me and Sagan as arrogant, knee-jerk defenders of establishment science. I did make a blunder when I attached Noah's name to a flood, but I'm apologizing for this in a reply to Prof Rose of Buffalo, who naturally caught me out on it. I'd forgotten that V puts the Deluge of Noah at 9,000 years ago when Earth and Moon passed through a great cloud of water that had been ejected from Saturn! (George M. Price had a much more plausible Flood theory—one that is still being defended in the major books by the modern Creationists.)

Ho hum,
Martin

[Euclid Avenue]   13 July 1979
Dear Marcello:

Haven't seen the Wheeler thing in *Science*, but I can check that at local library. Nor have I seen the blast in *Psychic Entertainer's Bulletin*, but I'm not curious to see it because I know what it will say. Phil Goldstein is specially miffed by my *Scientific American* column on, ESP Cards, etc. I hope when you discuss all this you'll emphasize that the leading parapsychologists are more down on the professional mentalists than people like me and Randi.

I know nothing about the rumor that Levy is threatening things about Rhine.

An article in *Zetetic Scholar* on Edison would be very welcome. I have no details about Edison's machine to communicate with spirits—I imagine the Edison estate is making access to such notes as difficult as possible. Here are some references that won't tell you much but will give some information about Edison's "authentication" of the famous mentalist Bert Reese.

1. See my *Fads and Fallacies* (check index for Edison).

2. Martin Ebon, *They Knew the Unknown*, page 125.

3. Houdini and Conan Doyle: The Story of a Strange Friendship, Bernard Ernst and a co-author [Carrington]. (Marvelous book, by the way.) See page 116ff for material on Reese and Edison.

4. A lot of Reese methods are given in Annemann's *Jinx*, but I don't have the dates. (One of his nice ideas was using blank pages torn out of old books for his billet reading, because the soft paper didn't "talk" when he secretly unfolded it after a switch.)

5. I enclose a piece by our old friend Bascom Jones from *Fate*, which I see I neglected to date.

6. See Mulholland, *Beware Familiar Spirits*. Check index for both Edison and Reese. Mulholland had lots of anecdotes about Reese, but alas, I didn't keep a record of them.

Best,
Martin

[Euclid Avenue]   15 July 1979
Dear Marcello,

Thanks for the advance peek, at your *Fate* interview. I could be wrong, but I suspect that granting this interview was unwise. *Fate* will gain, but your image will be tarnished. You have always carefully distinguished between sincere offbeat scientists and the charlatans, but surely it takes only a quick look at *Fate's* advertising to recognize the magazine as 98 percent rip-off. Bob Shadewald wanted to interview me, for *Fate* a couple of years back, and I at once refused. It is true I once did a pseudonymous article for *Fate* about Dr. Jaks, as a favor to a good friend who used it for publicity purposes (he was doing a mental act called "Curiosities of the Mind"), but this was at a time when mentalists were not confused with psychics. Even so, I'm now sorry I wrote it.

Yes, I think your memory of events is accurate. But some of your remarks puzzle me. Let me express my puzzlement in the form of questions, some serious, some rhetorical:

1. "... neither a *fervent* believer..." Does this mean, as most readers' will assume, you are a *mild* believer in the paranormal? I assume what you meant is that you are like an agnostic with respect to God, you have no beliefs one way or the other.

2. On Hynek and his Center, "... truly scientific approach to UFOs..." Not "mystical." Are you quoted right? As you know, H. now buys Vallee's UFO approach. If Vallee's approach isn't "mystical" then we are in another bind over the meaning of a term.

3. "... not because I believe in astrology, at least not traditional astrology..." Does this mean, as the words imply, that you incline to believe in *non*traditional astrology? Gauquelin, I presume.

4. "I happen not to believe in .the possibility of a truly rational ethic..." Here I am just curious. There are four main possibilities:

1. The emotive theory of Russell, Carnap, Stevenson, etc.

2. The sociobiology theory of genetically-rooted ethics. Wilson, Prince Kropotkin, etc.

3. The revelation theory: Christianity, Judaism, Moslemism, etc.

4. The denial of ethical values: nihilism, extreme cultural relativism, etc.

5. "... highly responsible believers. I had the greatest respect for them ... Gordon Melton is a good example."

Gordon Melton! Have you read his *Fate* contributions over the past many years? If this is a correct quote, it measures the vast difference between our respective mind-sets. I find Melton on the same level as Rogo and Colin Wilson, and only a few cuts above Spraggett and Watson.

When I said recently that I was beginning to think you are an "old-fashioned Fortean" I meant this as a kind of compliment because neither Fort nor Thayer took Fort's "anomalies" seriously. As I put it in my old F&F book, part of the big joke was to pretend it wasn't a joke. Can it be you are becoming a new-fashioned Fortean who takes such anomalies *seriously*?

The central joke of the Baker Street Irregulars is to assume Holmes and Watson were real. How Fort would have roared with laughter if he could have guessed that *his* joke would deteriorate to the point at which persons calling themselves "Fortean" would be unable to distinguish the gigantic gap that separates, say, a possible anomaly like ball lightning from Rogo's belief that people have communicated with the dead by telephone.

We are, of course, back to our old problem of how best to label parts of continua. We agree that science information is on a scale with 1 and 0 as the unreachable limits. Where we differ, apparently to a huge degree, is over how to estimate the probabilities, the degrees of credibility, without producing more confusion than light. Surely when you say (as you do in *Fate*) that there is "some" scientific evidence for astrology you don't mean that this evidence is stronger than, say, for phrenology or palmistry? Or maybe you do! That's my last question. Naturally you don't have to answer any of them.

Best,
Martin

CC: Ray

[Eastern Michigan University]   17 July 1979
Dear Martin,

Just a note of thanks for the various references and comments you sent me re both my PEA polling and the Edison business.

Thanks especially for the address for Gibson whom I had been meaning to contact for a long time but never knew exactly where to reach him. Re Rauscher, thanks for his phone number. If you can easily locate his address, I would very much like that since I would prefer to write him.

I plan to check with the Edison estate re his spiritualistic telephone. I suspect that you are right about their wanting to down play such matters. But we'll see.

Since I have the original Mulholland book, I confess to not having gotten the recent reprinting with your introduction. I presume your intro is not a long one, and since you mention your views are expressed in it on the ethical issues I am now researching, I wonder if you might be willing to xerox the intro for me. I am a bit reluctant to buy a second copy for your intro.

You are right about the Nelson catalog. I have it and bought a lot of his stuff when I first got into mentalism. I have found that a lot of "white readers" use his books, so his claim that non-entertainers used his books seems correct. I have been in touch with Micky Hades in the past about Nelson's poll he did for his last book (Hades has the original materials). I hope to eventually reanalyze Nelson's data from his customers.

Bill Nagler promised to find out about the $1000 mental trick for me at a convention he will be at this week. Will let you know what I find out when he writes me.

My PEA survey has now been collated (at least preliminarily) and the results are along the lines I indicated to you earlier. I hope to get more questionnaires soon to allow me some better internal analysis (differences by age, time in mentalism, etc.).

Surprised you know nothing of the Levy business. Randi told me that the science editor at *Time* told him of a call from Levy a while back threatening to tell all. I would have thought Randi would have told you about it.

Though you know my position is somewhere in between, I do disapprove of Ellenberger's use of ad hominem and have told him so several times! Nonetheless, I find him one of the most able of the defenders of Velikovsky and he actually is not really that much of a disciple of V's. He is more interested in the fairness question than he is in seeing V vindicated. He has a watts line and calls me just about every week with some bit of gossip or another. I've told him that I wish he would stop writing letters and turn his efforts to a book or article which would more likely prove effective than all the pot shots he takes writing letters to editors. I think he enjoys the "battle" of it all in some ways more than the issues. But I could be wrong. I only mention this since you brought him up. I do not intend to get into another Velikovsky discussion with you unless you initiate it. We have too many other things in common which we can communicate about without my irritating you on V.

You might be interested in knowing that Dennis Rawlins seems to think that you and he are in the same corner. He has offered to write an article on the Gauquelin-Kurtz retest mess (from which he has apparently dissociated himself) for *Fate*. He is apparently angry as hell with Kurtz and sees you as his close ally on the Committee. At least that's what he told my friend at *Fate*. I suspect that though you may be sympathetic to Dennis re his problems with Kurtz, I find it hard to believe that you are truly his ally. Politics may make strange bedfellows, but I suspect Dennis' perception is exaggerated.

Do you know if the Gauquelin retest article will be in the next issue of *Skeptical Inquirer*? I have been looking forward to seeing its final form. Actually, I thought *Skeptical Inquirer* would be out by now.

What ever happened with Paul Kurtz and the AHA? I see he is no longer listed on the editorial board of *The Humanist*. Has he severed all ties with the AHA?

I just got a copy of the list/addresses of the membership of the Parapsychological Association. I intend to run a survey on these. Any suggestions for questions you might like to see them answer? I am also in the midst of getting a liaison committee set up in the PEA to help out the PA in dealing with alleged psychics. This way, the PA can ask the PEA about, say, a Geller, and they can respond with "He's one of us, don't bother testing him for real psi." The PEA can also probably prove helpful in setting up controls or making a mentalist available to help out in other ways.

Best,
Marcello

P.S. Do you have a home phone number for Martin Ebon? I think I once got it from you and seem to have mislaid it.

[Euclid Avenue]   20 July 1979
Dear Marcello:

My last letter will probably bring a response, but so as not to confuse things, I'll here answer questions in yours of July 17.

Rauscher's address is [address withheld] (that's the rectory of Christ Episcopal Church).

I enclose a copy of the Mulholland book introduction.

I could be wrong, but my memory of the phone call to Leon, which both he and Randi did tell me about, was that it was a call from a reporter who was in touch with Levy and was trying to peddle something Levy had told him. Hence my reference to the "rumor." Randi and I periodically ask each other if we've heard any more about this, which we haven't.

Yes, Dennis's perception of my being on his side is exaggerated. Dennis is his own worst enemy, but I try to be polite in my exchanges with him. His tactics are terrible, although I mostly agree with his perceptions about parascience, and astrology and G in particular. Haven't any information about Kurtz and the AHA. He occasionally phones about one thing or another, but I hesitate to bring up what must be a sore episode for him.

Ebon's home and office phones are respectively [phone numbers withheld]. He and Christopher were on the Candy Jones show last week.

Don't know if you're interested in the Ira Einhorn case or not—he's the Philadelphia counterculture chap now in jail for the murder of his girlfriend whose body was found in a trunk in his apartment. Ira was active in the consciousness circuit, and was deep into parapsychology, especially its implications for psychic warfare. The current (July 23) *Village Voice* has a cover story on the case. A few odd facts come the light. Two of his friends were Ed Rosenfeld, articles editor at *Omni*, and Bill Whitehead, now at Dutton, but who edited one of Einhorn's crazy books for Random House. *Voice* pages are too big for my machine, but the local campus bookstore may carry the paper. Spraggett's trial is also coming up, in Toronto, which should be as interesting as the coming Einhorn trial.

Best,
Martin

P.S. I disapprove of Dennis doing anything for *Fate*, and will tell him so.

[Eastern Michigan University]   19 July 1979
Dear Martin,

Glad your objections to the *Fate* interview were so few. Let me respond to your questions and remarks one by one.

You may be correct about tarnishing my image, but since I did approve all of the copy except the headline and their biographical intro, it seemed worth the disadvantages for several reasons. First of all, nobody else asked me and I did feel it worth while to get myself on the record re the various matters surrounding my leaving the Committee. If nothing else, it will keep me from having to write it in more letters to more people who have inquired. I hope it will clear the air. But the main reason—aside from no better publication having asked me for such an interview—is that I want to publicize *Zetetic Scholar* to potential subscribers. Unlike the Committee, my resources are very limited and I simply need to get more subscribers if the journal is to continue publication. I really have little choice since, you will recall, I could not even get the mailing list from the Committee. I realize that a lot of crazies read *Fate*, but so do a lot of serious people. In fact, I remind you that *Fate* has run articles by Klass, Kusche, de Camp, Oberg and Schaeffer. So why should I end up more tarnished than they? One reason I wanted an interview instead of an article on me was because

I had prior agreement that I could approve the final text of what I said. And an interview, in my view, is actually less of an association with *Fate* than my writing an article for them since they clearly had to come to me and not vice-versa.

Re the question of "neither a fervent believer": I am indeed pretty much an agnostic re the paranormal; however, as I have consistently defined it, the paranormal includes far more than psi (cf., my taxonomy in *Zetetic Scholar*, #2). In regard to some issues, I probably am what you would call a mild believer. For example, the correlations (as opposed to the causalities) claimed by Gauquelin would be classified as paranormal in that we have no immediate and/or conclusive scientific explanation for them. I believe his work is probably valid. That is, a real anomaly may be present. I also am inclined to think that something anomalous is going on in Loch Ness. I think the evidence thus far shows that acupuncture is effective beyond its purely psychological (suggestibility) effects (as demonstrated by veterinary acupuncture). These would all be classified as paranormal phenomena by many and by my own definition. As for psi, I do not feel that there has yet been adequate accounting for all psi evidence by critics. I am inclined to disbelieve in psi, but I must confess the issue remains quite open; and here I am truly more agnostic than I was a few years ago (when I really agreed more with Hansel than I do now). In any case, the reference to "fervent believer" is their description of me, not language that I used. I can not be held accountable for what they say of me. But even so, their statement seems to me to be quite defensible since the rest of their sentence goes on to say "nor a hostile critic, he occupies a middle position..."

Re the Hynek quote. I stick with it. You seem to be unaware of the gulf that now divides Hynek and Vallee. Vallee does indeed espouse a mystical approach. Hynek has repudiated that position some while ago. In any case, my reference was to his Center more than to Hynek. Al Hendry seems to represent the more consistently responsible position of the Center. I will grant that Hynek may personally say irresponsible things. But the Center has its goals and public positions which should not be confused any more than Paul Kurtz's statements should consistently be taken to speak for the Committee (I at least hope you would agree with me there). Hynek, on the other hand, should be judged by his major two books rather than the unfortunate book of conversation between him and Vallee. Hynek's current public posture is very much an agnostic position on UFOs. That is, he merely contends that UFOs are indeed unidentified (that is a significant but small percentage of total reports) and constitute a legitimate anomaly. He does not anywhere indicate that he knows what they are and, in fact, has come to be very skeptical about the extraterrestrial hypothesis. His position, as I read him, is that a genuine puzzle exists and that Klass et al.

have not yet accounted for the evidence thus far put forward. On this, I would agree with him. Again, you may wish to consider this as a belief in the paranormal on my part.

Re the astrology statement: As you surely must have appreciated from our earlier correspondence on Gauquelin, I do "believe" in such "nontraditional astrology" to the extent that I recognize what seems to be legitimate evidence (correlations) supporting astrobiological claims like those of Gauquelin, Eysenck, etc.. As with all science, once their evidence has been put forward, the burden shifts to the critics to account for the discrepant, anomalous findings. This has not yet really been done. I have no reason to think that Gauquelin and Eysenck (unlike most others associated with neoastrology) are either incompetent or frauds. (On the other hand, I have good reason to criticize the Kurtz-Zelen-Abell test, as does Rawlins.)

Re the question of a rational ethic and the AHA: This certainly is unclear, but here's my position. As I understand it from my interactions with the Humanist movement (as a teenage Unitarian I used to read *The Humanist* quite faithfully), the AHA seems to believe that it is possible to build an ethic that could be *universally* accepted on rational grounds. I think this is nonsense since differential human interests create natural conflicts that make any universal ethic probably impossible on purely rational grounds. That is, in a world in which everyone is practicing the golden rule or being honest, a deceitful person has a distinct advantage. People and institutions have vested interests of all sorts, and what is rational for one person's sense of justice, as well as profit, is simply not necessarily another's. The AHA has a more positive view of mankind than I have, I fear, so that was a major reason I did not join the AHA. You can not get everyone to act virtuously by appealing to reason alone. That's all I meant. I am not an extreme cultural relativist but enough of one to preclude my being an AHA regular (as I understand their position; it may have changed in a more liberal direction unknown to me).

Re Melton: I consider Melton honest, sincere, and learned. We disagree on a great deal. The point is that I consider him responsible. He does not make up facts and data. In certain areas (e.g., the varieties of contemporary religious groups and cults, upon which he has published a couple of important books) he is highly knowledgeable. I have always found him very reasonable in discussions with me. In some areas, he is a fine debunker, too. I will be publishing a piece debunking Edgar Cayce by Melton. You say he is on the same level as Rogo or Colin Wilson. You may be right. I barely know Rogo and don't really care for him very much based upon our brief interactions. But Wilson I know much better and like him. Wilson is a sloppy scholar and Melton is far less so. But Wilson is in many ways a brilliant fellow and is, I think, sincere though very wrong. The problem seems to me to boil down to what we mean when we say we respect someone. You mean this more narrowly than I do respect the honesty and integrity of Melton (as I do your own). It does not mean I agree with you or him on many matters. But I think Melton is trying to do a good job in dealing with the paranormal evidence. Unlike Melton, there are hordes of irresponsible people on the paranormal. For example, I do not consider Puthoff or Targ as responsible as Melton. Melton has beliefs I do not have (based partly on his being a minister and theist) but I think he would not fudge data or block free inquiry. In short, he is knowledgeable, a serious scholar (of church history etc.) and honest. I wish I could say that of everyone involved on either side of the paranormal debate.

Re the matter of being a Fortean: As far as the real Forteana goes (that is the largely unvalidated reportage of alleged anomalies), I do not take it seriously but enjoy it. In that sense I agree with you that I am a Fortean. And I would also blast the new Forteans who take it all seriously since they pretend to a scientific posture. But I also think that there are valid and validated anomalies and that these are extremely important for social change within science. And the basic issues surround the role of anomalies in science, I take very seriously.

Re the question of scientific evidence for astrology being better or worse than for palmistry or phrenology: I do not think the evidence is at all equivalent. Astrology has been rather carefully examined through empirical work. The evidence for traditional astrology is almost entirely negative. The evidence for astrobiology (which is astrology in the sense of broadly considering possible interactions between the heavens and man's behavior) is slim but largely (if not mostly) positive. The studies are very limited (mostly Gauquelin's), and negative replications are lacking. On the evidence for palmistry, I have never been able to find a single serious empirical study done (people refer to such studies having been done, but I have not been able to trace areal one down yet). Thus, serious evidence for palmistry (that is experimental rather than "clinical" evidence) seems nonexistent. If tomorrow someone went to mortuaries and got hand prints for an experiment to test palmists, that would be most welcome; but so far no one has done it. There is some stuff on hand shape and feeblemindedness (by a Sheldonian psychologist), but I would not call that neopalmistry the way I term Gauquelin's stuff neoastrology. As for phrenology, the literature is full of negative studies demonstrating its error. At least that is the case for popular phrenology. There may be evidence for non-popular phrenology) but I am really not familiar with it (I presume that there was probably such evidence during the period of its acceptance by psychologists; and there is some revival of interest among anthropologists for some elements of phrenology).

Though I am sure my answers may not prove satisfactory, I hope they are clear enough.

In thinking further about the matter of "respect" I have for people, I think it ultimately boils down to whether or not I think I can learn something from the persons. If the guy is suspected by me of being a fraud, I might recognize his learning but would distrust learning from him since he might mislead me intentionally. If a person has knowledge, intelligence, and is honest with me, I guess that's a large part of what will gain my respect. Melton and Colin Wilson both have my respect, but I would not necessarily characterize Wilson as responsible since I think he is too sloppy a scholar. Of course, I therefore can not respect Wilson as a scholar, but I do respect him as a person who seems to want to "play fair" and who has intellectual capacities I admire. You, I am sure, would use the word "respect" less broadly, I suppose.

Best,
Marcello Truzzi

[Euclid Avenue]   23 July 1979
Dear Marcello:

Thanks for such a careful and responsive letter. I have learned more about your attitudes from it than from all our previous conversations and correspondence. Naturally I differ with you on your evaluations of astrology, Gauquelin, Loch Ness, UFOs, Hynek, Melton, Wilson, and even acupuncture, but the differences are, as usual, differences of degree and emphasis, and I can't see that more debate on such questions would be fruitful.

I write now to ask if you would mind if I sent copies of our exchange to Ken Frazier with the suggestion that he consider publishing them as a letter exchange? If so, we should each have the privilege of a bit of polishing, after which we could exchange and okay our final copies. In my case I am thinking mainly of such trivial things as adding first names here and there (Scott Rogo, Stanley Jaks, etc), changing Baker Streeters to Baker Street Irregulars, and in general making the letter more readable.

If you prefer not to have our exchange published, do you have any objection to my copying your letter for people like Ray and Persi who would be especially interested? Maybe you sent a copy to Ray already.

Best,
Martin

[Zetetic Scholar]   28 July 1979
Dear Martin,

Thanks for the intro to the Mullholland book and the phone numbers for Ebon and address for Rauscher.

Re the Levy rumor, it seems a somewhat curious business. Randi told me that the science editor at *Time* (presumably not Jaroff) told him that Levy called him to offer to spill the beans. Obviously something has gotten distorted somewhere along the line somehow. I am going to get Levy's address from Barry Poss and try to find out directly from Levy (Barry knew nothing about this new development). Apparently, there has long been a question about Rhine's exact role in the affair. Rhine was indeed warned about Levy in advance, but from what Barry tells me, there was so much jealousy of Levy among people at FRNM that Rhine took that into account and tended to support Levy thinking the early warnings were just symptoms of the jealousy that usually surrounded whoever was heir apparent to Rhine's mantle at FRNM. Anyway, I will let you know if I learn anything significant.

Thanks for the information on the Einhorn case. Martin Ebon sent me the *Village Voice* piece on the affair. Wild stuff.

Just reading Walter Prince's old gem *The Enchanted Boundary*. He does an especially nice job tearing apart both Houdini and Conan Doyle. I knew that *A Magician Among The Spirits* was full of errors, but Prince does a nice cataloging of them. It is remarkable how little the role of the conjuror re psychical research has changed over the decades. I think I have the makings of a nice piece on the whole business. I really think the main point about the role of a magician in psi investigation has been missed. The usual argument is that magicians should be included to catch trickery. Obviously, this may be useful but guarantees nothing at all since magicians fool one another all the time. The reason an investigator into psi needs to study magic is not so much to catch fraud per se; it is to make the investigator realize how easily anyone can be fooled including the trained magician. Magic teaches you to be humble about your own potential for being imperceptive. The problem with most psi investigators (Taylor is the best example) is that they are so arrogant that they believe they are too smart to be fooled. They think it impossible that an illiterate or child could mislead someone as sharp as they think they are. To me, that's the real message of familiarity with conjuring: all of us can be deceived by people we think might be less clever than we are. The true lesson of magic should be an increase in our humility. Instead we think it is some form of inoculation against being tricked. That's why the psychic researcher who know a bit about conjuring (e.g., Targ) may end up more arrogant than he was before he studied conjuring at all. He thinks he now knows it all. This is all the more reason why Randi and others who duplicate psychic effects should not expose them since you need wonder to fight wonder and thereby impress the witness with the fact that he can be fooled. If you tell them the method, they will just rationalize that either the psychic did it the real way or that what you did was somehow not exactly what the psychic did. You can not win that way. You need to baffle the witness equally and leave him with his befuddlement while you simply say that what you did was a trick and force him to show that there is an effective difference between what you did and the psychic did. There maybe some extremist who will insist that you too are psychic (like Conan Doyle insisted Houdini dematerialized over Houdini's denials), but nothing will convince such persons anyway. And others will normally buy your argument and at least become newly skeptical about the supposed real psychic's effects. Does my argument seem sensible to you?

Got the new issue of *Skeptical Inquirer* which I generally found less offensive than the last issue. One interesting thing, however: I see that when a letter writer writes a stupid attempt at debunking (such as the ones on Kurtz's article claiming that bad runs are not reported by parapsychologists, a practice generally not so and in any case mostly irrelevant if the odds are tremendously against chance as in some experiments), there is no correction by the editor as in the case of critical pro-paranormal letters which elicit replies. Shouldn't error be corrected whether or not it comes from pro or con sides? Since I know letters have been sent giving manifestly decent arguments against some things published in *Skeptical Inquirer* and these letters are not being published, it also says something about selectivity in that is being printed; that is, a bias obviously exists toward publishing debunking letters even if they are irresponsible ones (irresponsible in the sense of presenting stupid arguments that no serious student of the area should not immediately see through). But as you recently put it: "ho hum."

Best,
Marcello Truzzi

P.S. Are you aware of: Edward W. Karnes and Ellen Pennes Susman, "Remote Viewing: A Response Bias Interpretation," *Psychological Reports*, 44 (1979), 471-479. I can send you a copy if you want one. Very similar to the replication study of T&P I had been planning to do earlier.

Dear Martin,

I see our letters are crossing one another in the mails. Re your letter of July 23rd, presumably commenting about my letter of the 19th answering your questions of the 15th, I am glad you found my responses clarifying but I really must say that I am surprised in the sense that I thought most of the points could be gotten or inferred from our past correspondence. But I am glad if my position is somehow clearer to you. Please never hesitate to raise future questions that I might further clarify anything that might disturb you in my answers. As you say, most of our disagreement concerns emphasis and degree, but these differences have very real practical consequences; and since I have such high regard for you, I very much wish our positions could move closer towards one another.

Always feel perfectly free to let any of our mutual friends see my letters to you unless I specifically say something about confidentiality. So certainly do send copies to Persi and Ray (along with your own letter with the questions, presumably) if you want to. I don't think I sent any of this to either of them (I normally note copies sent on the bottom of the last page of my letters to you).

As to publishing the exchange(s) in *Skeptical Inquirer*, I appreciate your intentions but must decline at this time. There are many reasons. The most obvious include words of derogation about Rogo and similar things not really meant for all eyes. Far more important, however, is the lack of context for the exchange which was, after all, written for you in light of all our past correspondence. People like Ray and Persi know my positions about as well (or better) than you do, but the typical *Skeptical Inquirer* reader would not have that context. If you want to put forward some public questions to me and publish my responses written for a general audience, that might be more practical. I also should say that I would much prefer to have such an exchange in *Zetetic Scholar* than in *Skeptical Inquirer*. I really have serious reservations about publishing anything at all in *Skeptical Inquirer* given all matters that have transpired. Naturally, if they ever attack me, I will be forced to answer and hope they will publish it, but I'd rather not initiate anything for *Skeptical Inquirer*. After all, they could have done an interview with me but felt that unnecessary. In fact, as you know, they pretty much ignored my departure all together. Perhaps, if you think the matter worth pursuing, we could find some more neutral territory for publication. I could understand your possible reluctance to publish in *Zetetic Scholar* for reasons that to a degree parallel mine for *Skeptical Inquirer*, but there might be some external forum which both of us might find more comfortable. I would be quite flattered to publish any exchange with you, so I am reluctant to give you a flat "no" on your suggestion.

It really is a shame about the copyright law since so many of my exchanges with you and others in the arena debating about paranormal matters would be highly enlightening to others. I think anyone reading, for example, my exchanges with Phil Klass would soon get a very different outlook on the debunking of UFOs. I think many readers would be surprised at how some people seen as "opponents" (e.g., Ray and I) are almost in complete agreement, while some people supposedly on the same side are actually radically different in their views. Many people reveal themselves as far more open minded or, in some cases, dogmatic than one would have thought. If you ever get through here, you should take a look at some of the more interesting exchanges. In a strange way, I am somewhat grateful to the factors causing my departure from the Committee since my thus being publicly branded a "neutral" has put me in the position of being able to get various otherwise extremist persons to write me relatively revealing and less defensive statements about their views. At first, everyone seemed to be trying to win me over, but that has now stopped and I think I am getting the reputation of a fair-minded (if naive, possibly) person really willing to weigh the evidence. I think I can do a much better job with that reputation than I could if I were firmly allied with either camp. In any case, time will tell.

I trust you saw the *Time* piece on J. Gordon Melton's new book on America's approximately 1800 religions. This should support my comments to you about him.

By the way, I was very amused to see that *Time* was one of the sponsors of what was probably the worst paranormal "documentary" movie shown about a week ago. Undoubtedly the worst such film I have ever seen on a national TV network in prime time.

Have you yet seen Allan Hendry's *The UFO Handbook*? I must say that in many ways it is an impressive job. I may try my hand at doing a comparative piece on the older apparitions research and contemporary ufology. Very interesting parallels exist.

Best,
Marcello Truzzi

[Euclid Avenue]   2 August 1979
Dear Marcello:

    Yours of July 28 to hand. Yes, I pretty much agree with your attitude
toward magic and psychic research. There are, however, exceptions. I
think my pamphlet on Uriah Fuller, for example, played an important role
in the rapid decline of interest in Geller on the part of parapsychologists
who really didn't know anything at *all* about how he bent their key. The
possibility that Shipi stooged for Uri seems not to have occurred to many
Gellerites until Randi and I and others pointed out this possibility. The
many exposures of spirit mediums that produced physical phenomena, by
Houdini, Carrington, and so many others, also played a major role, I think,
in the decline of this type of seance. But it's a very complicated situation,
with many subtle angles. I thought the exposing of the clipboard, etc., in
the book by Rauscher and Spraggett went much too far, especially in view
of Rauscher's total silence about Uri's methods, or Kulagina's—in short,
all the two authors did was expose the secrets of a professional magician,
and to hint that he was gay!

    Yes, I'd like a copy of the Karnes and Susman paper on remote view-
ing. Have you seen the eye-opening piece on Kubler-Ross, "Afterword of
Entities," by Kate Coleman, in *New West*, July 30, pp. 43-50? If not, I
can send you this in exchange for the above. It's all about the lady's current obsession with Spiritualism as practiced by Jay Barham, in California
(naturally).

    Yes, there surely is a bias in the selection of letters for S.I., but S.I. has
never pretended to be "objective" in the sense that you try to be. I have
no objection to biases in biased magazines. *New Realities* selects letters
with an even stronger bias. When I wrote short and polite letter pointing
out huge errors in Wilbur Franklin's article many issues back, the editor
thanked me with a note saying he would be happy to run the letter, but he
never did. And I don't hold that against him. I suspect that here we are
back to one of our old disagreements over whether a periodical should, in
certain areas, take a strong stand one way or another, and not even attempt
to be "objective." If, for example, I were running a magazine that has as
its major purpose to defend libertarianism (or neo-conservatism), such as
Irving Kristol's periodical, I would be highly selective in what letters I
print. (I'm thinking of *The Public Interest*. Maybe not a good example
because Kristol tries to be fairer than, say, fellow conservatives like Wm
Buckley. A better example would be a magazine attacking, say, neo-fascism
in Germany. I view the "occult explosion," especially as it is now being
promoted by television and movie producers, as morally offensive, whereas

you can afford to be a fencesitter because you don't see it in moral terms at all. Merv Griffin's entire show yesterday was on the Amityville Horror, not a note of skepticism—all of the guests were gung ho for poltergeists. Bill Bogg's, two days earlier, devoted his entire children's show to psychic phenomena, with Alex Tanous telling all about how he helps police solve crimes. The kiddies loved every minute.) Naturally I agree with you about pointing out any recognizable errors in any letter that printed.

Did I recently say "ho hum" about something? Maybe you are thinking of a letter from Puthoff about Randi's recent reply to their demand that he pay them a thousand dollars because of an error he had made in an earlier document. Puthoff said his only comment was "ho hum." It's a funny way to put it; maybe Puthoff got the phrase from me?

Best,
Martin

[Zetetic Scholar]   10 August 1979
Dear Martin,

Just got back in town from visiting my wife's relatives in Tennessee, so I'm running a bit later that usual in answering your letter. Enclosed is the remote viewing article. I got to the office this morning to find I left it home, so I sent you only the second half of the *Fate* interview then. I'll mail this out in the morning.

The Kubler-Ross article sounds most interesting. Don't bother copying it for me, though, since I can get *New West* at the public library this weekend, I think. Let you know if I fail to get it.

I of course agree with your comments about *Skeptical Inquirer*, but how can you possibly write "but *Skeptical Inquirer* has never pretended to be 'objective' in the sense that you try to be." I fully understand that you never wanted SI (even when I edited it) to be a skeptical but neutral publication. I have always respected you for what I took to be your lack of any hypocrisy. You wanted a debunking publication and simply said so. But do you really think that the readers of the journal who presume it is arm of the Committee whose goal statement on the back of the journal clearly states that claims will not be approached with any bias and which explicitly states (as did Kurtz in one of his lead articles in SI, too, after I left) that no judgments will be made antecedent to inquiry, will agree with you that bias is to be expected? Surely you can not really mean that. As I even point out in my *Fate* interview, if the Committee and its journal were to take the advocacy position of defenders of scientific orthodoxy, that is perfectly reasonable and legitimate. I have absolutely nothing against the presence of a strong debunking magazine. My objection is that SI is acting like one while pretending to be open-minded, etc. I quite agree with you that biased magazines can publish biased stuff quite properly. But the Committee was set up, and its goal statement still states (as does Paul) that the Committee is not supposed to be biased. Its debunking functions should simply emerge from its honest and unbiased attempts to find the truth re claims of the paranormal. That's pure hypocrisy and bull shit. I can understand Paul's trying to have his cake and eat it too. He's just being opportunistic (or just plain stupid). But I have never understood why you and Randi (and to some degree Ray, too) have stood for the hypocrisy of publicly presenting the Committee and journal as unbiased. I had thought that maybe you did not really see it as biased somehow, but your letter of 2 August clearly states that you do see it as biased and properly so. How do you resolve this?

You may recall that at the meeting where I resigned as editor, I indicated that I was not basically opposed to a debunking publication; I just didn't want to be the editor of one. If the Committee had chosen (as you and Randi wanted) to more publicly announce itself as the defender of the anti-paranormal forces in science, that would be open and above board and the honest approach to what is now actually happening. Such a role is needed in science and I could support that. But Paul always wanted to pretend the whole operation was philosophically open-minded and would never pre-judge evidence. Yet I think you would agree with me that this has in fact been done. In fact, if Rawlins is right—and in this case I think he is—Kurtz has even gone so far as to actively bias the sampling against Gauquelin in the U.S. test of G's astrobiological claims. Whatever you may think of G's claims, how can you justify what Kurtz seems to be doing? How much can be justified in the name of some higher good? Do the ends of the Committee (fighting the irrational forces of paranormality and its advocates) really justify the means of pretending open mindedness and—if I am right about Kurtz on the Gauquelin stuff—even doing bad and biased research to falsify such claims?

I had thought that after I left the scene and after Kurtz lost the support of the AHA, you and the other debunkers would force out the hypocrisy of claiming non-bias and come out openly as debunkers. I still hope that will emerge as the case. Otherwise, I am afraid that the Committee is, in my view, becoming rather dishonorable an enterprise.

Do we really disagree so much? Dammit, Martin, I really don't think we do. Neither of us is convinced by the evidence put forward for the paranormal by parapsychologists, astrobiologists, etc. Though I am rather more "agnostic" than "atheistic" than you, my "agnosticism" really does amount to a practical "atheism" since the burden of proof is on the claimant. As a practical and scientific matter, therefore, I presume the negative, too. But where we seem to so strongly differ is that you seem willing to antagonize the claimants more than I do, and you seem more concerned about maintaining the boundaries of orthodox science than I am. I suspect that this has something to do with your being a science writer rather than a practicing scientist (no derogation intended), for I think that in general it is science writers (e.g., Asimov, Sagan, etc.) who seem to be given the job of boundary maintenance for science re the general public. I think you are actually more worried about releasing the floodgates of crackpottery upon science than most scientists really are. There may be a parallel between this and mentalists lesser concern about real psi than magicians. I am not sure. In any case, since you have always been open (at least with me) about your biases and advocacy role, I see you as quite honorable; but I really do think you may have been partially blinded to those like Kurtz who claim to be quite unbiased truth seekers. All of this has put me in the position that I must try to reveal Kurtz for what he is doing while seeking not to offend the open debunkers, like you and Randi, with whom I really have no quarrel. As you must realize, I have been rather patient in waiting to see what direction things took with the Committee. I did not leap into the media with any attacks on it. I really hoped that you guys would eventually take over and muzzle Kurtz and his silly public statements. Since that has not happened, and it looks like it never will, I really have little choice any longer.

Randi's position in all this is somewhat different than yours. He gets special legitimacy from being on the Committee that may help him professionally and with the media. But I do not see how you gain a damned thing from your own involvement. And if things keep up the way they have been going, I think you ultimately will damage your own reputation from association with the Committee. But then you seem equally concerned about me damaging my reputation with things like associating myself in any way with *Fate*. We may even both be right and both get damaged. But whatever the outcome, I do want you to know that I appreciate the openness you have always displayed with me about your position, and I have tried to reciprocate by being open with you about my own views. I may accuse the Committee of hypocrisy, but I do not directly accuse you of any.

Two small points regarding the second part of the *Fate* interview. Their summary of my first half in the previous issue was not approved by me and I have some misgivings about the last part of that summary which I think is slightly misleading, though I think unintentionally so. The other point I might make is that the very last section of the interview in which I insist on fallibilism sounds too much like I am a relativist, which I am not. When

I said more-or-less (I don't have the article before me) "Why do we need to know everything." what I meant was "Why do we need to pretend to know everything when we do not." In other words, I am very concerned that scientists should admit their areas of ignorance and show a kind of tolerance stemming from their agnosticism rather than prematurely insist on denying unexamined claims.

Best,
Marcello Truzzi

P.S. I have decided to go to the Parapsychological Association meetings this coming weekend. Ray will be there, too. So I will finally get to meet many of the opposition. Also hope to get to meet Feyerabend at Berkeley while I am thereabouts.

One of these days you will have to see my file of correspondence with Phil Klass. I really had not realized what a fanatic he is. If you are interested— and I could understand that you might not want to give it the time—I can send you the last few months of our exchanges dealing with his incredible attacks on people. He really is fast becoming the Joseph McCarthy of ufology. I can now understand why most people in ufology don't want to interact with him and have tried to just ignore him. Of course, he sees their avoiding him as proof that his argumentation is so impregnable, but in fact it is largely quite untenable and his behavior "backstage" (in relation to publishers, employers, and people at foundations) can only be characterized as vicious by me. Somewhere along the line, I hope you get to read Paul McCarthy's doctoral dissertation on the politics of the UFO controversy (U. of Hawaii, political science) in which McCarthy nicely chronicles the way Klass went after McDonald. I knew that he had also tried to make trouble for Hynek, but until his more recent attacks, I really was a bit blind to how extreme Phil felt and acted. I am sure Phil is sincere and imagines himself motivated by high morality, but I can now really understand why people like Hynek and Sturrock simply want nothing to do with him. They view him as a crank just as you view people like Velikovsky. You may all be right.

[Euclid Avenue]   8 August 1979
Dear Marcello:

Thanks for yours of 30 July. I'll drop the notion of publishing any of our correspondence.

I did see the *Time* story about Melton's new book, but what this has to do with our disagreement about him I can't fathom. We were concerned with his gullibility and reliability as to his opinions of the paranormal field. I have the highest respect for Conan Doyle as a writer of fiction, but no respect at all for him as a psychic investigator. I think Bobby Fischer's religious views (he was a convert to Armstrong's Church of God) are infantile, but that doesn't prevent my admiring his ability to play chess. I am one of Koestler's strongest admirers when he writes about political topics, but not when he writes about Uri Geller.

Ellenberger has sent *NY Review of Books* his *third* long (unprintable) tirade vs. Sagan. (Not that I'm against ad hominem arguments. I think they have their place: e.g., pointing out that Puthoff was, maybe still is, a Scientologist.)

Best,
Martin

[Euclid Avenue]   9 August 1979
Dear Marcello,

Just read part 2. This time I think there *are* some factual mistakes. Ms. Cottrell told us she had been tested at Rhine's laboratory just before her session with us. Since Rhine has issued no report, nobody knows what he found. Back in Kansas, as you must know, she had been working under psychologists in giving ESP therapy of some sort to autistic children!

No one on the committee planned anything about Suzie until her father contacted Paul and requested that the committee test his daughter. Why? Randi and I finally concluded (long story) that Suzie has never leveled with her father, and he truly believes in her psi powers. Suzie came, I think, because of possible TV coverage, and maybe because she underestimated how much Randi and I knew about her techniques. I think her willingness to use autistic children for personal publicity is shameful, and if we killed her career in show biz, she deserved it.

You give the impression that persons like me look down on *Fate* because of ads for fortune tellers. Then you show that such ads are harmless. It would have been fairer, don't you think, to mention such ads as those for goggles that show the aura, pyramids that preserve food, water-witching devices, expensive biorhythm calculators, good luck charms—all with horrendous price markups. And there are the sleazy books advertised that even you would agree are on the lowest level of occult huckstering. Who cares about fortune tellers!

Your last paragraph brings up a topic on which we seem unable to come to grips. The doctrine of "fallibilism" (to use Peirce's term) is taken for granted by everybody, certainly by me. To say that views of the world are fallible is to speak an obviosity, not to say anything profound enough to justify your "dammit." But it is not a case of "This is so or it isn't so" as versus saying "'we don't know." It's a case of knowing with varying probability estimates based on the total relevant evidence. The probabilities range between the unreachable limits of 0 and 1. The fact that there are no precise measures of these probabilities is beside the point. Everyone rates his science beliefs roughly in a hierarchy of probabilities: e.g., that the earth is not flat has a probability close to 1, that light is a limiting speed for energy transfer is almost as close to 1. Black holes? I rate their reality as about $\frac{1}{2}$. Most astronomers would, I think, go to about 70% and some to 90%.

When you make causal remarks about such persons as Gauquelin and Velikovsky, your remarks are extremely vague because they are never couched in terms of probability estimates. For instance, you say here "I may be inclined to think Velikovsky wrong..." Most people would take "inclined" to mean that your probability estimate is a bit under $\frac{1}{2}$. I, along with Sagan and virtually the entire science community, would make the estimate less than 1 percent.

I truly believe that most of our differences about the sensible attitude to take toward debunking, and toward eccentric science, and the occult explosion, rest on wildly different ways in which you, compared to most scientists, estimate probabilities. It would help a lot if instead of saying such things as "I'm inclined to think there is something significant in Gauquelin's findings," you would say something like, "I think the probability, as I now see it, is better than 50% (or whatever your estimate is) that G's correlations are not statistical flukes, or the result of bias on his part, but are genuine correlations that science should now try to explain."

It is easy to understand why you would be influenced by Feyerabend with his "anything goes" approach to science. It ignores what lies at the very heart of scientific method and progress in knowledge—the assessment of varying degrees of credibility that can be attached to hypotheses in the light of the total evidence. As someone once said: extraordinary claims require extraordinary evidence. And if the claims are in a field in which one is not trained, don't you think one should have the humility of trusting in an overwhelming consensus of the experts (e.g., about V) rather than support confusion by saying "I'm *inclined* to think he's wrong." To me, this remark has almost exactly the same ring as if you had said, "I'm inclined to think the earth isn't hollow and open at the poles."

Fallibilism, by all means. But that doesn't mean you have to treat a V or a G with the same respect that you would treat, say, the far-out theories of a Hawking or a Penrose or a Wheeler.

Best,
Martin

[Zetetic Scholar] 21 August 1979

Dear Martin,

Just returned from California and the Parapsychological Association meetings. A remarkable experience in many ways and I think Ray also got a lot out of it. Very friendly and revealing exchanges between us all, and I wish you could have been there for it.

Finding your letter of 9 August with comments on part 2 of the *Fate* interview, let me comment on your remarks.

First, re Suzie Cottrell. Though the central point I made that she had not been endorsed by any prominent parapsychologist still stands, I must admit that I did not know that she had ever been tested by Rhine at all. However, you will note that I said "So far as I know..." and that was certainly true. I did not know of the Rhine testings. Rhine has not endorsed her, as you point out, and there seems to be no evidence he was about to. I very much do approve of Randi debunking Cottrell. My point is not that this was wrong but that such debunkings should not be the *central* concern of a committee supposedly interested in dealing with the truth claims of the parapsychologists. My point remains constant: Any serious interest in debunking parapsychology must deal with the best cases offered by the most respectable parapsychologists. Debunking Cottrell is more a matter of public relations and protecting the public, a perfectly proper thing to do but something which is a far cry from what the Committee was expected to do in terms of its goal statements and general publicity.

Now I do indeed think Cottrell needed to be debunked, but I wonder about what evidence you have that she "used" her role with autistic children as you mentioned in an earlier letter. If she indeed misleads parents into thinking that she has communicated with their autistic kids by phoney means, that is most despicable indeed. But is this really the case? Do you have evidence of this? I have read nothing aside from your letter to this effect. I also call your attention to the terrible past things that have been done with autistic children by the "legitimate" psychiatrists. Until recently, it was quite common to medically treat the kids by actual physical beatings (all in the name of modern psychotherapy!) and Bruno Bettelheim (whom I consider even worse than Cottrell) for many years blamed the parents for the condition of their children in both autistic and schizophrenic cases. Laying such an unwarranted guilt on the parents (when we know have much reason to believe physiological factors are central in these cases) is even worse than the false hope that someone like Cottrell may be giving them.

As I thought I mentioned in my last letter to you, the comments I made in the article re fallibilism and Feyerabend easily gave the misimpression of more relativism on my part than was actually the case. Your remarks seem to ignore these qualifications I made in my letter. Let me seek to clarify this.

Feyerabend and Lakatos are generally today considered on somewhat opposite sides of the spectrum re the issue of the role of rationality present in overall scientific method. I still side more with Lakatos than with Feyerabend but consider them both an improvement on straight, especially earlier versions, of Karl Popper's views. I make a strong distinction between science as idealized and as institutionally practiced. I remain firmly committed to basic rationality as far as idealized science, that is, scientific method, is concerned. Here I think that Lakatos is correct. But science as institutionally practiced (as a social institution) is incredibly less rational a process. This is especially true in (a) the social sciences and medicine, and (b) those areas of the "hard" sciences, which get highly abstract, conjectural, and far removed from any operational data base in the empirical world (this includes much of modern physics and certainly includes much astronomy, especially such things as speculations on the history of the universe).

My remarks in *Fate* make it sound like I am willing to settle for "I don't know" much more than I do. But I consider most institutional scientists as far too arrogant in their public posturing as "experts" that should be trusted to know what is most true about matters.

I very much agree with you that the consensus of experts goes against people like Velikovsky and Gauquelin (though we might argue a bit about degrees of such consensus). But in these two particular cases, the simple fact is that most of this consensus is actually based on a lack of familiarity with the evidence and a trust in one another's judgments that exceeds any evidence. For example, re Velikovsky, few astronomers have actually read any pro-Velikovsky rebuttals to those like Sagan and simply do not realize that Sagan and other critics have actually distorted what Velikovsky is saying. I can only say that I have read both Sagan and the rebuttals and I no longer find Sagan so convincing as I once did. It is not that I think Velikovsky is right, but I do agree that Sagan actually did a very bad job in scientifically responding to Velikovsky. There is still merit in Sagan's charges, and I have independent reasons for thinking Velikovsky is wrong (e.g., V's ideas about collective amnesia), but I am no longer so willing to trust to the "experts" like Sagan to do my thinking for me. And, like Feyerabend, I am convinced that the reaction to Velikovsky has been far more reminiscent of priests attacking heresy than it has been rational dialogue with an extremist theory. I left religion years ago in objection to its authoritarianism; I do not tolerate similar authoritarianism in the *name* of science.

It is not that I want to treat far out theories with equal respect. Of course Wheeler and Penrose deserve more consideration than Velikovsky or Gauquelin. But *all* such theories that are presented as scientific ones and which do not violate basic rules of scientific discourse (like making up facts, etc., like a von Däniken) need a certain minimum level of respect and courtesy. Priority for discourse should be given to those theories judged most probable to contain pay dirt for us, so—as I have said repeatedly—I do not object to you or anyone simply choosing to ignore a Velikovsky or Gauquelin. But once you choose to engage them in scientific discourse, there is a minimum level of respect and courtesy that should be present in any scientific discourse based on the recognition of fallibilism so clearly demonstrated throughout the history of science. (Though his defenders, like Ellenberger, may show excesses, it should be noted that Velikovsky has been very restrained towards his critics and has *not* called them blind, stupid, irrational, pigheaded, etc., the way these critics have felt free to call him crazy, a crackpot, etc.).

Martin, when it comes to complicated technical matters, I must, like all of us defer to experts. But in the cases of Velikovsky and Gauquelin, such matters are not present. The major arguments against Velikovsky are relatively simple and presumably have been stated quite clearly by Sagan et al. It is not a matter of differences about the outcome of technical equations that the layman can not understand. When I read Sagan quoting Velikovsky and then his critics pointing out that Sagan has misquoted Velikovsky, I can simply go look at what Velikovsky actually said; and it seems that Sagan has indeed misrepresented Velikovsky. I don't have to be an astronomer to understand this. It does not make Velikovsky correct, but it certainly shows that his true position is not what has been "demolished" by Sagan. It seems to me that you and I share a common faith in science in terms of its method as a way of finding truth. But where we obviously part company is in the faith you obviously have in its practitioners to properly follow that method. I share your faith in science (in the abstract) but not in scientists. Where I must have faith in scientists, I prefer it to placing my faith in priests and others; but I would prefer to *minimize* the amount of faith I must place in anyone else who must mediate between me and the facts. Most people in most occupations are simply barely competent at what they do. I don't think scientists are very much different. And modern science in its currently institutionalized forms (with great vested interests and economic and other interconnections with society) is in great jeopardy of replacing the old priesthood with a new one.

Finally, your suggestion that I voice explicit probability statements rather than say something like "I am inclined to believe X is wrong" may be worthwhile. But it is not that easy to specify since many of the ideas I refer to are quite multidimensional. For example, the probability that Velikovsky is 100% right is probably nearly zero. But the probability that he is right about anything at all on which critics disagree with him is about 90%. Such probability estimates might be very different depending upon whether we are discussing the historical or astronomical elements in his theories. In the case of Gauquelin, I think the odds are about 90% that his evidence was properly arrived at (e.g., no data fudging, etc.) and therefore represent a legitimate anomaly. But the odds of his correlations being due to direct causal linkage between Mars and occupational choice is probably only about .01%.

Don't know if this letter makes things any clearer or better re your understanding of my views, but I hope it will help.

Cordially,
Marcello Truzzi

P.S. E. H. Walker was at the PA meetings and Ray and I met him and talked at some length. He was very different than either of us expected (good humored and remarkably pleasant about matters). He insisted that he thought he was largely chiding you and meant much of his attack upon you as humorous rather than really abusive as interpreted. He said he intended to write you directly about the matter. Not having seen the correspondence, Ray and I could not comment; but I must say that this would be consistent with the personality he displayed. Unlike John Taylor, for example, I found Walker able to show self-deprecating humor and similar signs of a lack of rigidity in his thinking. I had expected a real curmudgeon and he was nothing at all like that.

[No letterhead]   23 August 1979
Dear Marcello,

Many thanks for your last two letters. I think the giving of rough probability estimates for our respective beliefs is going a long way to clarify where we agree and differ.

Let me try to summarize where I assume we agree:

1. There is an external world, independent of you and me, that has a structure.

2. "Truth" in science is a correspondence between the meaning of an assertion in science and the world's structure. This is Tarski's "semantic truth" which all major philosophers of science now accept. Even John

Dewey would agree. His "pragmatic truth" was repeatedly stated by Dewey to be a correspondence theory. He disagreed with critics like Russell on details about the meaning of "correspondence," but this is irrelevant to our main issues here.

3. No statement about the world can be known to be absolutely true or false.

4. In the light of all relevant known evidence, joined with mathematics and logic, statements about the world can be given rough probability values. Carnap called it "degree of confirmation," .Russell called it "degree of credibility," Dewey called it "degree of warranted assertability." All amount to the same idea.

5. Science has no adequate method for assigning precise probability values to statements.

6. On the continuum between 1 and 0, extremes can be identified as almost certainly true (e.g., world is round) or almost certainly false (e.g., world is flat).

7. Scientists make rough probability estimates which underlie their subjective beliefs.

8. When a man is judged by almost all his peers to be ignorant of basic science, and when he defends with great dogmatism a theory to which all his peers assign an extremely low probability of truth (say under 1 percent), the man is called a crank and most of his peers are uninterested in either attacking him in print or even in reading what they consider his worthless books. (Note: By definition, all cranks are sincere; we are not considering the charlatans. Of course some cranks can be partly charlatans, but such in between things defeat all definitions except in mathematics.)

So far, I think you are with me on all the above.

Now we come to the all-important matter of personal judgments. I, and virtually all astronomers, physicists, and geologists (historians also) consider Velikovsky the very model of a crank. To me this is an informed, unbiased, objective evaluation. Because I consider V a crank, I don't want to waste valuable time studying his considerable output of writing. Of course one might say my strong opinion about V is "biased," but in that sense of the word I could say that you, who write about V as if his central dogma has a probability of, say, more than 5% of being true, are incredibly "biased" (incredible in the distance of your view from that of the experts) in the opposite direction.

Now if one is convinced, as I am, that V is a crank, it would be highly misleading to the public if I wrote about him as if he presented a genuine challenge to science in the Kuhnian sense. (A deplorable example of treating V in the Kuhnian context, is the chapter about him in Damon Knight's book on Fort. You and Knight are in extremely close agreement.) By the same token, assuming V is a crank, your treatment of him in your magazine as if he were a reputable scientist whose theories should be seriously heard and debated, serves to further confuse a public already enormously confused about how to distinguish good from bad science. You have said that you think my willingness to label V a crank (and the debunking tone I use in writing about him) reflects my being a writer and not a scientist. I think not, because all the scientists I know uniformly express themselves about V in much *stronger* terms. Not being writers, they don't write about such things. I in turn regard your tolerance of such as V as springing from the fact that you are a sociologist, inclined toward cultural relativism and toward overemphasizing the influence of cultural matters on scientific progress. (Kuhn does the same, but to a less degree than you (or maybe more!). Did you know that he wrote a paper insisting that at the time science abandoned the Ptolemaic theory for the Copernican, it was *entirely* a matter of aesthetic preference? But that's a long story ...)

Let's look at the later controversy over evolution. Huxley and Gladstone engaged in numerous and hilarious debates over whether the earth was ever totally covered with water in historic times. Huxley debunked Gladstone with strong rhetoric, and Gladstone debunked Huxley with equally strong rhetoric. Looking back on the debate, we see that Huxley had all the facts on his side, and Gladstone comes through as an ignoramus. At the time of this debate there were many Marcello Truzzis of the day who said: We don't know for sure who is right. Probably Huxley, but we mustn't be dogmatic. We haven't enough evidence yet. We just *don't know.* Let's not get carried away and treat Gladstone as a crank. After all, he is intelligent, sincere, etc. Maybe he'll turn out to be right. Science isn't infallible. Etc, etc.

The effect of this fence-sitting position was that it delayed the acceptance of evolution on the part of the public. It gave a spurious respectability to Gladstone's views. Huxley's efforts had a positive educational effect. Gladstone's had a negative one. The fence-sitters aided Gladstone, not Huxley. (I am assuming, of course, that at the time, the evidence for the fact of evolution was overwhelming among all those who understood the evidence.)

Velikovsky not only believes that the earth was totally covered with water in historic times, he thinks the moon was also. All the cratering we see on the moon, V believes, took place *after* the Noachian flood!

In your last letter you write that in your view the probability of V being right about *anything* at all on which he disagrees with his critics is about 90 percent. When I read that, I couldn't believe you had written it. Perhaps you mean that on about 90 percent of all the matters on which V and his critics disagree, V is right and his critics wrong. I find this almost, not quite, as horrendous a misjudgment. 90 percent? Surely you don't mean it I Say it ain't so, Marcello!

I, of course, think the probability (in both interpretations), is well under 1 percent, so you see, we do have an extremely radical difference of opinion with respect to our evaluations of specific theories. As for your conceding that V is not 100 percent right, I find this no concession at all, since it applies to *any* theory. For example, I think the probability of Einstein being right about quantum mechanics has a probability much *lower* than 90 percent. Here is something very curious indeed. I am *more* sceptical of the rightness of Einstein's views about QM than you are about V's central scenario!

A word about Suzie Cottrell. When she spoke to the assembled newsmen at Buffalo she went on at considerable length about her work with autistic children at the Biomedical Synergistic Institute (whatever that is) in Wichita, under psychologist Hugh Riordan. He may be a psychiatrist. I have never heard of him before. (Riordan, I was told, has since decided Suzie is a phoney. As I heard it, she and her father approached him on a deal by which he would cooperate and they would all make a lot of money out of Suzie's career as a psychic.) At the close of our session, when Suzie left in great anger, her parting words were, in effect: Whatever you think of me, if you have an autistic child, bring him to Wichita and I'll try to help him. I enclose three news clippings in which Suzie's remarks about autistic children are reported, and one in which I am quoted as calling it "shameful." Note how she applied her one trick (the only one she knows!) to letter cards. She would force an autistic child to take cards that spelled DOG, then someone would bring in a puppy!

A final comment about your surprise on finding Walker a pussy cat. When I was working on *Fads and Fallacies*, Gershon Legman said to me one day: "In case you worry about getting punched on the nose by someone you attack in your book, forget it. Crackpots are almost always meek, affable, friendly people in person." He was exactly right. I've met numerous cranks in my day, and invariably they are charming, polite, often humble about their views, but exactly the opposite in correspondence and in their writings. I sometimes think that you, not having been previously aware of this correlation, tend to be taken in by a crank's appearance of humility, which serves as a disguise for a total inflexibility of mind. Sagan told me that one day in conversation with V he decided to stick to one simple point. If electromagnetic forces play such a major role in determining the orbits of planets, how is it that astronomers, over the centuries, have been so extremely accurate in predicting planetary paths without at all considering their extremely weak magnetic fields? V, said Sagan, paused, stared at him a moment, and said: Let me make a prediction. Sagan waited breathlessly.

I predict, said V that ten years from now you will discover that I am right. At that point, said Sagan, he realized that rational argument with V was hopeless.

But I am wandering from the point. The contrast between verbal written abuse and affability in personal conversation has also been noticed as characteristic of extreme radicals. No one sounded more violent on paper that the Communist literary critic Mike Gold. Nobody was more of a pussy cat in person. Ditto for most of the anarchists who advocated violence. I could cite dozens of examples. Wilhelm Reich was savage in print toward his "fascist" enemies. In person, a charming, seemingly modest man. It is part of the cranks paranoid mind, knowing himself to be a great genius, that he can play the role of great humility in his personal contacts. Eisenbud is another good example. In person he is witty, charming, seemingly open-minded about everything.

Could you send me sometime, to further my understanding of your way of evaluating offbeat science, the names of three living persons, not charlatans, whom you consider outstanding cranks? (I assume we agree that there *are* such things as scientific cranks.) Persons you would rate as much farther out on the spectrum of low credibility than, say, V and Gauquelin.

You estimate the odds that Gauquelin's correlations are genuine (no data fudging of any sort, conscious or unconscious) as about 90 percent, and you add that the probability the correlations are explainable by "direct causal linkage" is only about .01 percent. If they are genuine correlations, am I correct in assuming you would rate the probability of "*indirect* causal linkage" to be about 90 percent also?

Incidentally, I don't consider Taylor a crank at all—just a blend of extreme egotism and extreme gullibility combined with the personality of a ham actor. Nor do I consider any of the eminent scientists who were duped by mediums to be cranks—Lodge, et al.

Best,
Martin

CC: Ray

PS: I regard my question about your nomination of the three greatest living cranks then as an important one, so let me focus it a bit more sharply. We rule out charlatans like von Daniken and L. Ron Hubbard. Let's confine it to the physical and biological sciences where there is the most solid core of agreed-upon beliefs among experts. (It is difficult to get a consensus on most theories in the social sciences, especially economics. Cultural anthropologists can't even agree on whether cannibalism is fact or mythology!)

And let's leave out psychiatry for the same reason—lack of consensus on what is highly confirmed. Also, when I say "great crank" I mean someone sufficiently intelligent and sufficiently persuasive to have attracted a considerable following of intelligent people. Thus I would not consider the head of California's flat earth society a major crank. He has written very little, and so far as I know has no followers. Nor should we include religious cult leaders like Herbert Armstrong who attack evolution. Again, his literature is too flimsy and unimpressive to make him a crank *scientist*. Price, on the other hand, is a model crank even though he was rationalizing Seventh Day Adventist doctrine—he produced a massive textbook (*The New Geology*) that is still *the* basic source of contemporary fundamentalist attacks on evolution.

From my perspective, Velikovsky is the greatest living crank. I believe he will go down in history as much greater than Ignatius Donnelly, with whom he has so much in common. My next two choices would be Gauquelin and Hynek. Obviously you do not consider any of these three men "cranks." (If you did, you would be debunking them with humor and satire) So— whom do you regard as the three greatest living cranks?
[Three articles about Suzie Cottrell]

[Eastern Michigan University]   30 August 1979
Dear Martin,

This is to comment on your letter of August 23rd.

First, let me clarify what may be a misunderstanding somehow. You seem shocked that I said that I felt the probability of Velikovsky being right about anything at all on which he disagrees with critics seems to have been badly stated by me if I read you correctly. By "anything," I meant any single thing. That is, if V and his critics disagree about, say, 200 different things, I think he is probably right about one of these points of difference with about 90% probability. In other words, I think he is very likely right about at least one point wherein he differs with his critics. I think you read my earlier statement to mean that I thought he was 90% probably right about any randomly picked difference between himself and his critics, which of course I do not. If you read my earlier paragraph over with the intent I meant—as hopefully clarified by this paragraph—I think you will see what I was getting at. I think V is wrong about most things but probably right about some small minority of his claims (which number a great many).

Now let me get to the earlier part of your letter and tackle things sequentially. I do indeed agree with you on the first 7 points you express. I have somewhat mixed feelings about your 8th point. It is not so much that I manifestly disagree with it, but it greatly oversimplifies matters and this leads to some later distortion, I think.

If a man is judged by scientific peers to be ignorant of basic science and defends his theory dogmatically, he is labeled a crank. The trouble with this is that it tends to work the other way around. That is, someone may be called a crank with the intent of convincing others that the "crank" is ignorant and dogmatic. The ignorance and dogmatism you seem to attribute to Velikovsky is the very thing at issue about his position. To call him a crank can mean prejudgment. Now I realize that you honestly believe you have examined his views and statements and believe he has displayed ignorance and dogmatism. But I and many others would simply disagree with you on this. You apparently want to consider all those who write for *Kronos*, *SIS Review*, etc., as other cranks despite their credentials, etc. I would prefer to see them as a minority view within science, which is clearly more than just a bunch of nuts. Now the debate about these matters in *Kronos*, etc., is mainly over the issue of whether or not V is in fact, as you insist, ignorant of basic science. He is certainly unorthodox in his views, but to call him simply ignorant is clearly incorrect because the simple fact is that he had the consultation of proper scientists with whom he corresponded and got comments and advice throughout his books. Given the broad range of his ideas, V has in fact sought expert opinion from a great many persons throughout his writing. He is simply not some sort of isolated writer sitting in a library spinning webs without ever seeking help in areas where he is ignorant. All this of course does not make V correct about anything, but this is not at all a case of some pontificating egomaniac screaming his eccentric theories at the world. Even if he indeed were guilty of the ignorance you and others like Asimov ascribe to him, the public record clearly shows that he does not fit your description of a crank in terms of dogmatism. He has been not only willing but welcoming of dialogue about his views. He has not ignored criticism at all. He has clearly sought to sincerely respond to his critics according to the regular ground rules of scientific discourse. He has not called others names. He has not ranted and raved. He has been remarkably calm and quiet in the face of much derogation. In your book *Fads And Fallacies* (as I recall) you define a crank as having paranoid views. This clearly does not describe the actions of Velikovsky's reactions to his critics. He does, of course, believe himself a great and revolutionary thinker, but that is far from what it takes to be a paranoid.

You described a conversation Sagan said he had with Velikovsky. As you described it, it does indeed sound like he's ignoring of criticism and acting irrational. But I would need more than Sagan's description of this conversation to make any such judgment. I have listened to the tapes of the AAAS symposium and I do not see Sagan as the calm rationalist and Velikovsky as the excited irrational megalomaniac. As I have mentioned

before, the criticisms of Sagan's critique of V are, in my view, largely well
founded. Contrary to your impressions, I do not think Sagan has really
played fair with Velikovsky at all. I dislike many of V's defenders, too, but
I must acknowledge their point that Sagan's critique is not the death blow
to V that Abell and others have claimed. It is not that I think V is right;
it is that I do not think that the case is closed by any means.

Now let me go back to the other element in your point 8, that your
"crank" is then ignored by the scientists who thus label him. As I have
written you in the past, if you think the probability of someone's being
correct is extremely low, you have a perfect right to give his ideas very
low priority and ignore him. But that is not what has been done with
Velikovsky. He has not been ignored at all. Instead we see spurious attacks
levied against him, which distort his views, misrepresent his words, impute
false motives to him, etc. My contention is that once you choose not to
ignore the man you think is a crank, you must argue against him with
sound arguments and not appeal to the authority of fellow scientists to
simply label him wrong and irrational. You can ignore him. You can even
simply state that most scientists think he is wrong and not worth answering.
But once you answer him, you better do a proper job of it. And that, I
would contend, has not been done.

Let me take some specifics relevant to yourself. You have several times
indicated your concern about Velikovsky having religious motivations di-
recting his work. Yet, in *Penseé*, May 1972, he was asked "To what extent
do you believe the Bible is an accurate document of historical record?" He
replied: "I'm not a fundamentalist and I oppose fundamentalism. I consider
any work written by a fundamentalist—say geology or paleontology—as of
reduced value (even though it may have some interesting facts brought together) because there is an axe to grind. You cannot approach the Bible
differently than you approach any other source." He later comments: "The
Bible in my work plays no other role than the Vedas, Upanishads, Icelandic
Edda, etc." I see no reason to doubt this from my reading of his works.

In your letter now you mention that V believes all the cratering on the
moon took place after the Noachian flood. Now though this also seems
doubtful to me, I must say that after reading the discussions of this in
Penseé and *Kronos*, including especially the debates on this with his critics,
the matter is not so clear as all that. And it certainly does not reflect a
dogmatic position on V's part. Martin, I simply see little in your comments
that goes beyond appeals to orthodoxy and authority. If you have in fact
read V and his defenders' answers to the critics and criticisms you raise, I
can not glean it from what you have thus far told me.

This brings me to the big issue, your request for my top 3 cranks.
As I have mentioned in my earlier letters to you, this presents a major

problem in that we don't entirely agree on definition. If we mean by a crank someone who tenaciously clings to a theory in the face of a majority of his peers criticisms and disagreement, that definition would apply to all sorts of people, and in no way suggests any irrationality at all. But you seem to want to include an irrational element in your definition of a crank judging from the way you use the word in your writings. Your crank must be ignorant and dogmatic. But how do we determine what constitutes ignorance and dogmatism? We obviously disagree about these parameters for Velikovsky. And it is obvious to me that this largely because you see him as irrational about it while I do not. I could just as easily call B. F. Skinner or Carl Jung cranks by my own perception of what constitutes ignorance and dogmatic thinking. But I would prefer to recognize that it would be unjust and improper for me to accuse these persons—with whom I strongly disagree and who I think "should know better" given the basic facts of science and common observation—of being irrational. So that is why I do not want to call people cranks at all. If we could use the term simply to mean obstinate and tenacious theorists, without the connotation of irrationality I would not object. Many scientists could properly be called "cranky" about their ideas and theories. But such pejorative terms simply don't belong in scientific description. If the person is truly irrational and paranoid—as you suggest Velikovsky and pseudo-scientists are, in *Fads And Fallacies*—then call him a crackpot, call him a nut, or whatever. But it you don't mean to imply irrationality, then why use the term crank? To me, an appeal in science to authority rather than to the arguments of the authorities is improper science and irrational within science. Since that is what I see some critics like Asimov doing, then it does not seem inappropriate to call Asimov a crank. But again, I would not do so for the reasons already given: such pejorative terms just don't belong in scientific discourse. If all you mean is that a person's ideas have a low probability of being right, not that they are irrationally derived and held, then call that person a poor scientist or a far-out scientist, but don't suggest he is not a scientist by labeling him a crank and thereby branding him as a heretic or illegitimate.

Now, if you ask me who I think is doing low-probability-of-truth type of science, and also clings to it tenaciously against what I see as reasonable criticism—this is as close as I will come to giving you 3 "cranks"—I would name: (1) those in the L5 Society who think it is really a solution for man's earthly problems to move us to outer space (and Asimov would seemingly be among these); (2) Thelma Moss for her work on Kirlian photography; and (3) Gerald Hawkins for his stuff on decoding Stonehenge. I mention these because we would probably agree on them. But I could also mention many leading economists, sociologists, psychologists, and others who put forward non-falsifiable forms of what they call "science" but which I would call pseudo-scientific.

****

On the matter of Suzie Cottrell, thanks for sending the newspaper stories, but they really don't shed that much light on the use of her deceptions re autistic children. You say in your letter that she forced DOG on the kids, but (a) there is certainly no direct evidence of this from the newspaper stories you sent, and (b) she does not even describe anything quite like you describe (that is, she says the dog is brought in after they "pick" DOG; even *if* forced on the kids, which is not demonstrated by you, the whole procedure might have some effectiveness for the kids). You speak of her "exploiting" autistic children, but how exactly does she do this. Does she take fat fees from the parents? As I mentioned to you in my last letter on this, I would call beating autistic children and making their parents feel guilty about them much worse (and maybe even more exploitative since psychiatrists charge fat fees) than giving some false hope to the parents (if that is all Suzie actually gives the parents from her work with the kids). Suzie may be as bad as you say, but I would certainly need more evidence than you have given me. It may even be the case that "fooling" the children by forcing DOG on them might somehow be therapeutic for them in the long run. What I would need to know before accusing her of exploitation is: (1) What is she getting out of it all, financially, etc.? and (2) What are the autistic kids and their parents getting from her attempts at therapy in *comparison* to more conventional and largely harmful or useless therapy by the "scientists"?

****

Finally, some comments on the personality of "cranks," and in particular my reaction to Walker. In general, I would agree with you about the personalities of charlatans. They are often among the most charming of people; it is definitely an asset if you are a charlatan, like good bed side manner for the quack doctor. I also agree with you that many crackpots (the truly irrational ones) are vitriolic in their writings and pussycats in person. But I don't think such generalization holds true for cranks whom I see as neither irrational or frauds. Cranks are merely extremists in science. Since, as I have tried to argue, being a crank has little to do directly with whether or not a person is correct, it can include all sorts of personalities except for strong intellectual conformists or authoritarians (who pretty much by definition would be ruled out since the crank necessarily does not want to follow the pack). Take Walker for example. When I met Walker, the thing that surprised me was his self-deprecating humor. This does not at all go with megalomania. I asked him about his correspondence with you which I had heard was strongly abusive. He told me that it was intended in large part to be joking and that he was baiting you, arguing that you should have had a thicker skin against his attacks given your own attacks on others. I said that this would be impossible for me to assess without seeing the actual correspondence. He just sent me the exchanges between you, and I can see from them now that he was indeed abusive but also that he indeed was putting you on in many places, especially in the early letters. To at least some degree, I think you may not have seen that was the case. This is not to say that you should not have been insulted by him. He was indeed insulting, and I do not excuse him for that. But there was also much humor intended and he was baiting you. (When I spoke with Walker at the PA meetings about your exchange, he indicated that he probably should write you some explanation. I urged him to do so. I don't know if he has done so.) He has now sent me a couple of letters and some of his articles. He signs his letters to me with such things under his name as "Mad Scientist." Such things clearly indicate to me that he has at least some psychological distance towards himself that eliminates any description of him as paranoid. It does not make him a nice person or correct in his sharp exchanges with you, but it does indicate to me that he is not so irrational as you probably see him.

****

In writing the above, I suddenly realized that I could have mentioned certain *schools* of chiropracty as among the cranks. Those who claim things like cancer cures with spinal manipulation (though these may really be conscious deceivers, so really charlatans). The area of medicine presents special problems re crank labeling. Linus Pauling is certainly an extremist re vitamin C's effects and goes beyond the known data in his claims in this area. He is in a minority and clings tenaciously to his claims. In the nonpejorative sense, he would be a crank. But he is not irrational about it and he may very well be ultimately proved correct about megavitamin doses of vitamin C curing everything from colds to cancer:

****

I've tried to take up all your points and hope this leads to more rather than less agreement between us.

Best,
Marcello

[Another Cottrell article]

[Euclid Avenue] 8 September 1979
Dear Marcello:

I am, of course, vastly relieved to learn that your 90 percent refers to V being right on *at least one* point. Ray had already guessed this was what you meant. I hadn't thought so because what I had asked you for was an estimate about V's central dogma: namely that about 1500 BC the earth suffered violent cataclysms because of close encounters with a comet that had erupted from Jupiter and later became Venus. *This* is the sole question with which Sagan is concerned in *Broca's Brain*, and about which I wanted your subjective estimate of credibility. It is important, because in my view such estimates are essential in defining those portions of the continua that define cranks. May I here ask for your estimate concerning V's central dogma?

I have of course no quarrel with your 90% except that I would make it 99+%. V raises literally hundreds of controversial questions on which experts themselves disagree. The probability seems to me fantastically high that he will continue to be shown right on more than one. So this estimate gives me no clue whatever about how you view the central dogma.

Megalomania. I do indeed believe that great cranks—by "great" I mean those who acquire world-wide followings—exhibit varying degrees of megalomania. The *sine qua non* of crankery is a persistent, unchangeable clinging to a set scientific beliefs that rated close to zero in credibility by the experts—note I say "experts." In the case of Galileo, for example, the rating of low credibility was not by other scientists but by churchmen, quite a different thing. Now when a man clings to a set of beliefs of near-zero credibility, and never budges in his convictions despite the evidence presented to him by his peers, it requires a psychological mind-set that clearly is megalomaniacal. This does not, however, take the form of ranting and raving, and screaming his views, etc.—*this* kind of paranoid behavior is characteristic of the minor cranks who spend their lives in mental institutions, working on perpetual motion machines, trying to convince others that they have found the secret of the universe, and so on. The great cranks in the history of science by and large have personalities that are mild and modest, often witty and charming and seemingly rational. Price is a marvelous example. If you look into his masterwork, *The New Geology*, you'll find it written without any rancor, or fuming at "enemies." In person he was a kindly man. Professionally, he was in charge of the buildings and grounds of a SDA college in Walla Walla. He had no training in geology. Yet he believed that all modern geology was totally mistaken, and that he, Price, had completely demolished the theory of evolution and proved by the record of the rocks that the earth had been created in seven days, and that all fossils were produced by the Great Flood. It is *this* kind of megalomania that is characteristic of the great cranks.

It is characteristic of V to a supreme degree. When I heard him speak in 1952 (see my book, p. 328) he said his greatest "fear" was that after his death his theory about Venus would become a new "dogma" that would be as hard to modify as the illusions of present-day astronomy. At that remark almost everyone in the audience did their best to conceal laughing. Example 2: After Sagan's famous AAAS attack, V told the gathering: "My work today is no longer theoretical. ...Nobody can change a single sentence in my books." (As reported in one of the news weeklies, V's followers gave him a standing ovation.) On this point Asimov shrewdly observed that he knew of no astronomy book that could last 10 years without major changes, V was talking about a 20-year old book. If this is not megalomania of a sort, then here again you and I have a very different understanding of how the word is used in common speech.

Great cranks, because of their ability at role playing, and their charisma, attract a following, mostly among nonscientists, on whom their megalomania becomes catching. Here is a quote from Sizemore, editor of *Kronos*: "To this date (1979) no one has brought up a single factor that would discredit V's theory on the origin of Venus..." Sizemore again: "I've seen all the evidence against him and it doesn't amount to a thimbleful." These are the remarks of a man who is supremely ignorant of the physical sciences. "Thimbleful"? Surely you can't agree that this is how responsible scientists talk.

Fundamentalism. This is a school of Protestant thought that emphasizes the literal accuracy of Biblical historical accounts. It is never applied to Catholics or Jews. When V said he was not a fundamentalist, he was merely disassociating himself from such Protestant fundamentalists as today's followers of Price, Billy Graham, and so on. But V is a devout believer in orthodox Judaism, who keeps a kosher household. On one occasion when I heard him talk, he spoke about how he believed that God had intended him to be a revolutionary figure in Judaism. (A "thorn in the side of Judaism" one of metaphors as I recall.) I do not have a record of his exact words, but I think it is accurate to summarize this "revolution" as follows. The majority of orthodox Jews believe that OT miracles are genuine miracles, performed by God in defiance of natural laws. Since God created and upholds the universe, he can cause any event he pleases. V's position is, that in dealing with his chosen people God did indeed cause seemingly miraculous events to take place—the Noachian Flood, the fall of manna, the stopping of the sun, the parting of the Red Sea, and so on—but that he accomplished these events by purely *natural* means. A kind of synchronicity, if you like. This same basic theme underlies V's attack on Darwinian evolution, and explains why, as I said in an earlier letter, that V's position is, in a sense, creationist. He believes that species appeared with great suddenness in geological history, as the result of massive and sudden mutations caused by catastrophes. This, by the way, has in the past been a popular approach with anti-Darwinians among both Catholics and Protestants (and still is today among "liberal" Catholics). You can see how "synchronicity" makes it hard to say whether the appearance of a new species is a creative act of God, or natural law. One believes that God, in his great Plan for life on earth, caused the various species to be "brought forth" as described in Genesis, yet he made use of natural calamities to bring about the sudden mutations. To ask if this is creationism is comparable to asking if it was God's plan that the earth be completely covered by water. Traditional Judaism views this as a miracle, as do Protestant fundamentalists. V's approach was to provide a scientific "explanation" for the Flood—i.e., the earth-moon system was enveloped by a great cloud of water that had earlier been ejected from, as I recall, Saturn. (I could check this by going to my V shelf, but is on another floor and is not worth the effort.) Price, by the way, also had a "scientific" explanation of the Flood, though quite different from V's. V also differed from Price in providing a modern "catastrophic theory" for the origin of species. (I am aware that catastrophes *played a role* in reputable evolutionary theory.)

Like Price, V would deny that his religious views played *any* role in his "science." But surely you can see how, in the case of both men, they had a powerful religious incentive to find "scientific" explanations that would be in accord with the OT account of God's dealing with man. When V writes that he treats the Bible the same way he treats other sacred books, he means no more than that "as a scientist" he is finding equal confirmation for his catastrophism in *all* ancient texts.

And V, like Price, *is* a hermit scientist even though he has a wide correspondence (as did Price) with hundreds of experts. V did his early "research" in the NY Public Library. He corresponds with other scientists only to find confirmation for his wild theories. There is no give and take of ideas in any sense comparable to that of genuine scientific discourse. Neither Price nor V budged an inch on any of their central dogmas. I think it would be salutary if you obtained Price's *New Geology* and read it as carefully as you have read V. As a work of crackpottery, I find it more impressive than V's first book.

Walker is another splendid example of a megalomania that is not visible in personal discourse with him. Check the close of his essay in Edgar Mitchell's big anthology. Walker thinks we are entering a religious revolution in which faith will be replaced by knowledge. He takes Copernicus as a symbol of science, Luther as a symbol of faith. "We are at a point in time for which certain knowledge, factual knowledge, can provide a basis for the God concept." If Walker believes that his QM theory of psi is the basis for the new revolution. "It is to be through efforts of this nature that the present acceptance of God, faith, will come to an end, and factual knowledge will become the basis of the new religion. This is to be the rock on which the new age is to be founded. This is the thesis I come to nail to your door."

Walker here likens himself to Luther nailing his thesis on the church door! If you do not see the megalomania in this, then we must be using the word in widely different ways. Naturally don't mean the extreme megalomania that puts people into asylums. It is the typical kind of megalomania that characterizes the scientific crank.

As to "crank," we continue to differ on what the word means to scientists in general. In your letter you deny that V is a crank, but suggest that Isaac Asimov is! And on the phone you said that upon reflection you decided that the outstanding living "crank" is Klass! But surely my letters and my old book make clear that scientists do not call a man crank unless he is stubbornly defending a body of doctrine that his peers agree is close to zero in credibility. What are Klass's scientific views that are remotely comparable in this way with V's? Just as you once confused "crank" with "charlatan," it seems to me you are now confusing "crank" with personality traits, or with the kind of rhetoric one adopts in private correspondence and writing for the general public. In your letter you cite Thelma Moss and Hawkins as nominees for two of the three greatest living cranks. Hawkins may indeed be a kind of crank, but surely a very "minor" one. As for Thelma, I hardly regard her as a crank at all since she has merely adopted the ready-made views of her mentors. Although I personally believe Pauling is wrong about the merits of vitamin C, by no stretch of the word would I call him a "crank." I agree with you that the *founders* of both chiropractic and osteopathy (you didn't mention the latter, but Andrew Still was certainly a very great crank. Osteopathy, of course, now rejects all his major doctrines.) But I asked for *living* cranks, not dead ones. In my view, V is the greatest living crank. You find him less crankish than Klass! We are back to an old argument about what Peirce called the "ethics of language." You cause only confusion when you adopt Humpty Dumpty's technique of making a word "mean just what I want it to mean." You have to respect the way it is commonly used, or you create endless confusion. To treat a crank like V as posing a genuine challenge to astronomy, and one that deserves to be treated in the same way one would treat, say, an unorthodox theory of black and white holes, is to mislead the already befuddled public and to do science a grave disservice.

No need to send the Mann mss on how Olaf may have done some of his card tricks. I find that Karl Fulves had sent me a copy earlier in the year. Incidentally, I wasn't much impressed by Mann's reconstructions. I think had he actually witnessed the effects, and not based his reconstructions on what nonmagicians wrote about them, he would have given simpler methods. Of course Mann realizes this. Even Christopher, whose knowledge of magic is vast, at first thought Geller was using a chemical that weakened spoons and keys!

Cheers,
Martin

CC: Ray

PS: Letter too long for me to write more about Suzie and what she told all of us about how she worked with autistic children (using alphabet cards), or what the psychiatrist told us on the phone. I agree with you that she probably did the children no damage, but I think you would agree that if this was merely a device to get publicity and advance what she hoped would be a show biz career (as I am convinced it was), then it was not a very admirable thing to do. Suzie and her father saw a lot of money in the sky! Unfortunately, her repertoire is limited to just one card trick (which she does superbly), though she may surface again with some new angles.

[Postcard]   14 September 1979
Dear Marcello:

Re postcard. I've never written a negative critique of Hynek and wouldn't if requested. The only time I had some critical remarks to make about Hynek was when I did a movie review of *Close Encounters*, where the remarks about H were asides in a long review of the film. I haven't written anything else about UFOs since 1952 (in *Fads and Fallacies*).

Best,
Martin

[Euclid Avenue]   13 September 1979
Dear Marcello:

Since I scribbled a postcard reply to your postcard, it suddenly occurred to me that my review of *Close Encounters* was presented by *NY Review of Books* as a review of three items, one a book by Hynek, so you naturally supposed I had accepted a request to review the book. Actually, what happened was this. *NY Review of Books* asked me by phone if I would go see the film and write a review, which I did (the *NY Review of Books* pays very well for such things). I had recently bought the Spielberg and Hynek books, so I worked references to both books into my film review, since they were closely tied into the film. *NY Review of Books* then decided to present the piece as a sort of triple review: one film, two books. I was intrigued by the project of reviewing a film (this is the only film review I've ever done). Had, NYR or any other publication asked me to review Hynek's book, or any other book on flying saucers, I would have declined. In fact, I decline *most* requests for reviews of what I consider crank books.

I recently looked at my chapter on flying saucers in *Fads and Fallacies* and although now dated, I find I have nothing in it I'd care to alter. The role of Ray Palmer in getting the mania off the ground (as you know, I see Ufology as simply another instance of a public mania, in the same category as medieval manias for seeing the Virgin Mary) has never been fully documented. He and Ken Arnold were good friends. I knew Palmer slightly when I lived in Chicago and he edited *Amazing Stories* there—a pleasant little man who enjoyed hoaxes (e.g., his Shaver hoax). He claimed his dwarfishness was due to an accident as a child, which may be true, but I was always skeptical of his somewhat conflicting accounts of what happened. He *looked* like a typical born-dwarf. As you know, he founded *Fate* magazine. There has never been a good biographical article about him, but his role in the paranormal explosion would make a colorful story. I have kept the early issues of *Fate* (Arnold had a piece on UFO's in the very first issue), and the UFO articles make fascinating reading. Palmer was always bitter about how little money he made out of the UFO flap.

I never knew anyone who knew Palmer *well*. Maybe Curtis Fuller? If so, he might be persuaded to write an article about him, if you think your readers would be interested.

Best,
Martin

[Eastern Michigan University]   17 September 1979
Dear Martin,

First, in regard to the "central question" in your letter of September 8th. You asked me what subjective probability I would give to Velikovsky's central thesis, that Venus came out of Jupiter, had a close encounter with the earth around 1500 BC and produced some events described in many historical records which most of us had thought to be purely mythological. I would give that whole episode a probability of being true something like 1% or possibly less. Looking at the part of the thesis that Venus was a comet coming close to the earth (aside from coming from Jupiter, etc.), I would give it something like 2% chance of being correct. Now that refers to my rough subjective probability and not really a carefully reasoned probability statement based on a really worked out probability (which would probably make me give it even less a chance of being true, probably something more like .05% given all the obstacles needing to be conquered as posited by astronomers). This reference to 1% is more-or-less my gut feeling about its chances.

My impression of your views is that it has only something like a .000000001% chance of being true. Is this correct?

But, as you point out, V's theory involves hundreds of lesser predictions and some of these, individually, may have a higher potential for being true. My own concern, however, is not really over V's likelihood of being right. What concerns me is the outrageous reactions against his views, reactions that continually misrepresent details of his views and include ad hominem attacks.

Somehow, we seem to have gotten a bit off the track in our communication, I think. You speak in your new letter of megalomania whereas you earlier spoke of *paranoid* behavior. Megalomania may be a necessary condition for being what you call a crank, but it certainly, is not a sufficient condition. Normal science has frequently been full of megalomaniacs who we have not called cranks.

You refer in your letter to megalomania as clinging to the scientific beliefs in the face of near zero credibility from experts. Yet you do not give me clear criteria for what an expert in this area is. You say that churchmen (in the case of Galileo) don't count (even though that line is not an easy one to draw between scientists and churchmen in Galileo's day). You seem to persistently think that all experts denounced Velikovsky. Yet this clearly ignores those who have come to his defense, many of whom are not scientists (of the first rank) but many of whom are, especially outside the natural sciences but including some within natural science. Just look at the people who wrote in *Penseé*, and now write in *SIS Review* and in *Kronos* (though I actually think somewhat less of the *Kronos* group).

I also find it amusing that you want to discount churchmen in the case of Galileo but seem to keep bringing up alleged Judaic orthodoxy as an important factor in discounting V's motivations. You may be quite right about V's motivations, but those motivations seem to have little bearing on his arguments themselves, and it is the arguments that need examination and/or acceptance/rejection. It is certainly clear that many who support

V's views (like Ellenberger, for example) neither support religious orthodoxy nor do they believe (as you do) that Velikovsky is either motivated by or arguing from such a posture.

You mention excessive rhetoric by Sizemore, and I agree it is extreme and somewhat silly language to speak of the evidence against V as a mere "thimbleful." But this is polemical writing, and similar polemical writing does indeed go on in responsible science elsewhere. And do you, on the other hand, consider Asimov writing an article about V as a "crackpot" (CP) to be responsible science writing?

On the matter of Fundamentalism, which you mention to explain away V's quotation, you may be right in your interpretation, but don't you agree, that many (if not most, as I would contend) of his defenders in science have accepted that quotation of his as dissociation? And you seem to ignore the remaining part of the quotation in which he explicitly talks about the status of biblical documentation as suspect from a scientific standpoint. I don't think his whole statement is really reconcilable with your interpretation. Also, to the degree that he seeks to give a naturalistic rather than supernaturalistic explanation of biblical and other sacred writings, he is not at all religiously orthodox and has been opposed by many fundamentalist/literalist types in religion.

But again, let me presume you are correct that he does indeed find a great incentive to find scientific explanations that fit his "OT account of God's dealing with man." So what? Motivations must be separated from arguments. If you believe him to be outrageously wrong, finding such motivations may explain his strange behavior for you, but his motives—which I as a nonbeliever find outrageous—can not be used to argue against his arguments. To do so is to commit ad hominem argumentation where it is not appropriate.

On Walker, I must decline judgment at this time. Your example of excessive rhetoric by Walker does suggest megalomania, but again, that is far from irrationality and paranoia.

You continuously refer to what "crank" means to scientists in general and point out that I use the word differently. But my whole point has from the beginning been that the word as used by scientists reflects a largely prejudgmental categorization that does not properly belong in science. The word as used by most people simply refers to those holding on to a minority view in a manner seemingly too tenacious. Based on this I am not at all surprised that V is called a crank by you and others. But I still must protest the word's use in so far as it suggest irrationality and an unscientific attitude by the so-called crank.

As I tried to show you in my last letters, if one spells out explicitly what you seem to mean by the use of the word crank, it can be argued that such irrational clinging to an idea against strong evidence can also refer to someone like Klass. You ask me what body of doctrine Klass clings to against his peers' views. That is easily answered. Klass clings to the view that there currently exist "normal" explanations for all significant UFO events. In other words, Klass seems to think that all UFOs as reported can be in fact identified or are IFOs. This is nonsense since it denies the existence of anything anomalous going on which is worth studying. And to make things worse, Klass persists in upholding the "explanation" of many particularly troublesome UFO cases, which he gave in his first book *UFOS IDENTIFIED*. There he insisted that they were some sort of strange new plasma previously unknown to science. His main opponent in all this was MacDonald who was an expert on plasmas. From what I can learn on this, Klass's theory of plasmas was generally considered ridiculous by plasma specialists. In this sense, Klass's theory was a kind of "paranormal plasma" explanation, which his peers rejected. His second book sticks with his plasma notions, and he and I have corresponded about this so I know he sticks with his plasma theory.

Now this issue in part brings up the question of who is an "expert" in the UFO realm. I think we might differ on the role of the Center for UFO Studies (Hynek's group). Please note that Hynek's first book got a very nice review in *Science*. It was well greeted by many scientists. Peter Sturrock is the head of the Plasma Institute at Stanford University and supports Hynek and rejects Klass's plasma ideas. I consider people like Allen Hendry and Hynek and Sturrock highly responsible "experts" in the UFO realm. Please note that Hynek has never expressed any commitment to any theory of UFO origins. He simply insists—and I agree with him—that there is a residue of unexplained reports, which constitutes a legitimate anomaly of probable worth for scientific investigation. Hynek (and I) remain agnostic about this residue of cases. They are just what the label says: *Unidentified* Flying Objects (I personally prefer the term Unexplained Aerial Phenomena). In no sense is Hynek's position on this that of a crank in my view (using your own definition). But Phil Klass continues to offer an explanation that is indeed rejected by the experts on plasma and in this sense, Phil's position is more radical than is Hynek's.

I also consider Phil a crank because of what I consider to be his irrational clinging to arguments that others have demolished and his clear evidence (to me) that he is incapable of confessing error (though he constantly says he would confess it if only someone would convince him he made one!). To make things even worse in the megalomaniacal sphere, Phil has frequently gone after people in what I would describe as a scientific vigilante seeking to destroy reputations in the name of some higher good which he wants to bring to science. Phil has done a number of things to people which I would characterize as vicious and these acts seem to be done because of what he views as his "Boy Scout ethics" but which I see as extreme megalomania of the kind you say characterizes cranks.

****

I appreciate your comments on the review of the UFO movie in *NY Review of Books*, but if I read your letter correctly, they made your article a review of the Hynek book (a) without your permission, and (b) without your even reading the Hynek book. If the latter is true (b), don't you think you should have written to inform the readers that this was the case? Otherwise it seemed you were knocking Hynek's book. Did you even write Hynek to explain all this? Certainly I thought you were knocking Hynek as well as the movie.

****

Re Cottrell, I ask you to reread your own letters. You began by condemning her primarily for her exploitation of autistic children. Now you admit she may have done them no harm. You do not really show that she even exploited them at all. If it was merely a publicity device on her part, I agree that it was not an admirable thing to do. But what real evidence do you have that it was merely a publicity device? You may have such evidence but you have not yet shown it to me. I think you are inferring such to be the case, possibly correctly, but hardly convincingly. I don't think much of Suzie, either, but my main criticism was that the Committee would do better to concentrate on the psychics endorsed by the parapsychologists. You then brought in her exploitation of children as apparently added reason why she should be debunked. But that hard evidence seems to be lacking and really is not very relevant to my main criticism anyway. I don't object to debunking Suzie, but I think this does more to educate the public against endorsements by Johnny Carson than it does anything to really further the debate between those for and against parapsychology's claims. I expect to see *Skeptical Inquirer* publish a debunking of Kreskin next (which it will probably take from Marks and Kammann's new book Paul Kurtz is publishing for Prometheus Press). My central point which started all this exchange, you will recall, was in my *Fate* interview comment criticizing the Committee for taking on the lightweights like Suzie as its apparent central purpose, which seems to be attacking media psychics rather than those put forward by the protoscientific community.

I am dashing this letter off to you rather hurriedly this evening, so hope it makes reasonable sense. Since I note you sent a copy of your last letter to Ray, I will send a copy of this one to him, too.

Best,
Marcello Truzzi

P.S. I got the enclosed letter recently from Krippner. I thought you might be interested in his comments on you and himself. I am also curious about your own reaction to these. To possibly clarify matters, I also enclose a copy of my own earlier letter to Krippner to which he is apparently responding in this letter from him.

[Euclid Avenue]   19 September 1979
Dear Ray:

Glad the phrenology stuff was useful.

I think we are in close agreement about ESP, though perhaps my probability estimate, necessarily fuzzy, is a trifle lower than yours. For me, the possibility that "something is going on" that we don't yet understand is about the same for ESP as it is for the UFOs. In both cases there is the residue of the unexplained. In both cases there just isn't enough adequate information to know which of which of the traditional explanations apply. In any case, my estimate of the probability that ESP is a genuine phenomenon is higher than I would assign to Velikovsky's central scenario about Venus, or to the genuineness of Gauquelin's correlations.

The book by A. R. Wallace that I mentioned on the phone is *Is Mars Inhabitable?* I discuss it briefly in an amusing exchange of letters with the astronomer Clyder Tombaugh that are reprinted in the last chapter of my *Sixth Book of Mathematical Games*. Tombaugh defends the canals, and I am the sceptic—one of those extremely rare cases of one side quickly winning a clear cut victory!

When you write "...we do not have ways to tell in advance if a particular belief system is correct or not," I assume you mean a belief system of substantial credibility, at a time before adequate evidence against it has accumulated. To take an extreme case: when the belief system that the earth is hollow and we are inside was advanced by Teed, science had ways at once for showing it to be incorrect to an almost certain degree. I think the same thing can be said today about V's theory of Venus. Another important point that I think is often forgotten is that sophisticated experimental designs in areas where the evidence is mainly statistical are relatively recent in the history of science. If you examine the actual supporting experiments in phrenology they are as crude as Rhine's early experiments. I believe that had the psychologists and medical men of Gall's day been as familiar as they are now with such things as double-blind techniques, etc., phrenology would not have flourished as it did among top scientists. There *were* ways to decide, but scientists didn't know about them.

Let me make the same point with Blondlot. The simple experiment that Wood used to discredit him was so simple that one would have expected

Blondlot to have thought of it. The fact that he didn't counts heavily against his competence. It does not make him deranged or insincere. An example of how far physics has progressed in understanding how to guard against bias from experimenter belief is provided by the story of Weber and his gravity waves. Nothing like the N-ray flap occurred, and I am convinced that one reason it didn't is that other scientists, taking the precautions that Weber did not, quickly disconfirmed his results and the community of establishment physicists accepted the disconfirmations. It is easy to forget that it is not just scientific knowledge that advances, but also scientific methods. This is not to fault scientists of the past who had beliefs, we now know to be wrong (e.g., Galileo never accepted elliptical orbits or the influence of the Moon on tides!), but only to point out that science does have ways to decide, in many cases, whether a new belief system is correct or not, and that the ways are steadily getting better.

I think it is a failure to make such distinctions that lands our friend Marcello in the lap of the "anything goes" or "almost anything goes" school of Feyerabend—a kind of Forteanism minus Fort's humor. On Feyerabend, by the way, see the exchange of letters between F and a critic in the October 11 issue of *NY Review of Books*. Same issue has a marvelous survey of new books about Egypt, including a good section on "Pyramidiocy."

Look forward to seeing you in NYC this fall.

Best,
Martin

CC: Marcello

[Two rejoinders in NYR with a handwritten note: Marcello: If Ellenberger had shorter letters and was less abusive of Sagan, NYR would have included one here. Compared to Ellenberger, Sagan and I are models of polite rhetoric.]

[Euclid Avenue]   22 September 1979
Dear Marcello,

Our letters are getting so lengthy that maybe we have reached a point of diminishing returns for both of us. I'll try to be brief.

1. Yes, I regard the probability that Venus was a comet before 1500 BC as having a probability *extremely* close to zero.

2. I don't think megalomania is either necessary or sufficient for crankery, only that it is a frequent accompaniment.

3. Can you name an astronomer or physicist of "first rank" who thinks Venus was a comet before 1500 BC? It's true that many scientists "came to

his defense," but that's quite a different thing—they defended his right to be heard and not censored, etc. I know of no first rank physical scientists who have endorsed V's central scenario.

4. Yes, I consider Asimov's writing about V as a "crackpot" to be responsible popular science journalism. Your unwillingness to call V a crank seems to me irresponsible. All over the science scene there are wild and wonderful theories being propounded, and among them may come genuinely new paradigms: new models of the universe, theories to replace quantum mechanics, theories of elementary particles, models of the brain, new angles on how evolution works, new theories of the origin of the solar system, theories about time reversal, Wheeler's geons and worm holes, etc, etc, etc. But V is totally *irrelevant* to astronomy and physics. To treat him as if he has issued a genuine challenge to the establishment, worthy of serious dialog between V and establishment scientists, is not to educate the public on how to tell good science from bad, but to confuse them.

5. I have never accused Walker of paranoia. I used such terms as megalomania and paranoid, both of which can vary from mild to severe.

6. Nor have I ever said V was "irrational." I consider him woefully un-informed, but not irrational. Indeed, rational arguments are characteristic of the great cranks. Grant their assumptions, and the rest of their belief system is worked out with incredible ingenuity. Cyrus Teed's books are filled with rational and good arguments, as well as "experiments" he per-formed to confirm his belief system. I, constantly receive angle trisections that are marvels of ingenious mathematical reasoning.

7. I completely agree with Klass (and the Condon Report) that nothing "anomalous" is going on over our heads that can't be fully explained in familiar, normal ways. (Naturally there will be that residue of unexplained, for the simple reason that there isn't enough evidence available.) I could, of course, be wrong—but my opinion is that the UFO craze is a typical mania, to be explained by sociologists and psychologists, and in no way comparable to reports of stones falling from the sky.

8. I do agree with you about Klass's plasma explanation. More than that, I am not convinced that ball lightning is a genuine phenomenon though here I give it a 50/50 probability. But this seems to me far too minor a thing to justify calling Klass a crank certainly not a "great crank," which is what I asked for. When I call V a crank and you respond with "How about Klass?" I find this comparable to my calling Stalin paranoid and having someone respond with "How about Ronald Reagan?"

9. Your statement that Klass's UFO views are more "radical" than Hynek's defeats me. Do you believe that the *Fate* interview (with H) dis-torted his statements?

10. Both assumptions about my *NY Review* piece on the movie are

incorrect. I gave permission to put Hynek's book at the top, along with Spielberg's novel. As I said in my letter, I owned both these books, and I certainly read both of them carefully. Hynek's book came into my review of the movie because Hynek, as you know, was the technical consultant for the film, and a major influence on Spielberg. You write "I thought you were knocking Hynek as well as the movie." Obviously! Of course I was knocking Hynek. When I said I would not have done a review of the Hynek book if a periodical had requested it, my refusal would not have been out of respect for Hynek but only because I would have considered the book not worth my time to review. I cannot buy your view that Hynek takes no sides on UFO explanations. He served as consultant to a movie, based in large part on his books, taking its title from Hynek, and Hynek even appeared in the movie. This is strange behavior for a scientist who does not wish to endorse the extraterrestrial view of UFOs! I don't doubt Hynek's sincerity, or his competence as an astronomer. (Have you ever asked Hynek what he thinks of V's Venus theories. Just curious.) To me, his case is a tragedy comparable to that of Dr. Andrew Ivy. In both cases, I think there are complex psychological factors involved, and about which we can only speculate.

11. Krippner. K is right in regarding my remark that he "well knows" that Nina is a charlatan as irresponsible. When I said that, I naively believed that K did know this. It was not until later that I realized how gullible K is—the full realization didn't hit me until I read his praise of the psychic motors in Czechoslovakia, and his description of how Pavlita had magnetized a wooden match. This, however, is no excuse, because I shouldn't have guessed, and for this I owe K. an apology.

12. Is Nina a stage performer? There are few nightclubs or vaudeville houses in Russia, and "psychic magicians" like Nina perform under the cloak of "lecture demonstrations." When the big flap about fingertip reading took place in Russia, Nina surfaced as the chief rival to Rosa Kulashova. Rosa, as you perhaps know, included standard card tricks in her stage shows, much as Uri Geller once did in Israel. Eventually she joined a circus and dropped out of the psi limelight. In the sixties, Ninel (Lenin backward) as she then called herself, was a popular psi lecturer and demonstrator, doing the equivalent of what magicians here call "club dates." After her release from prison, she has stuck pretty close to her apartment in Leningrad where she lives with her engineer husband, and performs mainly for visitors. I consider her the Uri Geller of Moscow.

I have bulging file on the lady, with translations of Russian news stories—including the famous case of her being tested by reputable psychologists at the Bekheerev Psychoneurological Scientific Research Institute, Leningrad, where she was caught cheating and branded a fraud.

This was in 1964 (or maybe earlier) and one reason why I assumed K knew she was a fraud. (The official psi line is, of course, that Nina was framed.) By the way, after her release from prison Nina began performing under the name of Nelya Mikhailova, but eventually went back to Nina Kulagina. (She may not have served time. This episode is obscure.) I will enclose a 1964 report in Rhine's bulletin that gives a fair account of what she was doing before her arrest (apparently for black marketeering).

An amusing news story in *Vecherni Leningrad,* 1968, reports that she had been sentenced to four years in 1964 for having hoaxed scientists under the name of Ninel Kulagina (nobody seems to know for sure just why she was sentenced), and accuses her of having "little magnets hidden in intimate places." The latter is strongly confirmed by the film in which she waves her hands over a compass to make it rotate wildly, and her large boobs flap wildly about just above the compass!

I hope you are not considering a "dialog" about Nina in your magazine in which there is serious discussion by both sides over whether she is a fraud or not! I find it sad that, now that Uri has been discredited, leading parapsychologists are still clinging to the genuineness of Nina, who is beyond the reach of those terrible US magicians! (I was surprised to hear from Martin Ebon, yesterday, that McConnell believes Nina to be genuine!)

Best,
Martin

CC: Ray

[Euclid Avenue]　　　　　　　　　　　　　　　　　18 January 1980
Dear Marcello:

Enjoyed your piece on juggling, and thanks too for the copy of the remote-viewing replication failure, and your contribution to the forthcoming anthology.

I have no comments on the AAAS paper, since all major points are clearly made by the authors themselves. There's nothing for a sceptic to add beyond saying that he's pleased to see these replication attempts made by sceptics. I would recommend that instead of seeking meaningless "I-told-you-so" responses from critics of parapsychology; you let the believers have their say in criticizing the replication, then give the four researchers a chance to reply to the criticism. It seems to me you'll only muddy up the water by soliciting comments from Randi, Ray, me, etc.

Professor Taylor has written an entire book about his new and total disenchantment! Perhaps you know about it. If not, try to snag a review copy from Dutton. Due out in May, I think, and titled *Science and the Supernatural*. One of the funniest books I ever read. I'm happy to welcome him to "our side" of course, but his reasons are as wild as his former reasons for belief, and he's too big an egotist to give anybody any credit for wising him up on the methods of Uri, etc. It's all the result of his own careful experiments, etc. Girard is in the book, too, but under the pseudonym of Alpha (Taylor is afraid of legal action!). (You say there has been no "smoking gun" expose of Girard, Uri, Ted S., etc. It all depends on how you define "smoking gun"—see Taylor's account of catching "Alpha" cheating. When you say Ted S. is "probably" not genuine, this hits my ears exactly as if you said the earth is "probably" not created 6000 years ago. One should be open-minded—nothing is *certain* outside logic and math—but do you have to be *that* open-minded?) NYR still hasn't run my Tart blast, but maybe soon. They've also taken a long review I did of the two new books attacking the talking-ape research—Terrace's *Nim*, and Sebeok's anthology, *Speaking of Apes*, but it probably won't run for a month or so.

> Best,
> Martin

[Euclid Avenue]   3 February 1980
Dear Marcello,

   Just when I think you and I have found something basic to agree on, the issue flares up again as though we had never discussed it. You write that you spoke of Ted Serios as "probably" a fake because, although you *believe* he is a fake, you don't "know" that he is a fake, and because he hasn't been "proved" to be a fake.

   I thought we had agreed that in science, or indeed in saying anything about the world (in contrast to pure logic and math), nothing is ever "proved." Nothing is ever "known" with certainty. Every scientific statement has a degree of confirmation. You close by saying that with respect to psychics, as in science, one "must demand proof, falsifiability, etc." But there is never proof, and falsifiability is as much a matter of probability as confirmation. Consider the familiar case of "All crows are black." Someone reports seeing a white crow. Does this falsify the hypothesis? Not at all! One can say the report is false, the person was lying or mistaken, etc. Suppose he brings you a white crow. Maybe someone caught a crow and bleached the feathers. Hypotheses in science are never completely falsified. The Ptolemaic theory went on for centuries with its epicycles. There are physicists today who dispute a basic tenet of relativity theory—that light does not partake of the motion of its source. We both know well enough how the failure to falsify applies to psychics. You cite the case of Palladino. But we can't *know* that Palladino always cheated, nor can we know that Palladino told the truth when she said she cheated. When you contrast Palladino with Serios I can only read what you say with amazement. The documents purporting to show Palladino had genuine powers, far superior in quantity and quality to those relating to Ted, who is a one-trick psychic (and with virtually a one-man investigator). Eisenbud, in his book, tells of catching Ted cheating on several occasions. Geller himself has confessed to cheating during his act in Israel. *All* of Geller's defenders now admit that he often cheats. Panati told me on the phone that he thinks Geller cheats 80 percent of the time. It is naive to look for "smoking guns" with respect to psychic charlatans. When Margaret Fox demonstrated her toe snaps and confessed, did the Spiritualists believe her? To this day Colin Wilson argues that her confession was false, and done solely to make money.

In view of the fallibility of all knowledge, it is a universal practice of scientists, when something is established to a very high degree of probability, not to be perpetually writing "probably," or "I am inclined to believe," and "it is likely true that..." and so on. Pick out any scientist you admire, show me any paper by him, and I'll find in it dozens of remarks in which he assumes the truth of something that, strictly speaking, is not "proved." An astronomer, writing about the craters on the moon, would sound naive if he expressed himself by saying that they "probably are older than 10,000 years." Yet he doesn't know that for sure, and indeed Velikovsky argues that the craters are all younger than 10,000 years, and came after the time when the moon was completely covered with water (at the time of Noah's flood). Now we can't be *certain* V is not right, but no astronomer or science writer worth his salt is going to say that on this matter V is "probably" wrong.

Consider Ted's one claim. He can see a photograph in a magazine, unconsciously remember it, then many years later, staring into a Polaroid camera, he projects the image, line for line, onto the film. Surely you can recognize that this claim is as outlandish, to me even more so, as a claim that one can walk on the water. It is a question of being sensible in the way one uses language. You don't say—to use Freud's illustration (thanks Ray!)—that the earth's core is "probably" not made of jello. You don't say that Ted Serios is "probably" not a genuine psychic. Strictly speaking, of course, the "probably" is justified, but to stick it into every such sentence would mean that one could hardly talk or write at all. You would have to say that Kreskin "probably" has no ESP ability (after all, where is the "smoking gun" with respect to Kreskin?).

As usual, we are not arguing over any substantive issue, only about how one should talk. The whole point is that on such matters you talk funny. And the funnier you talk, the more you cloud the issues, creating more confusion than light. We all take the fallibilism of science for granted, but that doesn't justify a bright fellow like you saying that Geller "probably" can't bend metal with his mind.

Best
Martin

CC: Ray

Dear Martin,

Just a quick reply to your letter of Feb. 3 and the points about Serios. (I still need to answer your earlier letter re the fundamental issues between us; I had expected to get a chance to review our earlier correspondence to do a proper job on that letter put still have not found the real time to do the job right. So that will have to wait a bit longer. I seem to have gotten myself very heavily committed to several writing projects lately and am now in the midst of a lot of administrative nonsense—budget proposals, etc. that are part of my job as sociology department head; but I will eventually get to the needed reply to you. Just be patient even if takes 'til this summer!)

Of course I agree that nothing is known with certainty (at least of an empirical kind and I agree with many of your points about even falsifiability being a matter of degree, etc. But I think you are missing the context of my paper which is intended as a possibly too serious consideration of Paul Kurtz's purportedly scientific analysis.

We obviously agree that the probability is extremely high that Serios is a complete phoney. I certainly do believe that to be the case. And I agree with you that the case for Palladino is basically better than the case favoring Serios. But Kurtz stated that Serios has been proven to be false in a sense that I think goes beyond even what you—a hard liner on the matter but one certainly better informed about Serios than Paul is—would consider to be the case against him. I know from past dealings with Paul that he simply dismisses the case for Serios as a non-expert on the matter simply based on what I think is his minimum information about the matter from Randi (and possibly yourself). It is one thing to say that the case might as well be closed on Serios given the evidence it is quite another to say that the case really is scientifically closed as though it were really a "smoking gun" situation. One of the background factors in my discussion of Paul's discussion is that I am at least personally convinced that Paul is really not that familiar with the case he is dismissing. In a debate, I think Eisenbud would make mince-meat of Paul. For that matter, I honestly think Randi has overstated the case against Serios, based on the total published record. I still believe Serios is a fraud, but I feel it is absolutely essential, that we distinguish between our strong convictions in such matters and the objective character of the evidence involved that it would take to convince a neutral party (I don't myself claim to be neutral and therefore think Serios is a fraud; this is largely based on inadmissible evidence in the scientific court of "due process" to which I try to be professionally committed and which I try to keep *Zetetic Scholar* attuned to.) All of this boils down to some of the points Ray raised in his AAAS paper on which I hope you will be commenting for *ZS* (I thus far have some very fine philosophers of science comments on the paper including Joe Agassi, Paul Feyerabend and Harold I. Brown and expect some others). In Ray's terms, Serios has been strongly discredited by "hit men" like Randi and others. This may have positive results in the short run by immunizing some young scientists against exploration of such nonsense. But such discrediting is not the same as disproving. And I agree with you that disproving is a matter of degree, but I think there is a distinct difference between the discrediting in this case and the disproving that needs quite a bit more careful fitting to the facts alleged about Serios. I think such a mistake of degree is very damaging to science in the long run if not in the short run. It expects people like Kurtz to dismiss ideas based on supposed authority of experts whose expertise is really dubious. It is very similar to the problem of having all those astronomers who know nothing about astrology saying there is no research supporting astrology's claims. As soon as the astrologers point to a few studies that do support them (even if the studies are terrible ones methodologically), they demonstrate the dogmatic character of their critics who simply denied the existence of the studies in the first place while claiming to be experts. In the end, the astrologers came out ahead from Kurtz's anti-astrology manifesto because they so easily demonstrated that (a) they were being misrepresented in their claims (e.g., astrologers do not claim gravitational influence from the planets so what is the point in astronomers demonstrating that there can be no gravitational effects that would allow for astrological effects?), and (b) that the astronomers really did not know the pro-astrological literature well enough to seriously criticize it anyway.

In the final analysis, the difference between us is that you obviously feel, like Ray's "hit men," that scientists simply can not afford to even weakly legitimize extreme unconventional claims by saying merely that the case has not been proved by those like Eisenbud. From my view, science assumes the negative about a claim with the burden on the claimant. If the claimant does not prove the claim (e.g., Serios's powers), I respond by simply saying "not proved" and either go on from there to something else or listen to further arguments of proof from the claimant. You don't seem to want to simply stop with saying "case not proved." You want to say "case disproved" even if you only have an unproved case plus some discrediting arguments (e.g., he was known to cheat in the past). I see no good reason to go beyond the "case unproved" condition and see potential harm in going *prematurely* to the "case disproved" position.

Now let me acknowledge one unexpected point you made. I frankly did not recall Eisenbud's having caught Serios cheating in the matter of picture production. I'm going back to the book to refresh my memory (it's been many years since I looked at the book), but I presume you are correct if you say so. This does put Serios more in a comparable position with Geller or Palladino. I may simply have been wrong on this categorization in the matter.

Finally, let me emphasize that our roles in this matter are quite different and our differences stem in part from them. I am trying to play the role of a kind of amicus curiae to the scientific community in these matters. It is essential that I have the respect—if not agreement—of all responsible parties in the dialogues I seek to promote. My saying that Serios is "probably" a fraud whereas you might have stated "is almost certainly a fraud" which I would agree with but did not state, is partly a function of my role. You are very much an advocate against Serios, and a needed and welcome one. I am not. And I honestly think that Paul—in light of the goal statement of the committee which he claims to adhere to—is supposed to be sharing something more like my role. Now I realize that you may disagree with me about where Paul should be in this matter. You have always wanted the Committee to be an advocacy body for orthodox science in an open and straightforward way that I disagreed with but which I respected. Paul, on the other hand, is busy trying to play totally objective and open-minded critic, which I don't think you really are. I have always seen you as certainly rational and honest but largely as polemical in your approach to these matters. It would be silly for you to say "probably a fraud" or maybe even silly for you to say "almost certainly a fraud" when you could easily just say "obviously a fraud." That's part of your proper role as a kind of defense attorney for established scientific beliefs re such things. But that just is not the same role I am in at all. And it is not the role I want.

Let me go a bit further. I probably could have said "almost certainly is a fraud" rather than merely "probably." Maybe I should have. I doubt that you would have complained about my language if I had said "almost certainly" or "highly probably a fraud." And I admit that that is what better would have matched what I think to be the case. But because of my role as a mediator between parapsychologists and critics, I simply did not give such a measure of degree of probability as I saw it because if I had I would have gotten a letter from psi enthusiasts complaining about my making such a judgment without supportive statements. I simply said "probably" because (a) I wanted to avoid the whole argument of degree (though you surprised me by bringing it up from your end of the matter), and (b) I wanted to emphasize the glibness of Paul Kurtz's mistaking discrediting for disproving evidence. In this sense, my piece is somewhat polemical relative to Paul's article and this is partially a result of that. In a sense, then, it was a debater's trick, but I don't think it was a cheap trick.

In a nutshell, then, you seem to be criticizing me for what I did not say rather than for what I did say. You are upset because my saying merely "probably" assigns no degree of probability showing that I am dubious about Serios. You think readers may ambiguously take me to be more open-minded on Serios than I really am or should be. Correct? You say that no decent science writer would normally refer to such wild things are merely "probably" wrong when they seem to be clearly wrong. And I agree with that and would agree that in a different context I should not say merely "probably" but should say something stronger. My doubts about Serios are no secret. I have challenged Eisenbud directly on the matter. But in the context in which I was writing—trying to emphasize the authoritarian stance Paul is really taking in his article—my mere and guarded "probably" remains appropriate.

All of this may tell you little you did not already know, but I hope it might shed some light on our differences which actually do remain small but, I fear, remain significant still in terms of how I feel I must behave. (Interestingly, it really does not have that much to do with how I think you should behave, for I think we probably agree about your advocacy role. Of course, I still think that you occasionally go too far for my own taste—as when you resigned from *ZS* over the Velikovsky dialogue which I hope you might now agree went reasonably well and with minimal obfuscation—but that, of course, remains your right and I respect that even if I disagree.)

Best,
Marcello Truzzi

CC: Ray

PS. I've been getting some pretty wild stuff from Sarfatti since I met him in at the AAAS meetings. Do you still hear from him? He seems pretty clearly a bit psychotic, but wouldn't it be funny if he had something.

[Euclid Avenue]   11 February 1980
Dear Marcello:

I think it best I not attempt any reply to your last letter. I'm content to let the letter stand as a difference in taste with respect to how a sceptic should talk.

I write mainly to enclose a copy of a letter to Rao. Franklin originally wrote me a very deceptive letter, saying that he was editing a book of essays on psi to be evenly divided to give the reader a balanced picture, etc. He did not specify who the editor was. His letter mentioned that Ray was a contributor, but when I phoned Ray, he had not been approached at all! My suspicions were further aroused by a copy of your contribution, which you sent me. I put two and two together, assuming this was for the same book, so I wrote to ask Franklin who the editor was. Franklin, as you surely know, was responsible for the line of psi books published by Scarecrow Press.

Note that the book consists of papers by sceptics followed by strong rebuttals, and that you are among the rebutters. I am sorry to see you pegged this way, but what can a scientist expect if (unlike all other scientists who are not parapsychologists), he makes remarks in his papers to the effect that Geller and Serios are probably fakes, and the verdict on astrology as a science is not yet in. In a strict sense of course it isn't in, but in exactly the same sense the verdict is not yet in on palmistry, numerology, and tea-leaf reading. Feyerabend is the only other writer who talks this way, but of course he is a philosopher and that gives him special privileges.

MIT is publishing one of the strangest books in years. I'll enclose a flier. You met Brams when you were here for a conference on deception.

                              Best,
                              Martin

[Euclid Avenue]   11 February 1980
Dear Rao:

Robert Franklin, in his letters, gave the strong impression that your anthology was to be objective in the sense of giving equal weight to pro and con arguments. It is possible that both you and he believe this is the case, but let me explain how the book appears to an outsider.

The number of pro papers is indeed balanced by the number of con, but in every case the sceptical paper is there as a clay pigeon to be shot down by a believer, whereas this works the other way around in only one case. Since I will be sending a copy of this letter to friends, let me give details:

1. Hansel's last chapter is followed by Martin's attack.

2. Girden's paper is followed by Murphy's rebuttal.

3. Girden's postscript is then followed by Pratt's rebuttal.

4. The Moss-Butler paper is followed by your rebuttal. (Here, as the sole exception, Moss and Butler are being allowed to reply.)

5. Kurtz's paper is followed by Truzzi's strongly biased attack, which is also a strong attack on our sceptical committee.

6. Gibson's paper is followed by Beloff's rebuttal.

7. The Gardner-Wheeler piece is to be followed by two attacks, one by Signer, one by Rockwell.

8. Hansel on Pearce-Pratt is followed by a rebuttal by Rhine and Pratt.

9. Hansel on Pratt-Woodruff is followed by a rebuttal by Pratt and Woodruff.

10. Medhurst and Scott are followed by Pratt's rebuttal.

11. Scott on Pratt-Woodruff is followed by two rebuttals by Pratt.

12. Diaconis is followed by Kelly's rebuttal, a letter from Puthoff, and unpublished letters by Child, Rush and Eisenbud!

13. The final Epilogue, by Mackenzie, is a strong pro piece as well as an attack on the irrationality of the sceptics. (I omitted, above, your introduction to the book.)

In sum, the sceptical articles are simply bowling pins to be knocked over. The result: a strongly loaded pro-psi volume. It is as if a Catholic bishop were asked to prepare an anthology of articles pro and con the Church. The bishop picked a dozen attacks on the church, then followed each with a stirring rebuttal by a Catholic. Hardly objective! Consider how you would react if, say, I edited an anthology that opened with Eldon Byrd's paper on Geller and Nitinol, followed by my *Humanist* paper attacking this paper. Next a chapter from Eisenbud's book on Serios, followed by the *Popular Photography* expose. Then a paper by Soal, followed by Betty Markwick's expose of Soal. And so on. I think you would be the first to raise a great cry of "unfair." I feared the worst, and it turns out to be worse than I feared. I do not want to be part of a book that pretends to be what it clearly is not.

Sincerely,
Martin Gardner

[Flier for *Biblical Games: A Strategic Analysis of Stories in the Old Testament*, by Steven J. Brams]

[Euclid Avenue]   14 June 1980
Dear Marcello:

Just returned from a month in Hendersonville, where we bought a house. We'll be moving early in 1981, and after a year of alternating columns with Doug Hofstadter, he'll take over permanently.

Don't know anything about the Carrington experiment on that finger lift. Of course at the start of the lift the weight would naturally increase, then if the lift continued upward, inertia would decrease the weight, and it would finally settle back to normal. So we need a precise definition of what is meant by saying the weight of the group "went down." A similar ambiguity is involved in Campbell's articles on the notorious "Dean machine" in *Astounding Science Fiction*, which was said to have "less weight" when operating on a scale. Since the inertia produced a fluctuation of weight, it was not, clear what "less weight" meant. Similar ambiguities are involved in Hasted's experiments with psychics who bounce up and down on a scale and lessen their weight, and TM meditators who levitate on scales. I didn't know "Hurling Bull" (as we used to call him) was still alive.

Did you see the article on "Ceremonial Occasions and Mortality" (on the "death-dip" hypothesis) in *American Psychologist*, March/80, 253ff, by Schulz and Bazerman? A splendid example of the trickiness of statistical correlations. Judith Tanur, sociologist at SUNY at Stony Brook, has an unpublished paper commenting on all this (you may know of her).

Thanks for copy of your letter to Puthoff. Did you hear what happened to Pribram? Washoe mangled his hand so severely that he had to have a finger amputated! (This ought to have some emotive effect on his attitude toward chimp language!)

I see that Gribbin has retracted his "Jupiter effect" in the current issue of *Omni*!

Best,
Martin

P.S. Feyerabend and Bunge "heavyweights"? That is of course their opinion of themselves, but hardly that of the majority of philosophers of science. (Certainly not mine!)

[Euclid Avenue]   20 June 1980
Dear Marcello:

When I questioned calling Feyerabend and Bunge heavyweights, I was comparing them to other philosophers of science. Neither Kurtz nor Hook have made *any* contributions to philosophy of science, and nobody thinks of them in those terms. Russell, Carnap and Reichenbach are of course dead, but I was thinking of such living philosophers of science as Popper, Quine, Hempel, Nelson Goodman, Grunbaum, Salmon, Nagel, Black, Putnam, and many others, all of whom in my opinion are more important than either F or B. Of course this is just my opinion. Both F and B are very prolific, especially Bunge. A typical list of references at the end of a Bunge article contains 10 references to his own papers, and a few by others. In my opinion, you over-value both men, especially F who is widely regarded by most philosophers of science as an eccentric egotist, with very little of substance in what he has to say. You like him, naturally, because of his (to me) extreme tolerance of eccentric science.

Have you heard about the fight brewing at Berkeley over giving Mishlove a Ph.D. degree? The two leaders of the committee, Scriven and Tart, got the two scientists who opposed the degree booted off the committee (with threats of lawsuits), so it looks like Mishlove, who couldn't pass an undergraduate course in statistics, is, going to get his doctorate after all. In my opinion it's a bigger scandal than the degree given to klassneda. Maybe Persi has sent you some documents. If not, write to Robert Pisani, Statistics, UC at Berkeley, for his memos to the president on this hilarious farce. He's so angry at Tart and Scriven (and I don't blame him) that he's anxious to have it all aired as soon as possible.

Did I mention in my last note that Washoe recently mangled Karl Pribram's hand and he had to have a finger amputated? (Don't know the details.)

Best,
Martin

Dear Martin,

Obviously, we have some semantics still confusing matters. I would categorize all those philosophers you listed as heavyweights also and I would place several of those you listed (Quine, Popper, Hempel and Nagel, for example) above both Feyerabend and Bunge. In a sense, one could speak of these as "champions" among the heavyweights, but—to mix metaphors—my point was centrally that I was getting "big league" players and I was explicitly contrasting them with Kurtz and Co. You may feel—and I would agree—that Kurtz and Co. have not made any contributions to the philosophy of science, but I don't think Paul would agree. In fact, he wrote at least one book explicitly dealing with the philosophy of social science. The funny thing to me about our exchange on this is that Bunge is an active consultant to your Committee and a rather hardliner towards matters paranormal. As for Feyerabend, you misperceive the reason for my admiration of F. I admired his work in *Against Method* (along with Lakatos's antagonistic but somewhat similar extensions of the Popperian view) for his emphasis on the social negotiation process in science. I find his views on some paranormal claims rather naive and suffering from a lack of real familiarity with the literature. I mainly admire his comments and analysis of Popper and Kuhn and their differences (in the Lakatos & Musgrave volume). I do, of course, agree with his criticism of the Committee and his support for the goals of *ZS*, but my evaluation of his philosophical acumen is not based on that. Our personal evaluations of F and B are, of course, largely subjective and unarguable, but your remark that "most philosophers of science" view F as "an eccentric egotist" may be true for "egotist" but not for "eccentric." This is not a subjective matter but can be ascertained from the literature. We apparently simply read different authors. F is, in fact, rather personally eccentric, I gather, but I do not think his philosophical views are truly thus described. And, I might add, several writers who have tried to deal with F have badly misrepresented his position (e.g., claiming he is an anarchist politically when he clearly is not).

I heard from Persi about the Mishlove battles, but I really don't quite understand the whole business. I thought Tart was at Davis and Scriven was also elsewhere than Berkeley. Whatever the case, it seems to me that the matter can hardly be compared to the Casteñada affair since, so far as I know, no outright fraud is involved. I will, as you suggest, write to Pisani.

Since I doubt that you get *SIS Review*, I enclose a copy of the review of *Broca's Brain* that might amuse you in its numerous references to you as well as Sagan.

I just got a copy of Bob Morris's review of the Marks & Kammann book, which will appear in the *Journal of the American Society for Psychical Research*. I am reviewing the same book for the *Journal of Parapsychology*. Though I think there are excellent things in the book, my overall assessment is similar to Morris's and in some ways a bit more negative. Will send you a copy of it when done.

Do you know anything about the causes of death of James Webb? I was very shocked to hear of it. I think Jamie was under 40.

Do you know Morris Goran and/or his books on science and pseudo-science? I've been belatedly reading them and am surprised at the lack of attention they have received.

My projected *Science 1980* paper has been killed. Probably will try to rewrite it and send it elsewhere. Pity since it was a very balanced piece, I thought, and Ray agreed.

ZS should be out at the end of this month—I hope. Hope you like it when you get the copy I will send you.

Best,
Marcello Truzzi

P.S. Do you still get Sarfatti's incredible mailings? When I met him at the AAAS meetings he seemed far more normal than I had expected, but some of his projects really make those like Walker look quite conservative.

[Euclid Avenue] 6 July 1980
Dear Judith Tanur,

The literature on the Gauquelin effect is so large and my photocopier so slow and small that I have decided to do as follows:

By separate mail I am sending you a duplicate copy I happened to have of a 1978 issue of our magazine that will introduce you (see page 118ff) to the Gauquelins and their books by way of a review they did in which they criticize statistical support of conventional astrology. The hilarious aspect of the Gauquelin claim is that they are strong *sceptics* of traditional astrology, but they claim that they have found significant correlations between various professions, and the positions of major planets in the sky at the time of birth! Thus all believers in astrology will throw the Gauquelin work at you to prove that the influence of the planets on human life has now been empirically established, even though the Gauquelin claims negate traditional astrology! The Gauqelins are quite sophisticated in statistics, and their correlations seem to follow if you accept their raw data. My own opinion is that the raw data is biased in subtle ways, but to check the sources is virtually impossible without an enormous expenditure of time in France, and obviously no statistician is going to attempt it without funding.

If you find this interesting, there are two later issues of the *Skeptical Inquirer* that you can obtain. Perhaps if you write directly to Paul Kurtz, telling him who you are and that you might want to jump into the fray, he will send you gratis copies or offprints. The issues are:

- Vol 4, No. 2, winter 79/80. A four-part report on the "Mars effect", pp. 19-63.

- Vol 4, No. 4, summer/80. Follow-up on above, by Gauquelin and by his critics, pp. 558-68.

Marcello Truzzi, who heads the sociology department at Eastern Michigan University, publishes a rival magazine called *Zetetic Scholar*, in which Gauquelin is defended!. Indeed, G is one of Marcello's "consulting editors." Marcello believes that the correlations are "genuine" (i.e., not the result of bias), but unlike G he does not believe there is any "direct" cause-and-effect relation involved, but perhaps some sort of indirect correlation, about which he is very obscure. Marcello resigned dramatically from our committee a few years ago, breaking with it over a question of whether the Committee should put out a "debunking" magazine (which it is) or serve as a periodical of "communication" between eccentric scientists and their critics. More colloquially, we believe in calling cranks cranks, whereas Marcello wants to be so open-minded that he is willing to treat Gauquelin (and Velikovsky!) not as cranks, but as scientists offering reputable challenges to orthodoxy, etc, etc. After lengthy correspondence with Marcello over whether the G correlations were "genuine," we each finally gave up on altering the other's viewpoint, though we remain good friends.

Best,
Martin Gardner

[Handwritten addendum] Marcello: Judith Tanur is a sociologist at SUNY, Stonybrook, who asked me for material on Gauquelin. You might want to send her your side of the "great" (?) debate!

Today's *NYT Book Review* reviews several books of interest to you: short review of a new blast at Freud; an *excellent* review of the new book claiming (in my view erroneously) that Darwin "stole" his theory from Wallace; and another good review by David Hawkins of Jensen's latest book.

[Euclid Avenue]   8 July 1980
Dear Marcello,

I can't understand why either of us wastes so much time arguing with the other over words. It must annoy you even more than it annoys me. I'm sorry now I said anything about Bunge and Feyerabend. My opinion of Bunge has nothing to do with his views on the paranormal; I merely find him not a heavyweight in the philosophy of science. As for F, we must indeed read different journals. I failed to come upon a single favorable review of *Against Method* in any of the philosophical journals that specialize in philosophy of science. I enclose the last page of the strongest attack, in the *British Journal for the Philosophy of Science*, which is of course the leading British journal devoted to the field. An almost equally strong attack appeared in Mind, but I'm too lazy to copy it. I suppose you saw the attack on it in *NY Review of Books*. Nagel has repeatedly spoken of F's views as eccentric, and I find another reference to F's "wild cavortings" in a footnote on a paper dealing with Hempel's paradox. Neither of us, of course, is going to spend weeks running down every review of F's books and making a statistical check on the terminology used by reviewers. I am only stating that my opinion, based on what I have read and my conversations with philosophers of science, is that the majority of the professionals in the field consider F's *views*, to be highly eccentric, and even malicious. I find his views essentially frivolous and largely irrelevant. Obviously you don't agree. But then I find *your* views mainly frivolous, just as you find mine hopelessly mired in dogmatic orthodoxy.

I'll try not to make any more offhand comments that will tempt us into frivolity.

Didn't know Webb had died. Don't know Goran's books. Sarfatti is, of course, mad, as he himself freely admits. His latest is a copy of a letter to Dutton, threatening legal action if they reprint his *Space-Time and Beyond* without letting him do all the revising. As you no doubt know, this is the book that poor Ira Einhorn agented for Sarfatti. I marvel at your ability not to let the psi scene turn your stomach. I manage to avoid this by laughing. I even think Sarfatti is funny.

                                        Best,
                                        Martin
CC: Ray (for no particular reason)

Dear Ms Tanur,

Martin Gardner sent me a copy of his letter to you suggesting I might get into touch with you.

Martin somewhat misrepresented my position to you, I think, in his letter. He seems to think that Gauquelin is defended in *Zetetic Scholar*. That is not quite the case. Gauquelin may be defended in *Zetetic Scholar* potentially, but thus far there has not been an article dealing with Gauquelin's work per se (though he and many others did indirectly defend his work, I guess, in a symposium review of another book reviewing experimental studies on astrology, a book by Geoffrey Dean). In any case, though just about every exotic claim may be defended in the pages of *Zetetic Scholar* by somebody, this is not the same thing as being defended *Zetetic Scholar*. I think you might easily read Martin's letter and get the impression that the journal itself takes some sort of editorial stance in defense of Gauquelin. That is not so. Also, though Gauquelin is on my editorial consulting board, so was Martin Gardner until he resigned because he did not approve of my publishing even an exchange of views on Velikovsky (an exchange in which he could have himself participated if he desired). However, Martin does state the difference between us clearly enough. He considers Gauquelin a crank along with Velikovsky and would prefer that work such as his not be given legitimacy by discussion in what should be—in Martin's view—a serious scientific publication. I on the contrary feel it is imperative that arguments and evidence like that of Gauquelin be publicly challenged and defended in serious scientific journals rather than in advocate journals of astrologers on the one side and debunking journals on the other. Martin assigns such a low probability to Gauquelin being right that he would just as soon ignore him altogether within the serious arena of science. He seems willing to discuss those like Gauquelin and Velikovsky only to the extent that a general public might be inoculated against their crank ideas lest they damage real science.

As for calling cranks cranks, here Martin and I disagree less than it may appear to you, also. If a crank is defined as merely someone who tenaciously clings to a minority viewpoint, I would also call Gauquelin and Velikovsky cranks. But I think Martin often seems to imply that they are also irrational and unwilling to discuss counter-evidence. Such irrationality would, for me, better be termed crackpottery. The line is a very significant one to me. I don't believe Gauquelin or the followers of Velikovsky are all incapable of reasoned dialogue. In fact, intransigence as been shown more by their critics—like Martin—than by the proponents of the exotic ideas of

Gauquelin and Velikovsky.

Martin also, I think, misrepresents Gauquelin and his evidence some-what. The Comite Para, which examined Gauquelin's ideas with their own sampling study, actually confirmed Gauquelin's central claims. In the U.S. studies replicating Gauquelin's work and adding new controls, the first (*Humanist*) study actually did confirm his predictions. Kurtz and Company's interpretation of the results was against Gauquelin but I think most impartial observers would conclude that the data are largely supportive of Gauquelin. Dennis Rawlins and Ray Hyman, both statistically trained members of Kurtz's Committee, would agree that this first study in the U.S. came out essentially favoring Gauquelin and Kurtz and Co. have been not so deftly avoiding admitting it. The newest study is also most inconclusive and certainly is not as simply represented by Kurtz and Co. in *Skeptical Inquirer*. I am told indirectly that Ken Frazier has decided to stop any further exchanges in *Skeptical Inquirer* on this Gauquelin test. I hope this is not so. They seem to have delayed publishing Gauquelin's response to their test publication until they could also publish a rejoinder in the same issue. This gives them the last word and I hope it is not true that they will simply deny Gauquelin any more space. If so, he is welcome to such space (as are Kurtz and Co.) in *Zetetic Scholar*.

More important, Martin seems to think that Gauquelin insists on a direct causality between planetary position and the occupations. I have spoken with Gauquelin at some length and he is quite receptive to any conjectures for intermediary causal connections. Gauquelin has indeed hypothesized a possible mechanism (a triggering phenomena for the fetus), but he is the first to admit this is merely hypothesis. Martin insists that Gauquelin is really an astrologer in pseudoscience clothing because his ideas lend support to the astrologers whom Martin seems to despise. But Gauquelin is very open (even at astrology conferences) in his denial of classical astrology. More important, to the degree that anyone has produced actual negative evidence against traditional astrology, Gauquelin is clearly the principal investigator. If Gauquelin had not added the strange finding he made among his researches, Martin would probably consider Gauquelin a hero of science giving deathblows to the astrologers. But, in my view, this makes Gauquelin merely "guilty by association" with the astrologers. And the irony is that Gauquelin wants to dissociate himself entirely from the astrologers but people like Martin won't take him seriously enough to allow him a normal scientific forum, so he is forced back into the astrological camp for his support.

I am also a bit amused that Martin seems to be bothered by the fact that I am somewhat obscure about just what the third factor that produces an indirect correlation might be. If Gauquelin's correlations are truly present in his data and the data are not spurious in even the subtle ways (about which Martin is equally somewhat obscure), it is incumbent on us critics to suggest possible other factors that might be intervening. I think it rather ironic that Gauquelin seems happy to discuss such possible intervening variables that would destroy his direct causality conjectures while Martin, who would benefit if such plausible third factors could be found, seems rather unwilling to discuss them seriously. I admit to some obscurity as to the mechanisms that might be involved, but I hope that dialogue might produce some fruitful ideas and eradicate the obscurity.

Martin and I have written long exchanges on all of this and somehow just seem to go past one another. Naturally, we view one another as somehow obstinate in refusing to see what each of probably think should now be quite clearly the correctness of our respective positions.

Obviously, I am in part taking this opportunity to write you as a means to once more—hopefully more clearly—to bring my differences up with Martin. I am sending him a copy of this letter and also will send a copy to Ray Hyman (just in case he disagrees with what I said about his views and also because he was privy to much of the earlier exchange between Martin and me). But I also hope my remarks might help you to see the differences between us in a somewhat different light than implied by Martin's letter to you (or at least what I inferred from his letter to you).

Unfortunately, Martin really did not tell me very much about what you were after yourself in asking him for Gauquelin materials. Perhaps I could be of some help to you. I presume you know Gauquelin is now in San Diego, and I trust you will contact him directly. I am sure he will be quite helpful to you as regards his own views and writings.

You might be further interested to know that Malcolm Dean, an astrologer and science writer, will soon be publishing his book *The Astrology Game* in which he goes into great detail about Gauquelin, the Kurtz Committee, etc. I am sure you will find that book of interest. I disagree much with what Dean concludes about astrology—he is a firm supporter but a very well read proponent who probably has tracked down just about everything available in this area—but much of his historical and factual discussion is most accurate and informative.

I enclose a flyer on *Zetetic Scholar* for your possible interest. Please note that although Martin is no longer on my board there are several members of Kurtz's Committee on the board and it is over-all quite balanced. Since Christopher Evans died recently, he has been replaced on my board by James Alcock, very much a critic and also the chairman of the Canadian branch of Kurtz's Committee.

Please don't hesitate to write to me if I can be of assistance to you. Like the Committee, I am very interested in having me and *Zetetic Scholar* act as a resource for those researching and writing in this area.

Sincerely,
Marcello Truzzi

Professor & Dept. Head
Editor, *Zetetic Scholar*

CC: Martin Gardner Ray Hyman

P.S. It is most unfortunate that Martin sees *ZS* as "rival" magazine to *Skeptical Inquirer*. I think I have gone way out of my way to avoid that image of it. I see it as complementary to both the proponent and debunking publications. On the other hand, I must admit that I think *Zetetic Scholar* lives up to the original goal statement of the Committee in favoring unbiased inquiry and insistence upon evidence than does *Skeptical Inquirer*.

[Eastern Michigan University] 12 July 1980
Dear Martin,

Obviously, we simply disagree about the status of Feyerabend. However, I can not help but wonder if you have actually read much of Feyerabend or whether you depend on the opinion of those like Gellner. In any case, I suspect you have not read Feyerabend's reply to Gellner and I think you would find it at least amusing if not convincing. I enclose a xerox of it. Paul answers some of his other critics in this book also. Until his new book is available *Problems of Empiricism* (Cambridge University Press), I certainly hesitate to really judge his efforts. I also enclose a xerox of a short recent piece of Feyerabend's that in effect outlines the argument of his *Science in a Free Society* volume. You might find it of interest. too. I don't expect you to be converted, but I hope to at least temper your opinion of Feyerabend. Some of the reviews of his work, including the piece in the *NY Review of Books*, have been terribly distorting. I might add that the ironic part of my role here in defending Feyerabend is that I am in many ways more impressed by his friend and critic Lakatos.

I have just finished reading the new biography of John B. Watson, and find it most instructive and relevant. Watson's principal intellectual rival, Titchner, was about the only colleague who remained loyal and friendly to Watson during his terrible crisis at Johns Hopkins University over his divorce (which drove him out of academia). I gather that Feyerabend and Lakatos also had a similar highly respectful view of one another despite their philosophical differences. If you have not yet read this new study of Watson (by David Cohen), I recommend it highly. It is also interesting how just about everyone distorted Watson's position into an extremist posture he never actually took. Relative to Feyerabend, I think it is a good lesson on the dangers of secondary academic sources to learn about someone's positions. It is incredible to me that it should take so long for Cohen to set the record straight on Watson's actual work and ideas about the character of behaviorism.

One small other thing. Whatever our own views of Feyerabend, there is a festschrift currently being published for him in Germany which includes many well-known scholars. Obviously, some others think better of Feyerabend than Nagel and company.

I hear Randi may soon confront Geller directly in San Diego. That should prove interesting. By the way, what is the status of our second Uriah Fuller book? Is it yet available?

You might be interested to know that I am currently reviewing the Marks & Kammann volume for the *Journal of Parapsychology*. I am afraid I do not share your complete enthusiasm for the book. Much in it of merit, but there are some, I think, serious problems. I am currently trying to get both sides of the story on some of the problems with the SRI chapter. Puthoff has been quite helpful, but I await word from Marks & Kammann in answer to several questions I have asked them. I gather they are not any longer on very good terms. A pity since a new edition of their book could correct the errors and it seems unlikely that they will be working together on such.

Have you seen the newest issue of *Pursuit*? I am very amused at the recent discovery that the universe is actually only about 9 billion years old while so many stars are supposedly 15 billion years old. The Forteans continue to have a good time.

I am off to the Psychic Entertainers Association meetings in St. Louis this next weekend. Really looking forward to it. I will be giving the preliminary report of my survey on their attitudes towards matter ethical and psychical.

You may have already heard this one: What is 10-9-8-7-6-5-4-3-2-1? Answer: Bo Derek growing older. Along similar lines, I presume you have heard the definition of a Greek 10. The back of a 3.

Best,
Marcello Truzzi

[Euclid Avenue]   20 July 1980
Dear Marcello:

Thanks for the Feyerabend papers. I am going to use your last letter, with its expression of stronger admiration for Lakatos, as a chance to set down some observations about both F and L, and how they relate to our differences.

I have subscribed to *British Journal for the Philosophy of Science* for more than 20 years, so I read F's reply to Gellner when it first appeared there. I had not seen his paper in *Inquiry*, but I have read so much of F that when I read this, I found nothing new. You wonder if I actually read F. I in turn wonder whether you actually have read the many long reviews attacking him, or just read F's replies.

My intent was not to get lost in details about specific points where I may or may not agree with F's critics, but only to support my contention that F is regarded as holding eccentric views by the majority of current philosophers of science. It is no rebuttal to send me papers by F! Naturally no philosopher agrees with his critics. It is as if I had had said that current philosophers (outside the Catholic Church) consider Mortimer Adler a "light weight" and do not take him seriously as a "philosopher," and you replied by sending me papers by Adler! The issue in question is not whether Adler, or L, is right or their critics are right, but only how they are regarded by their peers. If you had sent me some favorable reviews of F, from journals devoted to philosophy of science, I might have softened my original statement.

I turn to L. He is much more fun to read than F, and when his classic *Proofs and Refutations* appeared in the *British Journal for the Philosophy of Science* I read the installments carefully and with mixed feelings. Let me explain how I evaluate L, and why I think he fascinates you more than me.

L. was strongly influenced by the Hegelian backdrop of Marxism (you know, I am sure, that he was an active Communist until his arrest in 1950 and his three years in prison). By this I mean there is a strong subjective tinge to his understanding of scientific progress, and that this tinge extends to mathematics. He sees mathematics as a sequence of "conjectures," first put forth dogmatically, then modified by the discovery of "Monsters" (counterexamples), which demand a revised conjecture, and these theses and antitheses continue to alternate forever in the manner of the Hegelian dialectic.

Obviously all science is fallible, and it does proceed roughly in this fashion, but there is a qualitative difference between math and science in that math has a technique of deductive proof within a formal system that science totally lacks. This utterly commonplace distinction is strongly blurred by L's rhetoric. He sees the history of math as analogous to science. I agree with his critics who feel that L. pressed the analogy too far, that he distorts the history of math so violently that, as Holton put it in his book *The Scientific Imagination*, L' s history is really an "historical parody that makes one's hair stand on end."

Let me be specific by considering the central conjecture of L's *Proofs*: Euler's famous formula about the edges, corners and faces of a "polyhedron." In Euler's day it was tacitly assumed that a polyhedron was simply connected. Euler was fully aware that his formula did not apply to such "non-polyhedrons" as a solid with one or more holes, or two tetrahedrons touching at a corner, or any of the other monsters L slowly and slyly introduces in his parody. Euler was not a dummy. He knew his proof was valid (as it is) only when applied to what he defined as a polyhedron. But, as always in math, the process of generalization begins, and as topology developed, mathematicians began to modify the formula so it would apply to polyhedrons with n holes (toroidal polyhedra as they are now called), and to other solids. Along with such generalizations, key terms usually evolve. Today "polyhedron" is a broader term than in Euler's time.

Some analogies will clarify my point. For early geometers "polygon" did not include figures in which sides intersected, as in a pentagram, or two triangles touching at their corners. Today the pentagram is called a "crossed pentagon." Theorems about crossed polygons in no way "refute" Euclid's theorems about polygons. An early Greek mathematician could say "every number is either even or odd." This theorem is not "refuted" by expanding' the concept of number to include irrationals, because the earlier statement tacitly assumed that "number" means "integer." L's choice of Euler's formula could hardly have been more inept. A better instance (about which he also wrote) was the rise of nonstandard analysis, which restored the infinitesimals of the early calculus. But even here one can view the history of calculus through two kinds of lenses. From my point of view, which is also that of almost all mathematicians, nonstandard calculus no more refutes standard calculus than non-Euclidean geometry refutes Euclid. Nonstandard analysis became possible after set theory refined to the point at which it was possible to present calculus within a novel and new deductive framework. It represents a fruitful generalization, and genuine progress, not a "refutation" of standard calculus.

I suspect the main reason you are fond of L is that you see his "monsters" as analogous to white crows, or unicorns, or perhaps even centaurs in the history of science. If even mathematicians can dogmatically assert something, only to have it later overturned, this appeals to your Fortean side and makes it easier for you to entertain the notion (which I think L would have found absurd) that the white crows (*theoretical* crows) of V or the observational crows of Gauquelin somehow should be taken as serious challenges to orthodoxy, rather than viewed as crank notions. L was a fascinating and at times infuriating thinker who explored deep problems, but our differences are on a far more mundane level. We both once agreed that the universe has a structure independent of human minds (that is, independent of social processes), and that science can make genuine advances in better and better descriptions of that structure. (By "better" we mean, of course, better success in explaining and predicting.) We agreed that all science is fallible. Perhaps we disagree about mathematics. I would adopt the language in which one can say that math and logic give "certain" knowledge, true in all possible worlds, provided one recognizes two huge prices that must be paid: (1) the certainty presupposes the acceptance of a formal system that rests on unprovable axioms (accept the system and you must accept the theorems), and (2) the formal systems tell you nothing about the world unless you apply what Carnap called "correspondence rules" that connect the axioms with the world. Correspondence rules are, of course, fallible because they derive from observations of the world.

Where we differ is in our evaluations of the worth of specific instances of eccentric science. I think you are naive to suppose that V and G pose any challenge to science. I believe your considerable talents and energies are being frittered away in old-fashioned argumentation, on the Fortean level, when you could be grappling with the real white crows that pop up all over the place in the orthodox journals. It is a terrible waste of your time to try to establish meaningful dialog with Velikovskians, or with those who regard P and T, or Crussard, or Hasted, as competent psi investigators.

I am sure that the K and Marks book has defects, but if you focus on them in your review, magnify them, and counter them with unverifiable assertions by P and T, you will only inject confusion into the controversy. I would like to see you hop off the fence and start calling spades spades, and not forever be saying "on the one hand" and "on the other hand." I would like to see you stop reviewing books by, say, Rogo, praising them for their "scholarship" and damning them with a light wrist slap by adding that "sceptics will not be impressed." By trying to be "fair" to all sides, you end up alienating all sides. Maybe (like Ebon) you enjoy playing these games. Anyway, here is the heart of our disagreements. I would like to see you join in some battles (as you take sides on Casteñada) and not struggle to be forever viewing things from an imagined mountain peak from which all things look grey.

Best,
Martin

CC: Ray

Dear Martin,

I'm afraid you mistook my purpose in sending you the Feyerabend articles. I did not send them to you to buttress my argument that he had higher respectability and evaluation than you think he has. I sent them to you because (a) I thought you probably had not read Feyerabend's last book, and the *Inquiry* article is a kind of summary; (b) I had not noted that Feyerabend's reply which is in his book was in a journal which you read regularly, so I thought you might not have seen it? and (c) I think Feyerabend's reply does demonstrate that he was not properly read by Gellner in Gellner's own attack. None of this was to show anything about Feyerabend's status within the philosophy of science today. I'm not sure why you thought I was arguing that. However, I did earlier write you mentioning several places where he was published (e.g., the Boston Studies series, etc.), that he was dealt with at length by Suppes and others like Brown who have done surveys recently, and that a festschrift to him was coming out shortly. These obviously do represent peer judgments.

I found your comments on Lakatos most interesting but somewhat irrelevant to my views of him. Though I knew his work on math, I frankly did not ever try to wade through it because I am simply not competent to evaluate it. However his papers on general science history and philosophy (the other volume of his collected papers) I did find impressive, and I was particularly impressed by his work in the Lakatos & Musgrave volume on social change in science. I feel that he is a good compromise between Popper and Feyerabend in that he and I (and you) would probably agree about the cumulative character of scientific knowledge.

You may, of course, be right that it is a waste of my time to try to bring about fruitful dialogues between critics and those you consider incompetent. Only time will tell. Given the repetition I find of these arguments in this arena about 100 years ago, I am not as optimistic as you may think me. But I think you may miss my main purpose of late. I don't know if I can make the proponents put their case forward more rationally, but I do hope I can influence some of my fellow critics to deal with such claims more rationally and scientifically. I very much agree with Ray Hyman's characterization of many critical efforts (such as those by Randi) as more pathological than the claims being made. I also agree with Bunge that more damage has been done by dogmatists, rejecting or blocking ideas in the name of science but without adequate evidence or due process, than has been done by all the occultists put together. I very much object to the interference of the state or religion in scientific exploration, but I really see little to fear from the very disorganized occultists and am far more concerned about the institutionalized scientists who too often think they can act as a new priesthood (which is where I agree so strongly with Feyerabend). I feel science can stand and indeed thrives upon diversity and free inquiry. I sometimes think I have more faith in the scientific method ultimately adjudicating claims than you have. I guess in the final analysis, I just don't see the clear and present danger in occultisms that Paul Kurtz (and presumably you) and some others do. But I do see a clear and present danger in monolithic Establishment Science as exemplified by the AMA, government "experts" on nuclear power, etc. The one place in which I am an advocate of laissez faire is in science, I guess. And that is because I think I share with you a perspective that science is self-correcting when it works best. In terms of all this, I am really less interested in Gauquelin and Velikovsky than I am in the over-reactions to their works which, to me, say terrible things about the current state of the scientific community. I do not think that either G or V offer very serious challenges to science; but they are treated as though they did, and that is what I am concerned about. If they are indeed representative of a sickness in science, I think the cure so far offered may be a worse disease. As I have repeatedly emphasized in my correspondence with you, I have no objection to any scientist giving esoteric ideas low priority and ignoring them. But once you choose to do combat with these claims, it is incumbent that you set an exemplary model in fairly dealing with such claims. It is very much like civil liberties. I am anxious for due process in science even for those whose ideas may be repugnant to me and that includes the Shockleys, etc., with whom I might disagree. You seem more interested in protecting the lay public from what you see as charlatans or quacks. I can see that posture where life and health are concerned (e.g., psychic surgery) but I see no clear and present danger in V or G. And I don't see any in Targ and Puthoff either. In the case of Casteñada, I never condemned him simply based on de Mille's work; I called for UCLA to investigate based on a prima facie just cause basis since he seemed in clear enough violation of anthropological criteria for a Ph.D. If I could be shown the same for Mishlove, I might do the same, but so far I have heard little evidence of that kind put forward. I don't think anyone has suggested fraudulent data collection by him. So far as I can tell, it is a matter of judgments about competency and ultimately that must be left to his final Committee since any dissertation is only as competent as its Committee and there have been many incompetent dissertations written and accepted.

The outrageous thing about the Castenada affair is that UCLA seems to be ignoring it all and his committee seems to accept the charges of false ethnography while choosing to ignore them.

                                        Best,
                                        Marcello Truzzi
CC: Ray Hyman

[Euclid Avenue]   29 July 1980
Dear Marcello:

Thanks for the stuff from Tart and Puthoff. I love Tart's phrase "trans-temporal inhibition" which I assume means either precognition or time lag. Tart, Puthoff, Krippner, et al, judging from letters of theirs you have sent me, are obviously doing their best to butter you up, and for transparent reasons. I hope it doesn't influence your judgment in the direction of believing everything they say. I can give you strong evidence of deliberate deception on the part of both men; Tart's role in the recent Mishlove flap is almost beyond belief. But the whole psi scene has become so dreary that I have to cut down the amount of time I've been wasting on it.

                                        Best,
                                        Martin

Chapter 3

The Dissolution

These letters start with Martin asserting that the two are, in some ways, "in total agreement." That level of rapport does not last very long. The attacks become more personal.

In the 1983 letter to John Fuller, quoted earlier, Martin also said:

> [Marcello has] denounced Kurtz as a "scoundrel" in letters to Kurtz and others, and has done all he can to discredit the committee. Rawlins apologized to Truzzi for his former attacks on him, and there is now an uneasy truce between them: sharing as they do a mutual dislike of Kurtz and others but on opposite sides with respect to debunking. ... Marcello, of course, keeps playing up this unfortunate episode because of his continued, and to my mind childish, antagonism towards the committee and most of its members.

In a letter to Stanley Krippner (August 27, 1979) Marcello states:

> You seem to imply that you find him not merely irresponsible— which I could understand—but seem to suggest that he is also guilty of past lying. Martin and I disagree about a very great deal, and I consider him to be a very hard-line skeptic who often borders on dogmatic denial, but I've never known him to consciously distort the truth as he saw it. ... Both he and Randi are indeed extremists in their antagonism to psi claims, but I have always found them both to be remarkably honest.

The premise for this book is to illustrate, warts and all, the battle lines of the debate over the demarcation problem.

Dear Marcello,

When we discuss science on what I would call the high level, at which debate goes on between scientists, we are in total agreement. I fully agree that science "thrives on diversity and free inquiry." And I fully agree that the writings of such as Gauquelin and Velikovsky and Mishlove do less harm than scientific Priesthoods that in the past have stifled free inquiry. I agree that there is little danger from the work of P and T, or V and G. I can't even see any danger from Castenada, except that the affair lowered the prestige of UCLA, as I think the Mishlove affair will lower the prestige of UC at Berkeley.

But I find in your letter a failure to make a distinction between science as a growing body of knowledge, and areas where science overlaps with moral and political judgments. You mention the danger of "establishment" science such as AMA and government experts on nuclear energy. But we have to carefully distinguish the views of the AMA with respect to medical *knowledge* and the AMA as an organization to protect the income of doctors. I violently disagree with the AMA with respect to socialized medicine (I am, as you may know, a lifelong democratic socialist), but this has nothing to do with the AMA's views on say, Arigo, and the views put forth in Fuller's book. As for the big nuclear debates, these are not debates about scientific knowledge, but debates about what is prudent in the application of such knowledge. To take a striking recent example of a bitter conflict, Teller and Oppenheimer didn't disagree over the nature of the H bomb, and how it works. They disagreed over the government should finance research on it. All these things are, in my view, not relevant to our fundamental differences. They are, however, much more important than the trivialities that you and I disagree about.

Our differences are over how seriously a magazine such as yours should take such characters as G and V. To me, V and G are on a level so far removed from serious debates and challenges in science as to be much closer to, say, the debate between biologists and George Price, or between astronomers and flat earthers. On the upper levels of science there is vigorous debate going on, about a thousand things, and in many cases one can see one side of it as an "establishment" view in the sense that the majority of scientists hold to it. It is the establishment view that Black Holes exist. Phil Morrison thinks they probably don't. It is the establishment view that homosexuality is primarily if not exclusively environmental in cause. A vigorous minority (including, I would guess, most homosexuals) is that it is genetic. It is the establishment view that races cannot be ranked in order of statistically superior genetic intelligence. Shockley thinks otherwise, along with many biologists. It is on *this* level that there is rousing, bitter controversy, and free debate, and we are in total agreement that there should be complete freedom here of expression of opinion, and may the truth win out in the long run.

But your periodical doesn't deal with conflict on lofty level. It is down on the Fortean level where you are seeking dialog between those who think Geller (or Girard!) can bend a spoon and those who don't, and those who think the pattern of planets at birth influence one's choice of a profession, and those who think that all the craters on the moon were formed since Noah's flood! I agree with you completely when you write "I do not think that either G or V offer very serious challenges to science; but they are treated as though the did." That is what I have been tirelessly saying. Astronomers have never taken V as a serious challenge. He is so taken only by V himself, and by his loyal followers, and by magazines such as *Fate* and the *Zetetic Scholar*. I am trying to say that debate on *this* level isn't worth your time, energies and talents and insights. It is on the level of Doubt magazine, but without Tiffany Thayer's sense of humor and absurdity.

Naturally if a scientist wants to poke fun at V it would be best not to make mistakes, but surely the trivial little errors of Sagan are picayune compared to V's misunderstandings of science papers he read. I'm sure you realize that to write a long paper on the errors of, say, Reich's orgone research, to avoid making trivial errors a scientist might have to spend a year running down all of R's writings, including papers in his magazine, pamphlets, communications to co-workers, etc. No busy scientist has that much time, and of course that's why so few scientists care to engage in debate at all with the Velikovsky's of the world. Almost no top geologists wrote about Price's theories because they didn't have time to read all Price's books. But when for one reason or another they do write about a crank, as Sagan did about V, you listen to V's disciples, find out about minor errors (sometimes they are fancied errors), and by playing them up in *ZS* it gives the distorted impression that a genuine scientific controversy is going on. By using kid gloves on the cranks and their followers, and taking strong pokes with brass knuckles at the critics of cranks, you play a game that is unworthy of you. If you want to grapple with the real controversies, where the real paradigms are likely to occur, and real revolutions in science, I should think you would be eager to forget about the lunatic fringes where intelligent debate is almost impossible, and take on worthier topics. How about the American Psychoanalytic Association? How about Milton Friedman and his libertarian economics? (He's influenced Thatcher in England, and will have a similar influence on Reagan if he becomes president.) What about the big wave of anti-evolution views? If the fundamentalist side of this is too low level, then how about the biologists of considerable competence who are reviving Lamarckianism? Does modern quantum mechanics support a subjective epistemology in which reality is mind-dependent (here there are top physicists, including a Nobel prize winner, Josephson—and possibly Wigner, another Nobel winner—who take this view.)? These are debates worthy of *ZS*. Not V and G, and Big Foot, or whether Ingo Swann can influence a magnetometer.

Both *Time* and *Psychology Today*, by the way, may cover the Mishlove flap. Incidentally, you and I agree on the harm done by psychic surgery. If you have a copy of Mishlove's *Roots of Consciousness*, check him on the psychic surgeons and you can see the kind of harm his book can do. (In fact, the daughter of a neighbor of mine (Mrs. Zolotow) in Hastings, living in an Arkansas Commune, needed an internal operation. The girl read Mishlove's book and wanted to fly to the Philippines and have it done there. The mother phoned me about it, in tears, for advice. All I could suggest was send her a copy of Nolen's book, with its chapter on the Ph. "doctors.")

But now I'm wandering,

Best,
Martin

CC: Ray

[Eastern Michigan University]   4 August 1980
Dear Martin,

Thanks for your particularly interesting letter of July 30th. Once again it becomes apparent to me that we agree about far more than we disagree, yet somehow we seem to part in unexpected areas.

I am particularly struck by the distinction you make between the AMA as political entity and as a scientific body. I very much agree and have consistently made this distinction between science as practiced and idealized, that is, as institutionalized and as method and knowledge. But I think you may underestimate the AMA (and science institutions in the form of such things as the AAAS and even the Committee for SICOP) as historically opposing progress in scientific knowledge. Thus, the AMA has been opposed to just about every major advance in medicine since its history (from vaccination to innovative cancer experimentation). Like yourself, I am a longtime democratic socialist, but my opposition to the AMA is not merely political. On balance, I am convinced that when medicine is allowed to treat health as a commodity, it begins to become scientistic as a means of protecting its' economic position. I am very impressed by the recent empirical studies in the sociology of medicine that show the abuses of medicine not only in the political but also the scientific sphere. The simple fact is that although you and I might differentiate between medical knowledge and medical practices, the fact is that the practices are justified in the name of science. I find striking parallels in your words and those I have read by catholic apologists for the witchcraft trials. The Catholic Encyclopedia, for example, differentiates between the church's theology and the political abuses of the inquisition. I see you as trying to do the same sort of thing. Unlike Feyerabend, and like yourself, I share the distinction between hard scientific knowledge and the practices of people who happen to be scientists. But unlike yourself, and like Feyerabend, I think that there is far less hard scientific knowledge than is generally claimed by practicing scientists who want to use such merely alleged hard knowledge to justify their political and other acts that frequently block free inquiry by other scientists.

You refer to my treatments as being on a Fortean level. I presume that you mean two things by that: (1) Consideration of anomalies that may actually be myths and which have little serious attention from the scientific community. And (2) I take too frivolous an attitude towards anomalies, failing to differentiate the serious from the trivial. I am not sure that is an intellectual sin. You would prefer that I deal with the conflicts on a more lofty level. You cite various examples of such lofty issues. I would agree with you that these are of probable greater importance. But I also recognize that, for the most part, I have little to contribute to those issues except to support those people who are already fighting about these matters on the sides I sympathize with. I am not trained in astronomy or physics and have little I could possibly say about the black holes issue, for example.

On the other hand, however, I think that as a sociologist I may have something to contribute in those areas where I do have some expertise. In addition, I think that the level of discourse I am trying to bring to what you think the more trivial issues is a distinct elevation from the discussions now found in the Fortean literature or *Fate* magazine. I think this current issue of ZS demonstrates that well. I think the kinds of issues Ray raises in his article on pathological reactions to deviant claims in science are most serious issues. And I think the implications of that article—if correct— are very serious as regards the efforts of critics like yourself and Randi. Obviously, you do not think that seeking to inoculate the public against what you consider preposterous pseudoscience claims via *Skeptical Inquirer* is a waste of *your* time and talents. Yet I think it is abundantly clear and demonstrable that most of the contents of *Skeptical Inquirer* represent very poor scientific arguments against the serious claims found among the paranormalists. I think that *Skeptical Inquirer* represents very well the kind of pathological reaction that Ray Hyman has written about, and you and Randi are well cast as the "hit men" he describes. You have yourself made it abundantly clear to me that you do not disapprove of ridicule and ad hominem argumentation against those you consider pathological scientists, and one of the things I respect about you is that you have always been "up front" about this. (Given some of the similar tactics used by the claimants of the paranormal, I can have some sympathy for your views here.) But though I can see that such tactics are essentially political rather than scientific, I also—like Ray—think they may be promoting of scientism rather than science. And I do view the growth of scientism as a real danger to free inquiry by those of us really committed to a free science.

I really think you underestimate the context of the above in my thinking. Thus, you speak of my giving the false impression that a "real scientific controversy" is going on about the theories of Velikovsky and Gauquelin. You miss my point. The real controversy is not over their ideas as such, it is over the way such eccentric ideas should be dealt with. Since I am absolutely in favor of due process for any and all ideas within science, and I definitely see their ideas as within science even if at the outskirts, I must demonstrate my commitment by giving space to these ideas. This is not to say that you and I otherwise agree about the arguments involved in these two cases. We clearly do not, for I think that Sagan's errors were not trivial ones, and I think Gauquelin's claims are far more important (especially as concerns basic ideas within the philosophy of science about matters like action at a distance, the need for mechanisms in a proper theory, etc.) and far less extraordinary (that is, less astrology and merely correlation through possible other undiscovered but potentially interesting variables) than you find them.

You keep speaking of the "real paradigms" and "real revolutions" in science as though you have some special ability to predict these when I think the majority of philosophers and historians of science now would argue that there simply is no a priori way of knowing where the future breakthroughs will come from. You may, of course, be correct, but it is also the case that there are many current outlets for discussion of those areas you think are the real candidates for future paradigms. I have done nothing to discourage that. Yet you do seem to want to discourage what I am doing with *Zetetic Scholar*. I don't think you do this merely because you think I am wasting my time and intellect on such matters. You seem opposed to the small modicum of legitimation I may be giving to the cranks. But what is really wrong with that so long as it is done with at least equal attention to their critics? I am happy to publish solid debunking articles (e.g., the Karnes piece in this issue). Even the poll on Bigfoot and Nessie in this issue shows the general disinterest in these topics by the scientists. And, of course, the critical reactions to Beloff—especially by his fellow parapsychologists— make many valid points against his evidential cases (in fact, I think these criticisms are far better than those put forward by critics like Alcock and Randi).

In the final analysis, I think you and I simply differ about means and ends. To me the important end is scientific method itself. I want to promote the use of science in dealing with paranormal claims. Using authority or ad hominem or false arguments within science is to be condemned on whatever side it is to be found. You, on the other hand, seem to think that there are those who abuse the scientific method so badly that we need to protect the public from them; and if that means expedient non-scientific arguments against them by bringing in things like authoritarian "manifestos" like the anti-astrology statement of Kurtz & Co., that is a justified means towards a good end. I just can not go along with that, and I think this is a fair characterization of your stance (and I think Ray will agree that you have given this impression should you disagree with my characterization).

In some areas, where there may be a "clear and present danger," as with psychic surgery, I might agree with your attitude. But even here I have some reservations. I may oppose Philippine psychic surgeons, but I am not really against all faith healing and other forms of probably psychological means that may have curative force (here I think William Nolen would agree with me). I am not against seeking extraordinary therapies if all, normal means have been exhausted. If I were condemned to die of cancer by the M.D.s, I would certainly not consider it irrational of me to try just about anything that might work for me (the costs of which were not extraordinary and damaging to my family). And I also think you may not appreciate the degree of help we can get against such charlatans from paranormal promoters themselves. For example, I enclose a copy of a series of pieces attacking psychic surgery by David Hoy, whom I presume you know from his pre-psychic days. I think this piece by Hoy probably would be more effective in keeping business from these rascals than anything you or I might write for the likely audience of believers in psi who would otherwise go to the Philippines.

This last point is important in another sense. Since doing *Zetetic Scholar*, I have been in contact with many proponents who have offered me information damaging to paranormal claims that I think it is unlikely you or Ken Frazier would ever be given. Thus I plan to run a piece debunking Edgar Cayce in a future issue of *Zetetic Scholar*, and David Hoy has sent me a most interesting confidential memo on his observations on Geller's performances and some backstage stuff. I just wrote him asking if I could share that memo with you and await his reply. I think this is significant reason why my attemptedly fair-minded approach will pay off to the advantage of all of us interested in getting the facts before science. In the final analysis, that is really all I want; to get all the facts before science. What is really so objectionable to you about that? You seem to want to equate *Zetetic Scholar* with *Fate* magazine's approach, but I think any careful reading of the *Zetetic Scholar* issues demonstrates that is not the case. In a sense, I may in fact be lowering science to a Fortean level by concerning *Zetetic Scholar* with Fortean types of (to you trivial) claims. But I am also clearly trying to elevate the Fortean level of discourse by bringing in the critics, citing extensive serious bibliography to build expertise and avoid repetitious argumentation, and bringing in new and serious thinkers (professional philosophers, historians, etc.) into the arena. Surely I am doing as much elevating of the Fortean concerns as I am lowering what have been the traditional scientific concerns. What's really wrong with that? Why not allow the developing scientific interaction to adjudicate the value of the eccentric ideas? If we trust science, what are we to be afraid of? If you are correct, why won't that become apparent in due course given due process? So far, critics may be avoiding *Zetetic Scholar*, but that is surely the fault of the critics and not *Zetetic Scholar*.

I really hope you will thoroughly look over this current issue of *Zetetic Scholar*. I think it bolsters the arguments I am making here, and I hope still to win you back into participating in *Zetetic Scholar*.

Cordially,
Marcello Truzzi

CC: Ray

P.S. I sent you a copy of *Zetetic Scholar* by airmail yesterday, so I hope you have it by the time this letter gets to you. If it somehow does not arrive, let me know.

P.P.S. Almost forgot to mention: Tart called me yesterday partly to inquire if the Committee had anything to do with some problems re Mishlove. It seems that he was somehow removed from the graduation listings. I told him that I believed there was nothing done by the Committee and that it was unlikely that there was any involvement of any of its members in whatever was happening. I particularly stressed that I thought it unlikely that you were in any way involved for you had recently written me that you were amused by but basically disinterested in the affair.

[Euclid Avenue] 8 August 1980
Dear Marcello:

Thanks for the paper by Lemert which I indeed had not seen, and which I will read with interest.

Here is Charlie Reynolds address: 2 Grove Court NYC, NY 10014. His business address is Popular Photography, he where he is Picture Editor. *ZS* also arrived, but I haven't yet had time to go through it. I did glance through it. Somebody should tell Persi how to spell "sleight," unless these were a copying errors. Incidentally, my crazy novel is just out in paperback. I asked to have a few printer's errors corrected, and in so doing, three new errors were made in the corrected lines! The worst is when Peter recites the "Jabberwocky." It comes out "Twas brillig and the *slighty* toves...!" Maybe that was why I noticed Persi's "slights." Jungian synchronicity, of course!

Best,
Martin

[Euclid Avenue] 11 August 1980
Dear Marcello,

Your last letter covers so much ground that I must struggle to be brief. Obviously I have no method of sharply distinguishing "real" anomalies from the fake, but as I have so often said, one must label parts of continuums with words in order to talk at all. It's as if I mentioned casually that I bought a pair of brown shoes, and you wrote back to ask if I had a sharp way of distinguishing brown from white. *All* words have fuzzy boundaries. But I talk on a level of extremes in which one sees a qualitative difference between, say, Hawking's distinguished speculations about black holes and Velikovsky's crazy scenario. (No, I do *not* have sharp ways to define "distinguished" or "crazy"!)

You write that you don't know enough astronomy to publish controversies over black holes. Yet it takes considerable knowledge of astronomy to perceive V as a crank, and you were quite willing to treat *his* theories with respect. What I am trying (vainly) to do is to raise the level of discourse in *ZS*. Most of all I would like to persuade you to avoid the strong bias that pervades your short reviews. You habitually damn sceptical books with faint praise, and praise books by occult hacks (e.g., Rogo) with faint damnation. I attribute this to your desire to keep friendly with the hacks and to perpetuate the animosity of scientists.

I see no useful analog between the AMA. and the Inquisition. (I trust that if you ever have a major operation you will seek a doctor of conservative medical views.) Recently the AMA considerably softened its opposition to chiropractic. In my opinion the AMA went too far. How about a debate in *ZS* on chiropractic? Actually the AMA's PR committee on cranks is weak beyond belief. Many years ago I wrote to it for information on Arigo and was astonished to be told that it had no information, had published nothing, did not intend to, and had no interest in the Phillipine psychic surgeons. If you should publish Hoy's blast, you'll be acting far more vigorously vs. unorthodox medical views than the AMA. I assume you'll accompany Hoy's piece with a defense of psychic surgery by, say, James Cranshaw who did three articles about it for *Fate* in 1977, or a piece by a writer like Tom Valentine who wrote a pro book about it. Perhaps a middle-of-the-road, "objective" article should be included by Harold Sherman, far more famous a psychic than Hoy, whose book *Wonder Healers of the Philippines* was published in 1967. (You remember a Sherman of course. It was he and Ingo who made the famous OOBE trips to Jupiter and Mercury, so carefully monitored by P and T.)

Of course I'm kidding. I find the topic of psychic surgery to be suitable for ridicule in SI, but as far beneath the level on which I think you think you are operating as Edgar Cayce or Uri Geller. Uri? He's now a dead issue. You'd arouse far more interest among parapsychologists if you published articles on Girard.

Since you regard me as a "hit man" analogous to the Inquisitors, let me counter by an equally wild metaphor. I am pleased to hear that you are a democratic socialist. I assume you know that democratic socialists have been notorious in their attacks on fellow-travelers of the Soviet Union (especially during the thirties), and have taken strong and vigorous stands vs. Stalinism. Your stand on crank science and parapsychology seems to me to strongly resemble the acrobatics of the fellow travelers of the thirties, who howled with anger when they were accused of being Communists, but who were noted for their inability to perceive Stalin as paranoid (I have no sharp definition of paranoid) or to admit the existence of the labor camps and the millions who perished there until Khrushchev himself admitted it. The point is that when there are *extremes* of evil (and I have no sharp definition of evil) and *extremes* of crank science, it is unbecoming *not* to take a strong stand. Democratic socialism doesn't entail a "soft" approach to Stalinism. A belief in freedom of scientific inquiry doesn't entail a soft approach to Velikovsky and Gauquelin.

Best,
Martin

[Euclid Avenue]   14 August 1980
Dear Marcello,

Your postcard raises a question that I will happily try to answer because it interlocks so firmly with our basic disagreements.

I believe, along with almost all scientists, that there are beliefs so far out that one is justified in labeling them crank. Cranks do not raise serious problems for the scientific community. However, if lots of ordinary people take cranks seriously, the scientific community has an obligation (in my opinion; on this some scientists disagree) to point out the crank's major errors in popular publications. I see the SI as devoted mainly to such debunking. Since I, along with the astronomers, consider V the very model of a crank, I regard him as fair game for exposure and even ridicule. I did not know of Ken's planned series, but I assume it will be an attack on V, and not a debate. If it is the latter, I am indeed opposed to it.

Now about the *ZS*. I believe you regard it as a serious scientific journal, not devoted to debunking popular crank science, but to establishing dialog between scientists of orthodox views and those with unorthodox views. For example, I do not think you would consider running pro and con pieces on hollow-earth theories. Why? Because in *your* judgment you do not consider them serious enough challenges. You would not care to dignify such "cranks." Another example. In the latest Asimov SF magazine Milton Rothman has a good article in which he discusses the notorious Dean drive that so excited John Campbell. Since the notion of an inertial drive for a spaceship is almost on the level of angle trisection, I do not think you would care to run pro and con articles about it.

Our differences are those of judgment. I have no hesitation in calling V a crank. You have never been willing to do that. Indeed, in earlier letters, when I tried to find out who you *would* consider a great crank, the best name you could come up with was Phil Klass. Since *ZS* is not devoted to debunking, clearly you would not wish to run only an attack on V. What I objected to was your presenting V in such a way as to convey the impression that a serious debate was underway among scientists. The net result was both to dignify V and to denigrate Sagan.

I myself am now so bored by V that if a magazine offered me, say, $3,000 for a debunking article, I would turn it down. But so long as large numbers of people take V seriously (in my opinion most of those who do take him seriously do so because they feel he provides support for accepting O.T. miracles as historical), he is still fair game for a debunking magazine like SI, and for popular magazines like *Omni* and *Science Digest*.

Let me put it this way. The V issue symbolized our basic disagreement over how to evaluate positions along continuums. With so many stupendous unorthodox views all over the scene (not just in the physical sciences, but also in the biological and social sciences), it seems to me unbecoming of you to spend time on debating V, Bigfoot, and UFOs. On *that* level, which is the Fortean level, I believe one should take vigorous sides, debunk, and not sit on a fence and match every whopping error of a crank with some trivial error of a critic. Perhaps I should repeat again that I do *not* consider parapsychology on the Fortean level. Eisenbud and Thelma Moss, yes, but the claims of parapsychology are serious and far above the level of V and George M. Price.

The *NY Times* in a book review this morning gave a great title for a potential best-seller: *How I Lost Weight and Found God in the Bermuda Triangle.*

Best,
Martin

CC: Ray

[Euclid Avenue]   15 August 1980
Dear Martin,

You raise several different points in your letter of 11 August and I will respond to each. Or try to.

I do not insist on sharp distinctions to the degree that you think I do. We both recognize that we are dealing with continua and that boundaries are not easy to draw and probably need not be drawn for clarity of communication. We also agree that Hawking's speculations on black holes are in a category quite different from, say, Gauquelin's. Certainly we would agree on the acceptability of the former's speculations among credentialed scientists and on the unacceptability—for whatever reasons, good or bad—of Gauquelin's and Velikovsky's. But there are great gaps between Velikovsky and Gauquelin's views on this continuum of "distinguished"-to-"crazy" which we are speaking about. I think for the most part we might rank order novel conjectures in a more-or-less ordinal fashion, but it is the intervals between them that we would disagree about. It is bad enough to me that you want to lump them both together with the flat-earthers; you also seem ready to lump them together. (Oddly enough, I infer from your past letters that you probably would rank Velikovsky above Gauquelin in your "crazies" list because Velikovsky is so much more grandiose and attempts so much more that has originality to it; I would rank Gauquelin as far less a "crank" than Velikovsky because his ideas are really far more reasonable and far better supported by the evidence we have.)

I am particularly amused at your comments on my remarks about why I did not publish stuff on black holes in *Zetetic Scholar*. You say "It takes considerable knowledge of astronomy to perceive V as a crank." Yet in fact that is the opposite of what most of his critics—I thought including you— claimed. The usual charge against V is that he ignores fundamental and well-known processes in physics that should be apparent to anyone literate at all about popular physics and astronomy. Certainly, Sagan does not attack V on the basis of subtle or technical matters, and you and others seem to think Sagan is the best current attack on V's ideas. The fact is, most of you have criticized V for his alleged ignorance of basic physics, not for subtle misunderstanding. I can read Sagan and you and Asimov et al, with very little problems of comprehension. I certainly can not say that is true about criticisms of Hawking. Besides, there are plenty of places such latter criticisms can now be published in the orthodox press. If I thought I could make a contribution to those exchanges. I would do so. You also say you would "like to raise the level of discourse in *Zetetic Scholar*." (I presume this was in reaction to my comment about elevating the level of discourse about the paranormal.) I am delighted you say this. But surely the best way for you to elevate this level of discourse might be for you to join into the dialogues I am publishing in *Zetetic Scholar*. You and all critics are more than welcome. I remind you that I had quite a few critical commentators on V's theory responding to May's essay. I expect to have more (in fact, David Morrison is writing a response now). I enclose a copy of a letter from Sagan to show that he apparently does not share your view of the character of discourse in *Zetetic Scholar*. I still hope to get Sagan to write something for *Zetetic Scholar* when his schedule allows.

Look, Martin, I can see that you might feel that V is such a silly fool that he should now be ignored. But surely you must agree that if he is to be criticized that criticism should be as well done as possible and not merely propaganda. I don't see you objecting to *Skeptical Inquirer*'s giving V attention (both in the new series in the works and some items in past issues). Why then do you so object to *Zetetic Scholar* giving attention when it is obvious from reading *Zetetic Scholar* that critics are welcome, at least equally represented, and that I personally do not agree with Velikovsky? It is not as though I were keeping critics out. Quite the contrary, I have tried to get more in. I haven't even closed the pages of *Zetetic Scholar* to Dennis Rawlins! I assure you that I would more than welcome a good attack on V from you, Sagan, or anyone else willing to do the job well.

You comment on my faint damnation of books like Rogo's and faint praise for those like Marks & Kammann. I call your attention to many books I have commented upon very negatively. I try to review the books in my book notes in terms of their intentions and audience. Rogo's books, for the most part, do not claim to be scientific works by a scientist or published by a scientific press for a scientific audience. He is a hack and he knows it and we know it and so does everybody who probably reads *Zetetic Scholar*. You seem to also forget the market that *Zetetic Scholar* is going to. Unlike *Skeptical Inquirer*, I am not seeking a mass audience at all. I am almost exclusively subscribed to by academicians, most of whom seem to be scientists professionally. In general, I do try to go for the positive in my comments more than you might like. And I may say more negative about the critic's books than you might like—partly because such reactions are unlikely to be forthcoming in places like *Skeptical Inquirer* (though please note that I always say what it is that strikes me negatively; I don't merely condemn). In general, though these book notes are not intended as serious reviews but just meant to give my quick reactions to the books to tell the readers a bit more than just the titles. (I must admit that these are usually done far too hurriedly, but you must recall that I have no help at all in the way of staff, etc., in getting *ZS* out. It is very much a one-man operation and even then done in between my job commitments, etc. I wish I had more time to devote to doing the job right—including doing a better job of proofreading the copy.)

Re the psychic surgeons business, I thought I had indicated that I felt this was a somewhat different affair since there was a "clear and present danger" involved as demonstrated by the past history of persons spending great deals of money to no effect. But even so, I would not go quite as far as you in condemning such stuff absolutely. That is, I would not invite Sherman or Valentine to comment (though I might publish their comments in reaction to something said in *Zetetic Scholar*) but I might invite Krippner to comment. In fact, Dave Hoy just gave me permission to publish his piece on these "healers" in *Zetetic Scholar*, and I would expect Krippner to be likely to comment on it. On the other hand, I would also invite Nolen to comment—or anybody else responsibly interested in presenting data or ideas for us to consider.

I agree with your comment on the greater relevance of Girard to parapsychologists. I would be pleased to publish stuff on him if anyone submits it.

In regard to your comments on fellow travelers (which you acknowledged was a "wild metaphor"), I see your point but think it more off the wall than I think you realize. I am not a fellow traveler with the parapsychologists at all. I have openly criticized them and indicate that I am unconvinced by their evidence even in my article in the current *Journal of Parapsychology*. There is no question about my willingness to point out the paranoia present in those quarters or about the frauds that have been done and may now being done. But the proper parallel is between my position and that of the ACLU. My concern is with due process and responsible examination of evidence, on both sides of the issues. There were those who called the ACLU a "pinko" organization for its defense of some Stalinists. If you want to see me as a "fellow traveler," then that is the fault of your inability to distinguish between the issues of due process and the validity-invalidity of the evidence. But more to the point, you simply can not claim that there is a similar "clear and present danger" involved in the kind of tolerance I advocate. In political matters, where life and death is involved, it *may* be necessary to sometimes suspend a degree of fairness to avoid serious dangers. But where is the serious danger from parapsychology, Velikovsky and Gauquelin comparable even remotely to such political matters? Surely you don't buy the silly nonsense Paul Kurtz occasionally spouts about Western Civilization being gripped in a wave of irrationality that correlates the occult with Naziism?

I realize that you meant this as a wild metaphor intended to parallel my remarks about the rise of scientism and your "hit man" role. And I realize that comparing you to an inquisitor is far too harsh (but intended to evoke some not altogether inappropriate imagery). But this characterization (and I remind you that I borrow this terminology from Ray Hyman's paper, so it is not really my own idea) does largely fit your actions as I see them. It certainly would explain your apparent willingness to support one-sided (and often poor) attacks by *Skeptical Inquirer* on subjects which you seem opposed to my even presenting two-sided discussions of in *Zetetic Scholar*. Your comments to me—and I think to Ray, too—have in the past indicated that you are very much in favor of ridiculing and otherwise putting down the "crazies." You seem to do this to protect main-line science. Why is that really so very different from the well-meaning inquisitors defending Mother Church? To extend your political metaphors; it seems to me similar to those who confuse America with democracy and want to defend America by suspending the rules of democracy which are the real reasons why America should be worth defending in the first place. To me science means serious and careful adjudication of the evidence wherever it comes from. Outrageous ideas should have their day in court and will be judged on their merits. In a courtroom, even a criminal is treated with respect by the legal system. Yet you chide me for treating Velikovsky with respect. Why does it seem so difficult for you to grant that I can treat him with respect while still thinking him wrong? Surely he is not Hitler or someone who should properly evoke great hostility and emotion? I could understand your attitude if I had actually endorsed V somewhere along the line, but I have not. Are you aware that Greenberg, the editor of *Kronos*, considers me a member of the enemy (anti-V) camp? Why do you insist on seeing me as a fellow traveler? I remind you, too, that the same social democrats who opposed Stalinism were often described as fellow travelers by the right wing. How much "traveling" do you have to do to be a "fellow" traveler? Is my insistence on mere fairness enough? I suspect I am carrying your "wild metaphor" too far. I hope I am, but I still think that underneath it all you think I am some sort of naive dupe of the paranormalist forces, if not—as I think Paul thought—an actual closet occultist.

Back to the main point, about the sharpness of our definitions. You keep wanting to take about the *extremes* of crank science. I guess we just don't agree about how extreme some of these people actually are. To me, there is still a good sized jump from Josephson and Walker to Sarfatti. And that size jump is a truly significant one in terms of how I respond to the claims. If you can give me reasonable criteria to view someone as more or less extreme, I am willing to listen. But if you ask me to suspend the rules of fairness (and just plain good scientific method) in dealing with an outrageous claim, you will have to argue (some sort of clear and present danger factors (as I think can be done re psychic surgery). If I truly believe in the "righteousness" of the cause, I might be willing to be a hit man, too. But so far I have seen little offered to argue in this direction. And in the meantime, any suspension of the fairness of scientific discourse can only belittle the ultimate public reaction to science. You can not defend science in the long run by using poor or just plain bad (or ad hominem) arguments in the name of science, to defeat supposedly pathological science. (By the way, I have long been amused that the title of your book *In The Name Of Science* really could have applied to both sides of the controversy, but you seem consistently on the one side.)

Finally, re the matter of chiropractic, a debate within *Zetetic Scholar* might be quite welcome; but, again, there are a number of current outlets for such stuff and I have quite a bit in my files. Also, I believe I recently read somewhere that the British were doing much in the way of investigating chiropractic claims. From what I have read, there is quite a bit of difference between the original chiropractic claims (still prevalent in some places) to cure just about everything and the claims of the more "legitimate" (again, I would argue "protoscientific") groups that make rather modest claims. (There have, by the way, been some good sociology of medicine studies of chiropractors comparing them with both osteopaths and M.D.s.)

Well, all this probably clears up little between us, but perhaps...

                              Cordially,
                              Marcello Truzzi

CC: Ray Hyman

P.S. In looking at your letter again, I note this addition. You say that when there "are extremes of crank science, it is unbecoming not to take a strong stand." I guess we disagree about what "strong" means here. I am a public nonbeliever in most paranormal claims. I place the burden of proof on the claimants. I am also willing to state reasons why I might be predisposed towards disbelief, in those cases where it would apply. I consider this a strong stand. You obviously want to include such things as ridicule and other measures in what you term a strong stand. I see nothing that is productive for science in doing that. On the other hand, as long as my evidence demands are as great as yours before I am willing to say Gauquelin or Velikovsky have made a convincing case, I don't see how that can really be characterized as a "soft approach" to them. Obviously, it is "soft" in your sense of those words, but I don't think this refers to anything but the kind of rhetoric and has nothing to do with real scientific reasoning. More important, I think that your kind of "strong" approach when taken by those without good arguments (and please be aware that I really don't accuse you of not having good arguments, but I would accuse many people who have written for *Skeptical Inquirer* of having such poor arguments) is actually counterproductive for science. Thus, something like Kurtz's anti-astrology manifesto got headlines but resulted in much of the public viewing the scientists involved as representing scientism and authoritarianism (a point well noted by Sagan when he refused to sign it).

P.P.S. If time permits me to do so tomorrow, I will enclose a copy of Hoy's piece on Geller and perhaps also his piece on the psychic surgeons.

[Euclid Avenue]   23 August 1980
Dear Marcello:

Touché on my phrase "considerable knowledge." It was a mistake. I should have said that a moderate amount of knowledge of astronomy and physics and geology was needed to recognize V as a crank. I would estimate the amount as roughly equivalent to what one would get from college introductory courses in the three subjects.

I hope you print Hoy's piece on Uri, even though it is dated (Uri's act has changed some in the past four years); because of the marvelous story about the two ladies. It certainly puts Sherman in an immoral light—he owed it to his guests to publicize the hoax. Incidentally, I tend to regard Sherman, Ingo, and Hoy as "charlatans" rather than pseudoscientists, but I could be wrong. I base my attitude only on hearing them on talk Shows, etc. What is your opinion of Hoy's claims to be a genuine psychic? Is it his sincere belief, or is it a show-biz ploy? Did you know that Tarcher, of

Tarcher books (you reviewed several of them in the last *Zetetic Scholar*) is Shari Lewis' husband? Maybe you saw the interview with him in a recent issue of *New Realities*.

I'll pass this time on commenting on your last letter, and not burden Ray with a copy of this one. Am entering an enormously busy phase getting our house ready to sell, moving to NC, etc, etc. I see by today's *NY Times* that Reagan has announced "serious flaws" in the theory of evolution, and recommends teaching creationism alongside the "theory" of evolution in public schools. Now I wish some reporter would ask for Carter's opinion! It's hard to guess how he'd answer it, or sidestep it.

Best,
Martin

[Eastern Michigan University]   2 September 1980
Dear Martin,

Just a quick note in reply to your letter of August 23rd.

I doubt that I will be publishing the Hoy comments on Uri since as you noted, it is really only the last part that is really at all news. However, I will be publishing Hoy's comments on the psychic healers. I am about to write to him to get possible approval by him of using the piece as a dialogue initiator. If he agrees, I would hope you and Randi would want to participate. I don't intend to invite any of the non-scientist defenders (though they are welcome to enter the dialogue at the second stage), but I think I will invite Krippner and Meek and a couple of anthropologists who have written somewhat sympathetically about these claimants. And of course, I would invite those like Nolen. Mainly, I would like to get some of the people interested in psychosomatic explanations to contribute, especially someone like Jerome Frank, if I can entice them.

In regard to Hoy as a possible charlatan, you may be right but I still am not sure. My conversation and correspondence with him indicate that he really does believe in his own psychic (occasional) abilities. He seems truly impressed by the many testimonial letters he gets on his radio and television answers to phone calls asking him to locate lost objects. Certainly, he is not a scientist and thus not a pseudoscientist, but I am not really sure of his real views on his own abilities. Frankly, from what I have seen of such occult types, the reinforcement and personal validation elements are so strong around them that it would be hard for them not to be taken in by their own myths. I was very much impressed by this kind of (in my view) self-delusion among the mentalists at the Psychic Entertainers

Association meeting in St. Louis. Most of them, I am quite convinced, really do believe in their own psi abilities even though they consciously cheat on stage every night. And I must say that some of anecdotes they tell are quite impressive even if scientifically not worth much as evidence to anyone skeptical as I am. So, I can see that subjective validation could play very strongly on them to convince them. In this sense, they may lie the victims of their mentalism more than their audiences. But as for Hoy, I think he presents his mentalism very honestly as trickery to the audiences I have seen him work before. He separates his program into two parts. The first part is presented as conjuring and intended to show his audiences that they can not tell real from artificial psi. He then spends the second half talking about (but not truly demonstrating except in the sense of answering questions with intuitive answers) "real" psi. And, more important for me, he seems quite willing to debunk bad psi research and other psychics like Uri (and I really don't think it is just a matter of jealousy). My impression of him seem confirmed from others I have talked to about him who have known him for many years. Certainly he is a charlatan to some degree, but I really don't put him in the same unscrupulous class as Uri and Swann.

Only one comment on your letter of August 14th: You say "I have no hesitation in calling V, a crank. You have never been willing to do that." That is not exactly so. In my earlier correspondence with you, I distinguished between a crank and a charlatan and a crackpot. We agree V is not a charlatan, I think. I am not certain if we agree he is not a crackpot. I do not think he is irrational in his presentation or willingness to debate his critics. I don't even think he is paranoid (merely internally consistent). From my information, he was quite willing to try to deal with his critics in great detail and was willing to recognize points of error in his own views (despite some quotes of his that might suggest dogmatism to his critics). From some points in your letters of the past, I think you ultimately do think he was a crackpot, i.e., truly irrational. Now, if we discriminate a crank as being someone who is tenacious about his theory and even obstinate in his defenses even when almost everyone of reputation would disagree with him, V was indeed a crank. But by this definition, many great thinkers have been cranks. In other words, if we attach no particularly irrationality to being a crank, if we see at as a matter of character and personality rather than as a quality of the arguments being put forward, V is clearly a crank; and I have described him as a crank in this sense to many pro-V people with whom I have corresponded. But I think you use the word crank to suggest not only irrationality and a minority viewpoint, but I think you imply that the person is wrong or using improper arguments. There I do not agree. If you do mean that he argues illogically or irrationally, then I say call him a crackpot (as does Asimov). In terms of my definition—and I think this is not at all an unusual definition at all but corresponds to what most scientists would mean—V is certainly a crank.

It seems to me that I may have earlier written you (I don't now have time to wade through our large past correspondence) as I know I wrote Ray, about my typology of critical labels for deviant scientists. They were: crank, crackpot, incompetent and charlatan. The crank is simply obstinate and tenacious. The crackpot is irrational but sincere (honest). The incompetent simply makes scientific errors but is sincere (honest). The charlatan is a fraud and is dishonest. Critics tend to want to make exo-heretics (to use Asimov's term) higher on the scale (e.g., crank V is seen as a crackpot by them). At the same time, the endo-heretics are usually seen by others in the discipline as lower on the scale. Thus, a fraudulent scientist is more likely to be described as merely an incompetent. The important thing to me is that the labels crank and incompetent *primarily* speak to the character of the arguments being used. The crank's arguments are simply too weak and the incompetent's are truly erroneous. But the labels crackpot and charlatan refer to the character of the person. The crackpot is irrational and the charlatan is dishonest. The charges of crackpottery or charlatanry are serious violations of scientific norms and are not really excusable. But mere tenacity and even mere error can be quite forgivable and even a virtue if the person proves to be right in the end. (I realize these terms slop over into one another in actual use, but I think it important to be clear at least in my own use of them).

I realize you are very busy, so don't feel obligated to answer soon. Good luck with your moving, and do let me know your new address when you move.

Best,
Marcello Truzzi

P.S. I enclose some jokes I just sent someone else, which you might like. Also, I am not bothering to send Ray a copy of this letter either.

P.P.S. Martin, a question about something altogether different: A friend of mine is trying to find the original or early source for a magical effect. This is the situation. He thinks he read the effect somewhere and very likely in one of your books. The effect is the result of wetting your thumb and pressing it against the message on the pad where it was written by ball point pen. The message impression goes on the thumb and is then read by the mentalist. The central elements are wetting the thumb and fact that the ball point pen leaves what will make an impression. Does this appear in your publications or any other early place? Someone else claims to have invented this method and my friend insists he read it elsewhere much earlier. Don't know if you can help on this, but it would be appreciated if you could. My friend is being accused of stealing this, trick and can not now remember just where he first got it but only remembers it was not from his accuser.

[Euclid Avenue]   12 September 1980
Dear Marcello:

I am pleased that you are willing to call V a "crank," but I am more puzzled than ever by how you understand the term. You still seem to be defining it psychologically, since you write that a crank is "someone who is tenacious about his theory and even obstinate in his defenses even when almost everyone of reputation would disagree with him."

As you yourself mention, "By this definition, many great thinkers have been cranks." Yes, indeed. It makes Galileo a crank for insisting that the earth moved, when all the reputable thinkers of the time were against him. It makes Einstein a crank for believing in relativity for a short period when it was being ridiculed by big wheels, and there was even some evidence against it. It makes Darwin a crank for being tenacious about his theory when most geologists and biologists of the day were creationists.

Now here is the problem. Since no scientist, and virtually no laymen, would call Galileo a crank, or Einstein a crank, or Darwin a crank, and since all scientists (with only trivial exceptions) call V a crank—then are you not again indulging in private definitions that create confusion instead of light? Or put it this way. If you are willing to say that Einstein was not a crank, but that V was, pray tell me how you distinguish between the two? I can't imagine how you can do so without bringing in elements of my definition, in prior letters, that take into account the degree to which a man's opinions go counter to well-established areas of science. Please note that Einstein's relativity did not go counter to Newton—it included Newton. It did not counter major experiments, like the Michelson-Morley, but explained them. But if V's central scenario is true, virtually all of astronomy and physics has to be heavily revised at a hundred points.

We also are using "rational" and "irrational" in different ways—again as we hashed over earlier. If by rational you mean "logical" (as distinguished from respect for empirical evidence), then the crank is almost ways *extremely* rational. I find nothing illogical in V's reasoning. Price's arguments are models of logic and rationality. (By the way, I guess you know that V many times quotes Price or refers to him in *Worlds in Collision*—see index for references.) Cranks go to extreme lengths in being logical and rational. It is their assumptions that make them cranks, not their arguments, which are usually impeccable.

But the main question I would like for you to answer is: What criteria,

you use for calling V a crank and not Einstein? When you tell me, perhaps
we will agree.

Best,
Martin

P.S. Enjoyed the jokes. As for the wet thumb picking up an impression, I
can't recall if I ever mentioned it in a magic magazine or not. I would guess
it to be as old as ball-point pens, since before then a wet thumb was used to
pick up letters and numerals written in pencil, especially a soft lead pencil.
(This is the basis of the very old trick of the initials on a sugar cube that
is dropped in a glass of water, and person finds the initials on his palm.)
And it works better, I think on fountain pen writing than ball-points. I
just tried it with my ball-point pen and couldn't pick up anything. I'm
responsible for being the first to think (so far as I know) of pressing thumb
against a die with recessed holes to get an impression of the die face. (Used
this in a trick published in the old Phoenix magazine).

Time has a short piece on Mishlove in the latest (Sep 15) issue. Now that
Reagan and Anderson have made statements about evolution, *Scientific
American* is trying to get a statement from Carter. My guess is he'll sidestep
answering, but I could be wrong.

[Eastern Michigan University]   24 September 1980
Dear Martin,

Of course I persist in defining "crank" psychologically. I do not insist
that this be the only way to define it, and I do not think you have ever
limited your definition to that. But my point is only that I am willing to
call some of the people you call a crank "cranks" if I am allowed to limit to
such a psychological definition. Now if you insist—as you consistently have
done—upon using the term to include nonpsychological criteria, you are
obviously making judgments about the likelihood of the truth of the claims
"cranks" make. And here is where we clearly differ, for I do not think there
is currently anything like the clear criteria you obviously think exists for
making such judgments about deviant scientific ideas that are consistently
stated.

There is no set of absolutely clear criteria for calling Velikovsky a crank
while denying that same label to Einstein's contemporary critics who called
him a crank. Your point that Einstein encompassed Newton's views while
Velikovsky forces a drastic revisions in astronomy is largely right, but Velikovsky's partisans insist this is amuck exaggerated a claim and also argue

that V's ideas subsume some isolated facts and explain some anomalies. Also, Darwin's idea contradicted Kelvin's.

I do also note that you seem to want to make a distinction between cranks who are illogical (those I would term crackpots) and those who may think logically but have faulty premises. However, I would very much disagree with you that Velikovsky and Gauquelin actually have such faulty premises in the clear-cut way you suggest. In fact, I think they have striven rather hard to get criticism of their premises and (contrary to you) I think they would very likely accept such criticisms if demonstrably valid. Now I know you have written of cranks that they are usually paranoid and used Velikovsky as a supposed prime example of a crank. But I also think your case for Velikovsky fitting your own criteria for a crank (in *Fads And Fallacies*) simply falls apart upon closer examination. Let me hasten to add that I still think V is probably wrong, but I refer here to your criteria for a crank.

On the other side of the coin, I am particularly amused at your remark that no scientist would call Galileo, Einstein or Darwin cranks. Of course none would probably do so today, now that their arguments have been largely accepted. But the fact is absolutely clear that many contemporaries who were noteworthy scientists did call all three "giants" cranks and even crackpots. My whole point is that there is no really acceptable basis upon which we can prejudge whether some ideas are revolutionary or just crank ideas (in your meaning) ahead of time. Even Sprague de Camp has said as much in his article on cranks. It is because of this inability for us to accurately prejudge with assurance that I insist on limiting the term crank to my proposed psychological meaning. And if you do have rational basis upon which to predict that the ideas put forward are wrong as well as merely implausible, then you are really attacking their reasoning or evidence, and then we need to spell that argument out and use the terms crackpot or incompetent or even charlatan.

Of course I realize that most people use the term crank in the kind of fuzzy way you do. That is what I am objecting to. Science needs arguments and not merely pejorative labels to make its judgments about new and even crazy-sounding ideas.

Finally, a rather funny aside (at least funny to me). Totally independent of our correspondence, I just got an interesting note from Jerry Clark. He sent me a xerox of your section of *Fads And Fallacies* in which you outline the characteristics of a crank pseudoscientist. He underlined large parts of it and asked me if I did not agree that your description actually fit Phil Klass unusually well. I don't happen to fully agree with Jerry on this argument, but he had some good points. My point here is simply to indicate how very vague, really, even your own criteria seem to be when operationalized by others.

I am sure we will probably never convince one another and will continue
to agree to disagree. But I suspect that if we ever can really get together to
interact for a few hours on all this, we might actually get some persuading
done. I think both of us respect one another and I know I really rather
dislike being in such disagreement with you because of my regard for you.
I certainly could change my position after lengthy real discussion with you
if I found you more convincing than your letters have been to me. I hope
we eventually can spend an evening together comparing arguments more
closely and with less lag than our correspondence necessitates.

I came across a particularly interesting article last night, which includes
some of the points I have been trying to make to you, I think. I enclose a
copy for your probable interest.

Cordially,
Marcello Truzzi

P.S. In reading over your letter after writing the above, I see I failed to
directly respond to one point, You correctly point out that by my definition
of crank (the psychological definition), people like Einstein, etc., would be
called cranks. Of course that does follow, but it only follows if we agree
to use my psychologically descriptive definition as clearly stated. If so,
the term crank is not really a pejorative label anymore than is a word
like "stubborn." It is in fact because my use would point up the fact
that there have been correct past stubborn thinkers who offered ideas that
seemed implausible, that I would not object to calling Einstein a crank
anymore than I would mind hearing him called "cranky". (a term which
does have the common meaning of mere stubbornness and being difficult
rather than clearly wrong-headed). I have even toyed with the idea of
starting an overt Society of Cranks which people like me could join to help
take the pejorative sting out of a word that may indicate that someone may
be stubbornly correct as well as incorrect. I of course realize that people
commonly use the term either fuzzily or with your meaning. I very much
object to that meaning, however, because I do not think you or anyone can
show certain criteria for its operationalization (unless you too admit that
rarely your crank may turn out to be judged right by future analysts). If
calling someone a crank (your way) merely meant ignoring him, I might not
object so much, but it usually also includes condemnation and the strong
insistence that others should ignore his ideas, too. [I note that a similar
problem exists about the word "skeptic" which is used by many if not most
people (e.g., Kurtz & Co.) to mean deniers rather than doubters, confusing
disbelief with nonbelief.]

P.P.S. Thanks for the helpful comments on the thumb-impression with the pen. As a matter of fact, I knew that only certain kinds of ballpoint pens would work. I should have mentioned this to you.

[No letterhead]   8 December 1980
Dear Martin,

Thank you for *Further Confessions of a Psychic*. If anything I enjoyed it more than *Confessions*. Together, the two volumes constitute a master course in how to be a super-psychic. I was intrigued by the hints of a possible third volume. How about *Confessions of an Evangelist*? Outside of the movie "Marjoe", we have not really had any full fledged expose of the Godcons.

I assume you will not be attending, the CSICOP meeting this week in Los Angeles. It promised to be duller and mere futile than the preceding meetings until Dennis surfaced again. He has obviously been phoning you, me, Marcello and who knows. He says he is going to show up and make waves of some sort. He keeps phoning me at the oddest hours, typically waking me from sleep, and not even asking if he caught me at an inopportune moment. Before I have fully awakened, he is talking full speed ahead as if our last conversation had not been interrupted. Since you disappointed him in his last phone call to you, I think he has had to reluctantly admit that he is Dennis Rawlins, the last man of honor on Earth.

Danny Korem stopped by a week ago to give a magic lecture and then spend some time with Jerry [Andrus] and me. Jerry and he spent most of their time together debating religion. Danny calls himself a "Messianic Jew"—he believes in the divinity of Christ. He also has figured out beyond any doubt (in his mind), that the Bible has accurately predicted all sorts of events and now we are approaching the time of the "Rapture" when he and all other true-believers will, before our eyes, be wafted up to Heaven. He gave me a copy of his book "Fakers" in which he debunks much of what he calls the "pseudo-occult". He distinguished this from the true "occult" which is under direct control of demons and Satan. This in turn is distinguished from supernatural, which is paranormality under the control of God. Danny is very good natured and soft-sell about all this. But he had not the slightest doubt that he is correct. Once I got him off the religious bit, we had a wonderful session of magic and mentalism. The guy has a lot of very good ideas. Jerry is still shaking his head in wonderment.

I'll report anything of interest, if such occurs, at our meetings.

Best regards,
Ray H

[Euclid Avenue]   12 December 1980
Dear Marcello:

A few years ago Dennis would phone me periodically to talk at length about what a devious and awful person you were, and now he phones me periodically to tell me what a devious and awful person Paul is. Soon he'll be telling you what an awful person I am because I have stopped replying to his letters. You don't have to pass on any more of his remarks because I get them direct.

I know nothing about Paul's political views, but I know a lot of liberals who found Reagan the lesser of two evils. I share Mencken's opinion that all successful politicians fools or scoundrels or both. I consider, Reagan a fool and Carter a scoundrel, and since I held no high opinion of Anderson either, I didn't vote for president. I could be wrong, but my impression of Reagan is that he does indeed believe in both astrology and parapsychology, but that his evolution bit was designed solely to win votes from the rising tide of fundamentalism. In the new Uriah Fuller booklet I have Uriah predicting that the coming trend is a combination of ESP, PK and JC (not Jimmy Carter, but Jesus Christ). This has been marvelously confirmed by the new Millennium Foundation, which you surely know about, in which a million bucks has become available for psi funding, offered by a Texas oilman who ties it up with the Second Coming, and who found his oil with the help of Rev. Ireland, a psychic who (according to Sarfatti's latest report) does a blindfold act! Maybe you should apply for funds from the Millennium Foundation, or does the founder consider you one of the enemy?

Can you imagine how the mouths of P and T and Tart and all the others must be drooling with the thought of grabbing a hunk of Jerry Conser's oil money?

Best,
Martin

P.S. Did Sidney Hook actually support Reagan? If so, I would be grateful of a reference in this. It is hard to believe. Certainly his 1980 book *Philosophy and Public Policy* showed no signs of Hook moving away from Democratic Socialism toward Milton Friedman's libertarianism.

[Eastern Michigan University] 17 December 1980
Dear Martin,

1) I too think it likely that Kurtz, if he actually supported Reagan, did so as the lesser of two evils (and as someone put it, got the evil of two lessers), but the irony remains obvious in any case.

2) I wonder if my card miscommunicated one thing (I did not make a copy). I have no reason to think Reagan believes in psi. My reference to believing in psi referred to some interviews Kurtz gave (as well as comments he made to me about his telepathic link with his wife). Reagan merely believes in astrology, so far as I know. I wonder if perhaps the Russians have now hired an astrologer (a la the British in World War II re Hitler's supposed astrological advice) to foresee what sort of advice Carroll Righter might be giving Reagan. I can see it now: Righter's column says something like "Today is a bad day for big decisions" on the day the Russians invade Poland . . .

3) Re Sidney Hook. I did not mean to suggest that I knew that Hook supported Reagan. My reference was to the congruity of such support since Hook is at the Hoover Institute, which is Reagan's conservative think tank. Also, Hook's strong anti-communism, his anti-liberal views on such things as the invoking of the fifth amendment and on student freedom efforts in the 1960s make his views somewhat compatible with Reagan's. I did not mean to suggest that he had moved into the libertarian corner with Friedman, et al. I merely suggest that Hook might find much in Reagan commendable; so the idea of Kurtz (his protégé) going for Reagan has a ring of plausibility to me.

3) Re Rawlins. I still don't fully know what to make of him and his antics. I should say that he has not said anything really bad about you to me. He did express disappointment at your cynical attitude towards Kurtz and the Committee's handling of the Gauquelin business, but that seemed to me more surprise than disgust on his part. I see Dennis as a very frustrated idealist. He foolishly believed Kurtz was an honest man and a real truth seeker, and now sees him as an opportunist. He earlier believed I was an opportunist and a kind of closet occultist. He was largely led to believe this because of things Kurtz told him including a few things that were clear lies. He tells me now that he was wrong about me and Kurtz. I think he was carried away with his zeal earlier, and I am certainly still upset about what he then said about me. But he has indicated to me that he was wrong, and since no great damage was done, I can certainly understand if not completely forgive him. The irony of it all is that as far as I can ascertain, Dennis is basically right (re Kurtz) and has been

badly treated by Kurtz. So I can certainly sympathize with him. I also think he is clearly right about the terrible handling of the Gauquelin tests. Dennis is admittedly a bit of a wild man, but he does seem to have both truth and justice on his side as far as I can evaluate the limited facts available to me. Strangely enough, Dennis and I (as different as we are in our views about the likely value of claims of the paranormal) share a common interest in fair play and responsible criticism. As far as I can make out, you have always been a bit less so concerned. But you and Randi have also always been above board about your desire to ridicule the paranormalists and "fight fire with fire" in dealing with them. I don't agree with this approach, but I have always respected the fact that you and Randi are very clear about your views. There is no hypocrisy involved. Unlike Kurtz and Klass, both of whom pretend to be "boy scouts" (to use Klass's frequent phrase) in their ethical stance towards dealing with the claimants, I always felt I knew exactly where you and Randi were coming from. My shock was in discovering that Kurtz (and later Klass) were not so dedicated to scientific objectivity as they postured. In Dennis's case, I think he did not initially appreciate the difference between your views and Kurtz's. So your unwillingness to do battle with Kurtz in the name of scientific integrity came as a bit of a shock to Dennis who did not expect a cynical reaction from you.

Naturally, I wish you were more—in my view—fair-minded towards proponents of the paranormal. In the long run, I think it is the only way out of their errors, both for them and for all of us interested in such matters. You have always tended to characterize me as naive while I characterize you as too cynical. Perhaps we are both wrong. Time will tell, I guess.

Oddly enough, I now feel that the whole mess brought on in large part by Dennis was the best thing that could have happened to me. I feel far more comfortable as a would-be amicus curiae than I did as an advocate for orthodoxy. I also feel that my new marginal position gives me insights I never could have enjoyed while co-chairing the Committee. So, though I at first felt much anger at Dennis for his part in getting me to resign from the journal's editorship, I now feel things have come to a better end. I would not go so far as to say that I am grateful for his misdeeds, but I really can not be very angry about matters anymore. And, so far as I can see, he has been far more damaged by Kurtz's little power plays than I was since Dennis has no place else to go with his ideas as I did. He is too hard-line a skeptic to operate outside the Committee, but he is too honest to function as a member of it in light of his own values.

A question out of curiosity: Did you ever actually vote on a ballot from Kurtz to remove Dennis from the Committee as a Fellow? I don't refer to his removal from the Executive Council but from membership. At the

L.A. meeting, Kurtz spoke of such a ballot but Ray tells me he remembers none though he takes Kurtz's word that there was one since some other members of the Committee acknowledged such a balloting done earlier. I can not help but suspect that no such balloting actually took place (in light of my own experience with Kurtz and his ballots on the motions I made while on the Committee). It is also possible, of course, that Ray simply did not get his ballot somehow.

Happy Holidays, and a pleasant move to North Carolina.

Cordially,
Marcello Truzzi

[Euclid Avenue]   24 December 1980
Dear Marcello:

Happy to hear that there is no evidence Hook supports Reagan. He is at Hoover Institute as their "token socialist." Being a lifelong democratic socialist myself, I naturally share Hook's views about the USSR, and about the student rioting of the 60s, though on many counts I do not agree with Hook's philosophical views. One of the chapter's in my forthcoming book is titled "Truth: Why I am Not a Pragmatist." Since Tarski's famous semantic definition of truth, Sidney is almost the only philosopher around who still defends Dewey's attempt to redefine truth in terms of criteria. But this is a technical matter, and too complicated to deal with in a few sentences. I could no more imagine Hook supporting Reagan than I could have imagined Buckley supporting Carter.

I don't consider myself "cynical" about the committee—only practical. Dennis wants to fire Paul, fire Ken, and I can't imagine who would like to have in charge except himself. Since he now trusts nobody. As for the "balloting," I assume you mean a hand vote, because so far as I know all voting has been done that way. I can't recall voting on anything, but that may be because I missed the morning business session and arrived only for the afternoon one. If there was a vote to expel Dennis from the council, I would have voted for it, because Dennis's tactics make it impossible for any group to function with him on it. If there had been a vote to remove him as a fellow, I would have voted against it. I honestly don't know if there was such a vote or not.

I do believe in "fighting fire with fire," as you put it, if you mean the use of ridicule and rhetoric. I agree with Mencken that "one horse laugh is worth a thousand syllogisms," and with Dewey's remark somewhere that no one is ever convinced by argument to give up a childish belief, though he or she may outgrow it. But I certainly do not believe in argument by deception, and I have never said anything about anybody that I didn't believe to be true. I may have been mistaken, as I was about Krippner's knowing Kulagina to be a charlatan, but I certainly don't believe in the kind of lying that is characteristic of, say, Puthoff. If you have read many of my replies to letters in the *NY Review of Books*, you will know that I am quick to concede any mistake that I make because of inadequate information or haste or carelessness. By the way, do you know anything about when and where Hubert Pearce died? It was about six or seven years ago. I need the date for the anthology I'm pasting up of my 30 years of writing about eccentric science. Nobody seems to know, including Martin Ebon. I've written to Ian Stevenson about it—perhaps he will provide the information. Pearce, in case you've forgotten the name, was Rhine's top psychic in the early days at Duke. He became a Methodist minister in Arkansas. I wrote to the Public Library at the town where I thought he was a minister when he died, but they informed me that he died "outside the state," and they had no record of when or where. I once exchanged a number of letters with him, trying to get him to answer some questions about his famous run of 25 hits, but without success. I suppose I shouldn't print any of the letters without permission from his widow, whom I presume is still living.

Have you a copy of my Uriah Fuller booklet 2? I can't remember if I sent you one or not. You probably won't like it. Assume you saw Geller's interview in the latest issue of Marc Seifer's rag.

> Best,
> Martin

[Postcard]   28 December 1980
Dear Marcello:

I introduced the terms Catch 22, 23, 24, etc in my "Mathematical Games" department, *Scientific American*, October 1975, but there are references to the many "outs" in my chapter on Rhine in the old *Fads and Fallacies* book, I draw a blank on Eugene Rubini. Thanks for the advance look at your contribution to *Journal of Parapsychology*. After January 20 my address is [address withheld]. No phone yet.

> Best,
> Martin

[Postcard] 3 January 1981
Dear Marcello:

I have subscribed to *Fate* since Vol 1, No 1 (which I still have, and which contains Ken Arnold's first piece on flying saucers.) I do *not* agree with you and Clark about Klass, and I *do* agree with Klass about you and Clark. (We are hopelessly of different mindsets—let's stop trying to convert one another.) How you can admire a man who took Doyle's fairy photos seriously is beyond me. After January 20 my address is [address withheld].

Best,
Martin

[Postcard] 5 January 1981
Dear Marcello:

Do let me know if you learn anything about Hook and Reagan or about the date of death of Hubert Pearce. I have never written Krippner, as far as I can recall.

I would send you a copy of Uriah Fuller #2, except *all* my books are now packed. After January 20 my address is [address withheld]. No phone yet.

NYTimes reported yesterday (Jan 4) that the 6 top editors of *Quest* resigned because they refused to print an article by Herbert Arnstrong, publisher!

Best,
Martin

[Zetetic Scholar] 19 January 1981
Dear Martin,

Just for the record. I indicated that I agreed nearly entirely with Clark's article on Klass. I neither indicated I agreed entirely nor did I indicate I agreed with Clark in other areas. You say "How can you admire a man who took Doyle's fairy photos seriously is beyond me." I think you know full well that *I* never took those photos seriously except as a minor puzzle as to method and details. Why can I not agree with arguments in Clark's paper on Klass without that being transposed by you into *general* admiration?

In addition, you say that you don't agree with me and Clark about Klass. Fine. I had hoped you might in lieu of the evidence presented and because I did not know you were an admirer of Phil's first books on UFOs which Phil still sticks by in its entirety. But so be it. But then you say that "I do agree with Klass about you and Clark." I don't really know what that means to you. Klass has indicated to me that he considers me far worse than naive, has accused me of lying and general bad faith. I hope that is not what you agree with.

We do indeed have different mind sets, but I know that evidence still has an effect on my views. I had hoped the same was still true of you. I have no interest in converting you to anything, but I have tried to persuade you with arguments and thought you were doing the same towards me.

The major purpose of *Zetetic Scholar* for me personally is to expose myself to new arguments and evidence on either side of the issues covered. I will admit towards general movement towards greater agnosticism on many issues where I was once quick to dismiss many claims. If "we are hopelessly of different mind sets" in your view, I must insist that you speak only of yourself and not of me. I have no vested interest even in my own current agnosticism about the paranormal which you seem to (and Klass definitely does) translate into my somehow being "soft" on the "pseudoscientists"' (which in Klass's case frequently sounds—as Clark points out—more like McCarthy's attacks on those seen as soft on communists but were merely socialists). Don't misunderstand me, I don't accuse you of the extremism I see in Klass, but your card certainly indicates more tolerance of his outrageous attacks on character than I would have hoped for from you. I do not expect you to repudiate Klass, but I had hoped you might recognize his excesses. It is one thing to take a kind of hard-line "law and order" approach to conducting science (which you do) and the kind of vigilantism in the name of higher values that Klass engages in when he goes after someone like he went after McDonald. I still hope that our differences are not so great as those between me and Klass. If so, dialogue probably is hopeless; and that would be a great shame, for I continue to respect your erudition.

Best,
Marcello Truzzi

[Postcard]   5 March 1981
Dear Marcello:

Thanks for letter. Yes—we agree on the upper levels of *Theory; all* philosophers of science are "fallibilists" in Peirce's sense of the word. But on the level of decisions and behavior, I believe the probabilities often justify taking strong *sides*; and you incline to fence sit. It's probably a matter of temperament. Surely Huxley—"Darwin's Bulldog"—took strong sides (witness his rhetoric vs Wilberforce, and his numerous articles and book reviews slashing the creationists). Have you seen *The Math. Experience*, by Davis and Hersch (just pub.)? You'll like its chapter on Lakatos.

Best,
Martin

[Postcard]   16 March 1981
Dear Marcello:

I agree 100% with Huxley's quote. Do, sometime, check on his discussion of pseudoscience in his book on Hume, which I have recommended to you before (the discussion about *centaurs*).

Please note my new address. Your letter went to Hastings-on-H, and just reached me today.

Best,
Martin

[Eastern Michigan University]   19 March 1981
Dear Martin,

Thanks for reminding me about Huxley on Hume and pseudoscience. I will try to get the book this weekend if the U. of Michigan library has it.

I am a bit surprised that you say you agree with Huxley's quote 100% since it would seem to me that your actions towards those you call pseudoscientists suggests to me that you stress the first part of his injunction (rationalism) but minimize the latter part (fallibilism). Perhaps when I read more of Huxley as you have, I will be enlightened on this matter, but it seems to me that Huxley's stated reaction (ideal) to someone with a crank idea (2 + 2 = 5) is far more like my own posture than yours. He enjoins us to suffer fools gladly, but I don't see you doing that with Gauquelin, Velikovsky, etc. Do I just misinterpret your reactions to such people? When I "suffer fools gladly," you have labeled me naive rather than virtuous as would, I think, Huxley. Or is it just that I am willing—in your view—to suffer fools longer than I should? I guess what bothers me is that I really think you and I share most fundamental views in regard to paranormal claims, neither of us are dogmatists, and neither of us is stupid, and I certainly respect your learning. So why do we seem to disagree about the things we do? Sharing our letters with people like Ray and other friends does not seem to have helped clear things up, at least not much. I hope you realize that I sincerely am less interested in having you change your opinion and agree with me than I am in somehow having our views come more together. I really do respect your intellect and knowledge, and really wish that my possible errors in thinking about these matters could be corrected. I truly wish we could agree, for though I do enjoy debating with you in our letters, I regret that we are not on the same "side" in these matters and I wish that either one of us could win the other over. If you have any items you think I might read that would perhaps, in your view, help me "to see the light" please don't hesitate to recommend them to me. Whether you are right or wrong, you have more experience than I have relevant to these matters, and I recognize that I have and still can learn much from you.

I recently sent you a copy of an announcement about my new Center. God knows if it will get off the ground. But it seems worth a try. My study on psychic sleuths is coming along very well, and I think the report on them will be important. If you have any odd stuff on this topic, please let me know since I am running down everything I can find. The literature on the subject turns out to be very large and I have already got a bibliography (minus newspaper stories) of about 200 items. Some real surprises and I think a real pattern is emerging. If my hopes for negotiating a survey of police officials through one of their major publications works out, there should be some important disclosure possible.

I will be giving a talk to the ASPR next month. It should prove interesting since I have not met most of the NY crowd. I am supposed to go visit Honorton's psychophysical labs while there. I trust you got the satire on you and Randi by Honorton that I sent you.

The Clever Hans symposium papers are supposed to be out next month. It will be interesting to see the pieces in print to see what—if any—changes the authors have made since their oral presentations attacking one another.

Best,
Marcello

[Woods End]   31 March 1981
Dear Marcello:

Glad to get Wilson's pyramidiocy piece for my files. It's obvious nonsense, and just another (and rather crude) example of how easy it is to extract pseudocorrelations from a mass of complex data. Piazzi Smyth did it better.

Yes, I saw the Brinkley show. Don't know anything about Kokolov.

Sorry I forgot to thank you on Pearce. According to Ian Stevenson, who heard from his widow, he did not die in Arkansas. The date of death was January 17, 1973, and the place was Southwest City, Missouri. The public library in his home town of Wynne, Arkansas—or rather where he was last a minister (so far as I've been able to learn) wrote me that they had no obit since he had died "outside the state." A letter to the library at Southwest City was returned with a note on the envelope saying the town had no library. Nobody seems to know the widow's current address. With more effort I could probably track down exactly where and of what he died, but it seemed not worth the effort.

According to the current *Star*, Uri is shacked up with Hanna (Hanna Shtrang?) and they have produced a baby boy which Uri says has more psi power than he has.

Best,
Martin

[Zetetic Scholar] 1 April 1981
Dear Martin,

Of course our differences in temperament must play a role in our attitudes, but I think you miss a central point. Huxley chose sides on the Darwin matter, but Huxley's statement admits that he himself does not usually live up to his own prescription, and he apologies for such lapses. The problem comes down to: When is it necessary to chose sides? I think Huxley may properly have been compelled to choose given the strength of organized religion's attempt to then block inquiry. If, as Peirce argues, our main concern should be to do nothing that might block further inquiry, that can act as a decision criterion. I prefer to put things in terms of whether or not there is a clear and present danger present to further inquiry. By that criterion, I think I am quite consistent. Also, there may be a clear and present danger of other things that need to be taken into account. Thus, I can understand your concern with psychic surgeons ripping off people and milking families. But surely no such clear and present danger exists with most of the occult (I presume you agree with me and not Kurtz that occult interests are not the harbinger of fascism and political evils). Since psychiatry and the AMA have tremendous power in our society, their excesses surely do more harm quantitatively than a few unorthodox practitioners like fortune-tellers and chiropractors, etc. Surely the pendulum has now swung in regard to who is trying to block inquiry today. Surely it is the

debunkers (AMA, CSICOP, etc.) who want to discourage taking alternative ideas seriously enough to really encourage investigation. For example,
CSICOP has conducted investigations into Gauquelin (in principle, that is
commendable by me), but you felt that even looking seriously into such
preposterous claims would only legitimate them. Weren't you discouraging
inquiry by others? By mocking all parapsychology (a la Randi), don't you
discourage the sincere and honest inquirers as well? I agree with Sagan that
the only real argument against pseudoscience is more and better real science. And that means serious inquiry, which means taking the proponents
seriously in a way that I think you just are unwilling to do.

I could even understand your feeling that you had to chose sides if you
were a scientist working in an orthodox area that needed defending and simply wanted to fight off competing screwy ideas. It would be a pragmatic
matter then based on your scientific role as defender of a perspective that
needed to maximize social support (grants, etc.). But, dammit, you are a
science writer and supposedly without the vested interests of the everyday
scientist. Like me as a sociologist, you should have more rather than less
detachment about these matters. You should see the big picture, which
includes alternative ideas needing scientific due process. You are like an
officer of the court of science, yet you act like an advocate. Why is this?
You are not alone. I find science writers apparently more interested in
protecting the boundaries of science from "outsiders" than are most professional scientists. Why this attempted role as gatekeeper? Do we perceive
the role of science writer differently? It seems to me that you can "afford"
more tolerance than Huxley could. Why not exercise it, then? You don't
need to court the kudos of the orthodox scientists for whom you, in effect,
"run interference." Why can't science writers be more interested in keeping
science honest than in protecting scientists already "in power"?

Perhaps this makes our differences clearer.

Cordially,
Marcello Truzzi

[Postcard]   4 April 1981
Dear Marcello:

Naturally I agree with you in respect to what you call "alternative
ideas needing scientific due process." I could name several hundred such
unorthodox ideas for which I have utmost tolerance and respect. I just
don't consider Velikovsky's theory, or Gauquelin's work, or UFO contact
reports, or the garbage in *Fate* magazine in this realm. Or the writings

of Rogo (e.g., his book on phone communication with the dead). But for Pete's sake, let's stop wasting your time and mine rehashing what we've both said over and over again!

Best,
Martin

---

10 May 1981
Dear Marcello:

Here's a list of pieces I found in my USSR files that are not on your list. I have a copy machine, so I could copy any or all of them for you if you like, except for the special issue of *Psychic*, which you can obtain (I assume) from *New Realities*.

Do you have a copy of the Defense Intelligence Agency report (on your list)? I note that it has 45 references, many in English, some of which are not on your list, and which I do not have.

I assume you are *not* interested in references on the Russian ladies who read with their fingertips. The big flap about this followed Albert Rosenfeld's big article in *Life*. There are dozens of references, the most important of which are in my *Science* paper on "Dermo-Optical Perception: A Peek Down the Nose." Let me know if you want a list of what I have on this, but of course it is not government research, so it probably falls outside the area you want to cover. However, the literature does include pro articles in *Russia Today*, and articles on government investigations of Kulashova and Kulagina, which reported that both ladies were using fraud. And Russian parapsychologists like Vasiliev were very much involved. The line between official government research and research by individuals is, of course, blurry, and some of the references on your list belong more to the latter category than the first.

Best,
Martin

P.S. No. 5 on your list is by Jules Asher
[Attached is this bibliography of 17 references]

- A. Vardy, "Parapsychology Versus Marxism-Leninism," *Review of Soviet Medical Science*, 3:2, 1966, 59-62.

- Nikolai Khokhlov, "The Relationship of Parapsychology to Communism," *Parapsychology Today*, ed. by Rhine, NY: 1968.

- Martin Ebon, "Moscow: Behind the ESP Enigma," *New Realities*, 1:2, 1977, 34ff.

- Martin Ebon, "The Woman Who Could Read Unopened Letters," by Ludmilla Zielinski, in *The Psychic Reader*, Chapter 1.

- Martin Ebon, "Iron Curtain ESP," *Human Behavior*, November 1978, 38-41.

- National Enquirer pieces:

  - Paul Bannister, et al. "US Defense Department Documents Warn of Threat from Russian Psychic Experiments," p. 37 (failed to date it).
  - Douglas Dean, "Amazing Russian ESP Tests—I Saw Them with My Own Eyes," October 1, 1972, 47:5.
  - Douglas Dean, "I Witnessed Startling Advances in Russian ESP Research," 47:4, September 24, 1972.
  - William Dick, "Russians Perfecting ESP for Spying," 46:19, January 9, 1972.

- William K. Stukey, "Psychic Power: The Next Superweapon?" *New York Magazine*, Dec. 27, 1976, 47-55 (interview with Duane Elgin).

- Alexander Ivanov, "ESP in the USSR," *Fact*, No.2, Mar/Apr 1964, 41-43.

- Irene Agnew, "Parapsychology In Russia," *Science Digest*, July 1972, 69-71.

- John Wilhelm, "Psychic Spying?" *Outlook* (Wash Post Sunday Magazine) August 7, 1977, B1 ff.

- Anita Gregory, "Crackdown on Parapsychology," *New Scientist*, 13 February 1975, 397-398.

- Pamela Painter de Maigret, "PK Training in Russia," *Fate*, May 1976, 36-44.

- Dan Fisher, "Soviets Widely Preoccupied with Occult." *LA Times*, October 17, 1979.

- *Psychic*, 2:6, June 1971. A special issue devoted entirely to USSR psi research. Interview with Ullman, etc.

[Gardner's review of Polywater, by Felix Franks, with note: Marcello—Here's an example of what I consider an anomaly to be treated with respect, not ridicule. The book is a must for you.]

[Eastern Michigan University] 2 June 1981
Dear Martin,

I was just read a section of a Chicago story on COSICOP [sic] in which you were quoted saying that earlier I tried to place some "California witch" in a position of authority on the Committee. You will please recall the fact that what I then urged was that proponents of the paranormal—of all kinds—should be allowed membership of some sort on the Committee but *without a vote.* That is why I invented the term Fellow to label the important, voting members of the Committee. I even spoke of both Members and Associate Members for the Committee. I *never* tried to get any occultist as a Fellow nor did I ever want any occultists to have any power (in the form of a vote) at all. Initially Kurtz agreed with me and that is why we changed calling you guys Members to calling you Fellows. What followed was that we ended up with Fellows and no Members at all. And, of course, Kurtz also nicely arranged it so not even the Fellows had a vote, leaving that power to his Executive Council which he stacked giving himself three votes via including Nisbet and Zimmerman as his echoes. I'm sure Ray would verify my memory of all this. The reason you guys told me you did not want proponents even as Members was because you feared some sort of public walk-out that would garner publicity should they stage something like that.

I am amazed that you seem to think—if you are quoted correctly—that *Zetetic Scholar* is simply a frivolous enterprise. It should be obvious that *Zetetic Scholar* has not become an occult organ and one needs merely look at the participants in *Zetetic Scholar* to see the continually improving quality.

Perhaps you were misquoted or quoted out of context. I hope that is the case.

I realize that we have our differences in approach to paranormal claims, but surely by now you know that I am sincere in my skepticism. The quote from you in the Chicago paper's story clearly implies that I may say I am a nonbeliever but that you somehow think I am not really. My friend who read me the quotes felt I should be angry with you about them. Instead I am more hurt because I really hope we understand one another better than that.

I also realize that newspaper reporters often distort things or take them out of context, and I hope that's what happened here.

Sincerely,
Marcello Truzzi

[Woods End]   9 June 1981
Dear Marcello,

I've not seen the Chicago newspaper piece so I can't say if I was misquoted or not. If you'll send me a Xerox maybe I can tell. I recall speaking
informally on the phone with a reporter. It is possible he tape-recorded. If
so, I have no objection to his sending you a transcript. I myself would be
curious as to what I actually said.

As I recall our early discussion, when the committee was still in a planning stage, you argued strongly for allowing members, of the "opposition"
to become advisory members of some sort. There was then no terminology
for such categories. You maintained that this would keep lines of communication open. I and others objected on the grounds that we would likely soon
be outnumbered, and would lose heavily if the believers later staged a mass
walk-out to protest the committee's bias. Your distinction between members and associate members was considered. The associates, for a higher
fee, would receive official memos, press releases, attend meetings, and so on.
You strongly urged allowing such persons to become associates, including
your friend the lady witch, and also a man (you wanted me to meet him
some time) who ran a church of Satan in California. You may recall that
Puthoff was one of the first to apply for associate membership. I think we
were wise to abolish this category.

I discussed these things hurriedly with the reporter. If I spoke carelessly and gave the impression I was talking about "fellows" with "voting
privileges," then I humbly apologize. On the other hand he may have misinterpreted what I was saying. I do not recall saying anything about voting
rights or positions of "authority," but he may have assumed this is what I
meant. Incidentally, a good indication of your attitude on this is that your
own list of "consulting editors"—the closest thing you have to "fellows"—
includes seven firm believers in ESP and PK, some of whom I consider
cranks. So there is a very sharp disagreement between us on the value of
having such persons in positions of "authority"—whatever that means—on
a periodical devoted to scepticism.

Now about frivolity. If the article gives the impression that I think the
ZS frivolous, then it accurately reflects my views. This surely shouldn't
surprise you, since I've been saying this plainly in letters for many years. I
recall once likening your attitude toward cranks to that of a person fascinated by 10-in-1 shows.

I have no doubt about the sincerity of your scepticism, but that is
a vague word of many meanings. I do not doubt that Tiffany Thayer

was sincere in his scepticism; he differed from you mainly in allowing his periodical, *Doubt*, to be funny. I regard the Fortean movement as frivolous (see my *Fads and Fallacies*) and I consider *ZS* frivolous in a similar sense. Modern science swarms with enormous anomalies, many of which have a good chance of becoming Kuhnian paradigms. A good current example are the quasars that belch out gas that seems to move faster than light. One distinguished cosmologist is convinced that this proves the quasars are nearby, and that the universe is not expanding! He could be right. I doubt it, but I treat his theory with utmost respect.

The *ZS*, on the other hand, is on the Fortean level of being concerned with the Geller effect, star patterns that correlate with human choices of a profession, a comet from Jupiter that becomes the planet Venus, Bigfoot, deep mysteries behind UFO sightings, and so on. I consider these topics trivial and beneath your dignity. Moreover, I believe it to be tragic that you waste your time and talents on these things, treating them as somehow serious challenges to orthodoxy. In the long run, I think that this obsession with crankery will prevent you from making any genuine contributions to the sociology of knowledge.

Naturally I do not think you *believe* in any of these things, but I do think you would *like* to believe. I recall that when you were planning a big test of ESP, you said in a letter that you hoped that the experiment would demonstrate it, as seeing a unicorn would prove the existence of unicorns. Belief is as vague a word as scepticism. It all depends on subjective probability estimates. On most of the topics with which you are concerned, your subjective estimates vary widely from mine, and always in the direction of favoring the anomalies. I rate the probability of a mystery behind UFO sightings as near zero. You have expressed a much higher probability rating. And similarly with other topics. We both, of course, recognize all science as fallible, and in *that* sense we are both sceptics about *all* scientific claims. But belief and nonbelief are attitudes on a continuum, and on this continuum we obviously diverge.

This divergence is reflected in our way of writing. You will praise a book by a hack journalist of the occult, then give the book a light tap on the wrist. At the same time you will lambast a book by a scientist, attacking the paranormal, then add a light note of praise. (I mean the *ZS*, not you personally, although your short reviews reflect the same tendency.) We *incline* in opposite directions. You think I am too dogmatic, too hard-nosed, too blind to the open possibilities of science. I in turn think you cannot distinguish between genuine aspects of unorthodox science, genuinely open, and areas so close to zero in the probability of being confirmed that they can justly be declared nonsense. No, there are no hard-and-fast rules for making such a distinction any more than there is a sharp way to distinguish the ocean from its shore, or from a harbor, or even from the atmosphere. Because the *ZS* is preoccupied with matters that I believe to be unbecoming to your intelligence and humor, I consider it a frivolous, time-wasting enterprise—with one exception, your bibliographies.

    If you care to print this letter, I won't mind.

                              Best,
                              Martin
CC: Ray

[Eastern Michigan University]   15 June 1981
Dear Martin,

    I enclose a copy of the article in the Chicago Tribune that was earlier read to me [June 2, 1981, section 2 p.1-2]. You will note that you are quoted as saying of me: "I don't know if he's part believer or not," Gardner adds, "He claims not. But conflict developed when he wanted to bring into a position leadership people we regarded as cranks. Like some lady in California who's a professional witch."

    Obviously, the key phrase I found offensive here was "into a position of leadership." Both your letter of June 9th recalling my distinction between associate and regular membership, and my last letter to you, should agree that I never wanted anyone but the Fellows (so named and created especially) to be in any kind of position of leadership. Quite the contrary; in fact, Dennis Rawlins did not like my proposal because he thought I was trying to get the funds from such associate memberships under somewhat false pretenses if I did not also let them have some degree of power! But, as I recall the whole thing (and I think both Ray and Dennis would agree with my recollection here), the main reason we dropped such associate members (and note that this would really have meant three levels of membership: Fellows, Members and Associate Members way down on the powerless bottom) and even members entirely, was because there was some fear of a publicized walk-out by the believers should they not be happy with things.

    Now in your letter, you say "If I spoke, carelessly and gave the impression I was talking about 'fellows' with 'voting privileges,' then I humbly apologize." I presume that most people would equate "a position of leadership" with having a voting privilege. So, I presume you were misquoted.

    Let me comment now on the rest of your most interesting letter. I note that you mention that you see my Consulting Editors for *ZS* as "the closest thing I have to Fellows." That is wrong on two points. First of all, my consulting editors have no votes, are known to differ sharply in opinion so can obviously never be spoken for as a body; I hardly think they are in positions of authority. (Of course, since the Committee's fellows have no votes on the Committee itself, I guess you could counter that they have no authority on the Committee either. In fact, in the case of both the Fellows and my consulting editors, these people are well known persons who *bring* authority to the Committee and to ZS. They are not put into positions of authority at all. That could only be so if they had some sort of power in our respective organizations.) The second point of error overlooks that the "closest thing" I have to Fellows are the Senior Consultants to my Center for Scientific Anomalies Research.

Another point in this regard is the surprising remark you make that my consulting editors include "seven firm believers in ESP and PK." You obviously know more about them than I do, for I know this to be the case only for Martin Ebon, Bob Morris, John Palmer, and Charles Tart. Who do you imagine are the other three? Perhaps you put Harry Collins into that category, but you would be incorrect if that is the case. Please note that I have 24 consulting editors. I had two other, James Webb and Chris Evans who died (and who were firmly anti-ESP, though Evans was quite open minded), and yourself before you quit over the Velikovsky dialogue in ZS. So, out of 27 consultants, I have had only 4-to-7 whom you'd classify as "firm believers." That's not even 1/3! You also have a penchant for repeating statements to the effect that I want to give "equal time" to both sides. I have never argued for equal time, only for fair or equitable time. There is one hell of a big difference. My god, Martin, all you have to do is look at ZS to see that the bulk of each issue is devoted to critical stuff. You still keep wanting to define skepticism as denial rather than the raising of doubts. How can you read the literature and not see that there is abundant skepticism of psi research among psi investigators when one looks at what they have to say about each other's works.

You are generally quite right about the difference between our orientations as outlined in your letter of June 9th, but you are simply wrong in thinking that your orientation is either more scientifically proper or more sceptical than my own.

You refer to my book reviews praising an occult hack and giving it only a light tap while doing the reverse with a scientific work. Yes, that is the case to some degree, but I think you overlook my reason: The scientist claims to be doing science and needs to be held fully accountable. That is why I generally am very offended by bad parapsychology. But an occult writer usually does not claim to be doing science. He is not trained as a scientist. It would be silly to swat such a mosquito with a canon (as you would prefer I do). Unlike your journal *Skeptical Inquirer*, ZS is not written for a mass audience. I am not trying to protect the great unwashed public from the horrors of supernaturalism. Why beat the gnats to death? But when a book like Hansel's comes out which is scientifically and scholarly irresponsible while put out as an important scientific work aid obtaining praises from fellow critics, somebody needs to tell it like it is. I think I have generally been consistent about this by damning strongly the bad parapsychology that similarly gets praised by psi researchers (e.g., Targ & Puthoff's work).

I come now to the most important point in your letter. You say that you think *ZS* is frivolous because it seeks to seriously discuss a bunch of topics with nearly zero-point probability of being true. Yet all these same topics are discussed in *Skeptical Inquirer*. Does this mean that you are now willing to publicly admit that the SI does not intend to take these subjects seriously? That it is only all right to take them up if we debunk them and thereby perform a public service in inoculating the public against pseudoscience? If that is so, isn't that pre-judgment of the issues in a way that absolutely contradicts the public goal statements of the Committee? Now I do think you mean that. I don't think you ever took the alleged fair-play, no pre-judgment policy put forward by Kurtz and the Committee to really be the proper posture for the Committee. I think you always saw it as a debunking effort. I think your letter clearly implies that. For that reason, I may take you up on your offer to publish it in *ZS*. But could you clarify this? I don't want to misrepresent you on this.

In any case, glad you like the *ZS* bibliographies, I agree with you that they may be the most significant part of *ZS* in the long run.

Cordially,
Marcello

CC: Ray

[Woods End]   18 June 1981
Dear Marcello,

I was relieved, to see I was not quoted as saying you wanted believers to be "fellows." When I spoke of membership in the Committee I clearly meant your category of members on the middling level, between fellows and subscribers. I believe the words "position of leadership" were put in my mouth by the reporter. If I did say that, or anything like that, I was wrong. I don't think I did. Aside from this unfortunate phrase, I stand by the quotation. I don't think Lyon tape recorded. We just chatted while he probably took notes, and later wrote from memory.

I consider my remarks all quite fair and moderate compared to your calling the Committee a "knee-jerk mouthpiece for orthodox science," inflexible, and "choking off legitimate research." It seems to me you still have not made the useful distinction between "unorthodox science" and "crank science." (No, I must say again that there are no sharp lines or hard criteria.)

Among your consulting editors the three you left out and who I included among my seven, did indeed include Harry Collins. When he visited me he told me he believed the Geller effect to be genuine (if that isn't PK what is?) when he arranged his test of the children, and that the test was not designed to trap them, but to test Taylor's "shyness" theory. Has he since changed his mind? There was no indication of it in the letter he wrote me last January in which he explained why he had not revealed to me his beliefs when he came to see me. He repeats in his letter that he expected his test to *confirm* the Geller effect, and of his "surprise" when they caught the kids cheating. His "official position" he states, is one of "neutrality." But I spoke of his personal beliefs, not has official stance.

As always, we should sharpen this by probability estimates. I maintain that Collins's subjective estimate of the reality of ESP and PK is above 50 percent. This is strengthened by his defenses of Girard, and of Hasted's strain-gauge tests. (He was defending the strain-gauge tests when he came to see me as tests that were carefully controlled and could be definitive. That was when I suddenly became suspicious of his motives in coming to see me.) Do you wish to maintain that Collins 's subjective estimate of the reality of PK is *exactly* 50 percent? I would welcome any documentation you can supply on this.

My third person was Gauquelin. Instead, of speaking of belief in ESP and PK I should have been more general and called it belief in the paranormal. I really don't know what G's views are about ESP and PK. Do you? It would be interesting to know. If you correspond with him still, you might ask him sometime and tell me what he says.

The phrase "take seriously" is ambiguous. I take seriously all the topics discussed in SI in the sense that I think public infatuation with such nonsense (which I sense as fading among the Educated but growing among the uneducated) should be taken, seriously. I do not take them seriously as anomalies in the Kuhnian sense, or as serious challenges to science. I am sure that Kuhn himself does not. They are on much too low a level. I rate them as only slightly above hollow-earth theories (with the exception of parapsychology, which I take to be on a higher level with respect to a *very* small number of researchers). To me they are absurd challenges to be debunked. That is indeed my view. But please note that I speak here *only* for myself, not for the Committee or any other members.

I see the SI as primarily a periodical for enlightening editors, writers, etc., who until now have had nowhere to turn for the skeptical side of the current craze. Call it "debunking" if you like. I prefer to call it science education. (In no way can a "debunking" of V be considered a knee-jerk defense of "orthodoxy." It is a defense of science. And, no, there are no *sharp* criteria to distinguish science from pseudoscience.) I believe your are wasting your talents trying to establish dialogue between, say, Velikovsky supporters and detractors. It is like establishing dialogue with Rogo on the topic of communicating with the dead by telephone. There are levels of bogus science that are not deserving of the respect accorded to what I consider valuable unorthodoxy in science.

And I found hilarious Clark's description of himself as a "moderate proponent." Surely you don't consider *Fate*, of which he is an editor, a "moderate proponent" of openness in science! If you consider *Fate* and Clark as "moderates," then our mind-sets are much further apart than I have supposed!

Best,
Martin

CC: Ray

[Eastern Michigan University]   22 June 1981
Dear Martin,

When I, first wrote you, I had not seen the full text of the newspaper article but only had the sections quoting you read to me on the phone. When I got the article in full, I just sent it off to you without even giving any thought to the "quotes" attributed to me. As a matter of fact, I seem to be in the same situation you are in. I was not really quoted accurately but the general spirit of the comments was correct. I note that in my case the writer did not put quotation marks around my remarks as was done in your case. Actually, I gave the writer a more balanced picture than came out in the article. In fact, I was a bit surprised at the whole thing because the writer gave me the impression that he was far more negative on the Committee than I was.

Of course, we continue to disagree about the matter of whether or not the Committee is acting to choke off legitimate research. I very much think it does, but then you and I would disagree about what constitutes legitimate research (e.g., our differences over Velikovsky). I think that any reasonable person who has looked objectively at what has happened with the handling of Gauquelin's work by the Committee (and I realize you like Randi, may wish to somehow separate Kurtz and his Committee press releases from the

Committee; but until there is public dissociation of the Kurtz efforts from the Committee proper, I must equate the two) would have to agree that the handling of the matter has acted to choke off legitimate further inquiry by those who have read of the supposed discrediting work of the Committee which is actually abominable science.

In regard to the beliefs of my Consulting Editors re psi and PK, you refer to "three" people of whom you mention Collins and Gauquelin. You don't indicate the third person. Re Collins, I don't think you fully appreciate his position. Collins is probably the most extreme relativist writing in the sociology of science today. He is not a "believer" in the usual sense in anything connected with our debates on the paranormal. You are correct that he did originally expect confirmation of the metal bending with the kids he looked at and was surprised to find them cheating. I have not yet read Collins' manuscript, but his book dealing with Geller and the metal bending stuff will be out rather soon. It is in press for Routledge & Kegan Paul. From my conversations with Collins, he is not truly a believer in the paranormal but sees all reality as so socially negotiated that there is no clear answer (for him), to the differences between, say, you and Hasted. I don't agree with Collins (I believe in relativism in the sociology of knowledge only as a heuristic rather than a philosophical position). His extreme form of reality constructionism (very heavily influenced by his work with physicists and by Wittgenstein) does not neatly put him into the paranormal camp at all.

As for Gauquelin, I simply don't know what his views are on psi and PK and will ask him. However, I have strong reason to classify him as a non-believer since he has never tried to bring that sort of thinking into his conjectures about the source for the Mars Effect. I think that as a deviant psychologist, he is probably open minded about psi. I have no evidence to think he is a proponent. I will try to find out, however.

Of course, this really makes little difference since a few of the Fellows are known believers in psi, and even Kurtz privately believes he and his wife share telepathy (something which I have never claimed).

I am amused at your remarks about Jerry Clark. Jerry is indeed an associate editor for *Fate*, but you can hardly hold him responsible for the opinions of Curt Fuller, the editor. He is indeed a *moderate* proponent of psi. Jerry is personally very responsible for the debunking articles that have been appearing in *Fate* these last few years. In the next issue of *Zetetic Scholar*, I will be publishing a full list of all the debunking articles in *Fate* over the last five years. You may be surprised to see the list (nearly two articles per issue). Since *Fate* comes out every month and *Skeptical Inquirer* comes out only quarterly, the simple fact is that there are as many debunking articles published in *Fate* since the arrival of the Committee as published in the *Skeptical Inquirer*. And I think the quality is about the same if not better. More important, the debunking articles in *Fate* actually reach the believers who read *Fate*, whereas the *Skeptical Inquirer* pretty much preaches to the already converted. And, I might add, (a) *Fate* has frequently debunked the claims first (e.g., the Amityville Horror, the Bermuda Triangle, & von Daniken), and (b) many of *Fate*'s authors of these pieces are by contributors to the *Skeptical Inquirer*, too.

Now I know Jerry quite well and he not only is a moderate, but I find that he is getting more conservative just about every year. He has already retracted a number of his earlier views (e.g., in this last issue or *Zetetic Scholar* he explicitly rejects his earlier book on UFOs). I think you would be amazed at how much you and Jerry probably would agree about.

I continue to be amused at the way we obviously speak past one another. You are willing to take seriously paranormal claims only in the sense that they are nonsense hurting the public. Obviously, what you take seriously is not the claims but the effects of the claims. You merely want, as my earlier letter indicated, to inoculate the public, to, as you think, educate them to their errors. We of course differ on that. But I can only again respond by citing the goal statement of the Committee, which says something very different indeed. You may counter by pointing out that you speak only for yourself, but that is simply nonsense in any practical sense. You are not just any ordinary Fellow on the committee. You are one of the few Fellows voting on the Executive Council. You are possibly the best known member of that Council. And the Executive Council's majority clearly agrees with your position on all this rather than on the goal statement which vows to take seriously the claims and not just their effects as nonsense.

You end with "In no way can a 'debunking of V be considered a knee-jerk defense of orthodoxy.' It is a defense of science." Surely you don't mean that to apply to all debunkings of Velikovsky? I will certainly agree that not all debunkings of Velikovsky are knee-jerk reactions (I respect even if I do not agree with Sagan's debunking effort). But surely the early criticisms of V by those who did not even read his book were knee-jerk reactions. (Let me add that the term "knee-jerk" was not actually one I used in my interview for the column. I don't like the term; but I adopt it here because you used it in your note and I take issue with the meaning you make.) Are you implying that defenses of Velikovsky can not also be done in "defense of science"? Dammit, Martin, you seem to simply refuse to grant scientific legitimacy to anyone who wants to defend what you consider implausible nonsense. Most of the people respect who defend Velikovsky insist that V is completely right far less than I think you really want to see him as almost completely wrong. They are as upset about the absence of due process accorded V as I am (and I do think he is almost completely wrong). You seem on the one hand to believe in science as a self-correcting system while at the same time to want to give it little chance to debate "crazy" ideas. You think V is nonsense and don't think it legitimate for a scientific journal to seriously discuss his ideas (only correct the serious damage his whacky ideas may be causing among the public). So, you want him only debunked and not seriously examined in public dialogue by proponents and critics (as in *ZS*). Yet you insist that your attitude does not choke off research. But then you would probably add "legitimate research," even though you admit that there is no clear demarcation criteria that can be used *a priori* to distinguish real from pseudoscience. And I think you would even admit that V has been made into a martyr because of some of the incompetent critiques of him earlier. What the hell is so terrible about full and free debate on these matters? Even flat earth theories, should a scientific (protoscientific) group emerge that wants to seriously put a case forward? If the ideas are so damned stupid, why won't the critics make mince meat of them? I could understand your objection to one-sided presentations by proponents. These might mislead the public, I am as against one-sided presentation by proponents as you are. But what is there to fear from two-sided presentation? Do you think I would stack the cards against the critics? That I would get weak critics (as you critics have frequently picked weak proponents to attack)? Dammit, Martin, science is its method, not its current fashionable body of theory. What we need to teach layman and everyone is scientific method, not reverence for the dogmatic authority of people who happen to be professional scientists. Even professional scientists often do bad science. In fact, most people do what they do rather poorly. Scientists are not that different. I fully agree with you about the need for science education, but to me that means teaching the methods and critical view of science, not merely the pontifications of experts. Surely you must see that. You seem to want to limit the self-critical functions of science to the scientists themselves. That is an elite view of science I simply can not share. Our differences over the word "skeptical" bring this out so well. I am skeptical to all claims, not just those by the exoheretics. By your criteria, I am sure even Torquemada probably saw himself as a skeptic.

I fear that no new ground is being covered in this letter. The sad thing to me is that your perspective allows you to rationalize the evils being done by people like Kurtz in the name of your own version of "the True Faith." In the final analysis, our differences may be moral ones. I think we are both pragmatists, but we obviously define the situation very differently. Time will tell which of us has chosen the best strategy.

A final comment. I enjoyed *Polywater* and thank you for calling it to my attention. My impression of the book was somewhat different from your own. You see it as vindication for the neat self-correcting system against a pathology. My reading of the book's last chapter clearly differs from your own. I wrote the author and will eventually learn more on this matter. I hope.

                                        Best,
                                        Marcello Truzzi
CC: Ray

P.S. Collin and John Taylor have both agreed to write reviews of Hasted's book for *ZS*. Would you be interested in writing one too? You remain a welcome contributor to *ZS*, whatever our differences.

[Woods End]   27 June 1981
Dear Marcello,

Sorry about skipping my second person. When I retyped a very rough first draft of my letter, I inadvertently omitted an entire paragraph!

It was about Hynek. I asked if you had actually read his famous *Fate* interview. If so, do you maintain that Hynek has no private beliefs in the reality of such psi phenomena as ESP, PK and so on? Please let me know if you think I erred in including Hynek among the seven.

Remember, we are talking about subjective belief, not official pronouncements. Your description of Collins makes him out to be even stranger than I had imagined. Not to have a personal probability estimate, above or below .5, on something like PK seems to be as unlikely as a person not knowing whether he believes in the Incarnation, and rates the probability as .5. Based on both conversation with Collins, and a long letter, I simply do not believe that he is absolutely neutral in his subjective estimates. As a sociologist he obviously believes in maintaining a neutral attitude, and wishes to keep his beliefs out of his work, but that of course is not what we are arguing. As for Gauquelin, I await with interest what you can discover about his opinions on the evidence for ESP and PK.

I am compiling a dictionary of definitions in a curious language I call Truzzese. If you consider Clark a "moderate," could you supply me with three names of *believers*, who are prominent journalists of the paranormal, who you would call "immoderate"?

I enclose a copy of my forthcoming review of *Polywater*. The polywater claims are precisely in the category of claims that I consider legitimate, and which deserve to be treated with respect. (*Scientific American* ran a major article on "superdense water", by Deryagin himself, November/1970.) I have never considered it "pathological" science on the low level of, say,

Velikovsky and the creationist literature. Polywater is almost a paradigm of the type of unorthodox science you should be studying and debating—not Bigfoot and the Bermuda Triangle.

Best
Martin

CC: Ray

[Eastern Michigan University]   7 July 1981
Dear Martin,

1) Re Hynek: I am sure I read his interview a very long time ago and frankly don't remember its contents. I will have to go find it and read it again. Do you have the date handy?

It is quite possible that Hynek has private beliefs favoring psi. I see no special reason to be concerned about that if so. It seems to me that the old distinction between belief and knowledge is significant here (though I recognize this is rather a continuum). My point is that I know from conversations with Hynek that he is open to the claims of psi and may even have an experiential basis for belief in psi. But I think he would distinguish (as I do) between such a claim as experiential (subjectively validated) versus objective. I have found psi believers to be of two very different sorts: those who have had experiences and use this as the basis for thinking they have knowledge of psi, and those who have had experiences which they believe prove psi subjectively but who still fully appreciate that such subjective validation does not constitute scientific (intersubjective) evidence. It is like the fellow who has seen a sasquatch but realizes that it is reasonable for others to doubt his experience and even retains some degree of doubt about the interpretation he himself has given to the direct sighting experience. I think Hynek falls into this latter group. Like me, Hynek does not want to close the door on matters and recognizes that some legitimate experimental anomalies appear to exist in the parapsychological literature. Unlike me, he may have some personal experiences I know nothing about (and which he may discuss in the *Fate* interview you mention), but I think he would still be, like, me, an agnostic on the matter of psi as far as considering the scientific merits of the evidence. Unlike me, he may personally lean towards rather than away from belief in psi.

You say "remember, we are talking about subjective belief not official pronouncements." I was not aware of that. I don't really see much relevance to such subjective beliefs. Half the Committee for SICOP may have subjective beliefs in psi for all I know (as Kurtz definitely does). I am not really that interested in motives. The critical question is can the scientist differentiate between his own subjective proclivities (which are not the same as biases if they are recognized as not objective by the scientist) and the kind of evidence science requires? It is the latter which we are supposed to be interested in and the former only counts in so far as it confounds the latter. If we did not argue this way, skeptics would be considered too biased to do honest research just as would be believers. (I disagree with psi researchers who claim such nonsense.)

2) I would be happy to clarify what I mean by *moderate proponent*. A moderate proponent is one who thinks the evidence is strong towards the anomaly and is merely *inclined* to accept the claim. Jerry Clark is clearly that. A moderate proponent is also one who continues to have reservations and is relatively easily able to change his mind given new evidence. A moderate proponent tends to have little vested interest in the claim. Jerry had changed his mind about many matters and I have argued him away from many claims he took seriously earlier. Jerry has publicly rejected some of his earlier published views. I think he is very open to sound argument and tries to be reasonable. Now you ask me to name some immoderate proponents of the paranormal. That is easy. A clear example would be William Cox. Another would be Scott Rogo. Another would be Hans Holzer. You specifically asked for prominent journalists who were proponents and immoderate. I just mentioned Holzer and Rogo (though Holzer is really much much worse than Rogo, and Holzer is probably an outright charlatan). There are many outright charlatans writing proparanormal, I believe. These include people like Holzer, Brad Steiger, Allen Spraggett, Sybil Leek, Jess Stearn, etc. They are not only immoderate but of doubtful veracity or sincerity. I presume you are asking me for immoderate but sincere proponents. These would include, in addition to Rogo, Raymond Bayless, Fred Archer, Tom Bearden, Marilyn Ferguson, and about half the authors of articles in *Fate*. But I would classify as moderate proponents such people as Martin Ebon, and perhaps even Stan Krippner (though he is near the immoderate end in his earlier writings as compared with his current views).

But I also would distinguish moderate and immoderate "skeptics" and would classify Randi as immoderate and Ray Hyman as moderate. I see you as nearer Randi on some dimensions but I think you more towards center on the immoderate side. I see Hansel as immoderate, too.

It may be the case that even neutrality can be both moderate and immoderate. Intensity is a separate dimension from the valence of an attitude. You may view me as immoderately neutral. I see myself as moderately neutral while seeing Harry Collins as immoderately neutral since he is a total relativist philosophically.

I trust that the above has clearly responded to your "challenge" that I name three immoderate believers.

I hope it also demonstrates why I think you see what I would consider multidimensional problem in simplistic black-or-white terms. If this means you need a dictionary of Truzzese, I can only hope that my dictionary is consistent in its terminology and ultimately useful in its distinctions between categories apparently lumped in Gardnerese.

I'm not sure why you brought up the Polywater business again. I recognize that you distinguish some anomalies as legitimate scientific concerns while relegating Bigfoot and Velikovsky to the limbo of the flat-earthers. But I thought I had also told you earlier why I thought you were misguided in doing so. By the way, this will be taken up in a dialogue between me and Leonard Zusne in the upcoming issue of *ZS* (#8).

Zetetically yours,
Marcello Truzzi

CC: Ray Hyman

P.S. I am in the midst of reading Henri Ellenberger's *The Discovery of the Unconscious*. What a remarkable piece of scholarship. You might find the book worth reading. Much to disagree with but much to appreciate also.
P.P.S. I think Curtis Fuller is immoderate in his beliefs but is probably somewhat more moderate than Rogo. I might add that though I think both Rogo and Randi are immoderate on different ends of the spectrum, I think both are sincere "hit men" even though both call one another liars.

[Woods End] 10 July 1981
Dear Marcello,

I enclose the *Fate* interview with Hynek, and will be interested in knowing whether you think his remarks are those of a "moderate" believer. My main reason for asking you to cite immoderate believers (not charlatans) and journalists of the paranormal was to see if your views will be reflected in any of your future short reviews of their books. I was glad to see Rogo on your list, particularly since I recall your many reviews of Rogo books. There has been one in every issue of *ZS* except the last. Let me list your remarks:

1. *Mind Beyond the Body*. "A first rate collection presenting the experimental evidence for OBEs. Though I did not find the evidence here overwhelming. I was impressed by the intelligent level of the book. Recommended." (Of course Rogo only edited this. Your use of "overwhelming" here is interesting.)

2. *The Haunted Universe.* "A highly entertaining tour..." You don't buy it, but you also voice no wisp of criticism.

3. *Phone Calls from the Dead.* "A rather fascinating book. ...fun reading..." You go on to say the authors do not "claim to present a careful scientific case"' and you poke fun at the hypothesis, but nowhere do you indicate that Rogo may be irresponsible.

4. *The Poltergeist Experience.* "A fascinating excursion ... by a prolific writer on the paranormal!" Whatever one may think of the phenomena, you continue, "the narratives present entertaining enigmas, and Rogo's grasp of the historical materials is admirable."

5. *Earth's Secret Inhabitants.* This by Rogo and your "moderate" Clark. "Hardly a scientific work but great fun and generally well documented."

If you consider Rogo "immoderate," on the level of Cox and Spraggett, et al., I find it odd that this opinion nowhere creeps into your copy. You also cite Marilyn Ferguson as "immoderate." Here is what you said about her *Aquarian Conspiracy*: "A remarkable survey ... probably the best introduction to this vast area and well worth reading for the information contained aside from question of assessment." "Lots of information but uncritically and some indiscriminately presented." Again, praise with light wrist taps. No indication that you consider her "immoderate". Just below it I see a review praising Curtis Fuller's book on Ufology. "A major collection. ...Recommended highly." You do say that hard-line critics are not included, but you feel they have been "scathingly and honestly examined."

Compare, now, with the highly *unfavorable* review of the book by Kammann and Marks, with only slight praise. I hope you are right in believing that Clark is slowly becoming more moderate. I know he changed his views on the fairy pictures of Doyle, that was only after a "smoking gun" episode. Surely you must regard the evidence for the fakery of the photos as pretty overwhelming at the time Clark wrote his two-piece defense of the pictures. But, as you say, maybe he is slowly educating himself. I'll consider him a "moderate" when he severs his ties with *Fate* and its rip-off advertisements.

                                    Best,
                                    Martin

CC: Ray

P.S. Please let me know your reactions to Hynek's pronouncements in his interview. You once chastised me for being unfair to Hynek in my review of *Close Encounters*. Since all I said about him was taken directly from his own remarks, I am particularly interested if you still think I was unfair.

[Eastern Michigan University]   14 July 1981
Dear Martin,

Thanks for your letter of July 10 and the enclosed copy of the Hynek interview. Let me comment on the Hynek interview first. One of the most interesting features of what you sent me was what you chose to underline and mark in the margins. Hynek repeatedly says he does not know what UFOs are and that no answers currently exist. He emphasizes this over and over again but you think such statements of agnosticism are unimportant (not worth marking). Instead, you mark all his references to statements he makes about *possible* psychical explanations and his inclination towards a general connection between UFOs and other paranormal phenomena. Now, I certainly know that Hynek is quite willing to consider such alternatives quite seriously. You, on the other hand, would not really want to consider such seriously. From my past conversations with Hynek, it is apparent to me that he felt that he was being driven to consider such exotic alternatives because the less exotic alternatives (such as Klass's plasma notions) did not fit the evidence he felt had been reported correctly. It is also the case that Hynek was then very much being influenced by Jacques Vallee's psychic conjectures and ideas. Hynek has since moved away from Vallee's position rather substantially. Hynek is—and was—willing to entertain many exotic ideas as hypotheses. I know that he believes—as do I—that parapsychology contains anomalous research results still needing adequate explanations. I do not think he is a confirmed believer in psi. But I must admit that a couple of statements in the interview in *Fate* indicate that he may be more disposed towards belief in psi than I had thought he was. However, I remind you that this an interview with Fuller which presumably was edited by Fuller for *Fate* readers, and the context is likely to bring out any such pro-psi elements in Hynek's views. The critical thing is that (a) Hynek recognizes that his paranormal conjectures are not only conjectures but far-out conjectures, (b) Hynek would find mundane explanations more acceptable as a scientist, and (c) he insists only that UFOs constitute a legitimate research problem and one that has not yet been solved. It would appear that you resent his even seriously conjecturing about the paranormal in a positive way that might give aid and comfort to "the enemy, " i.e., the typical reader of *Fate*. And this grossly colored your review of Hynek's book, which you really did not review at all but instead of which you commented upon this interview. Where are all the paranormal conjectures in his book you reviewed?

As for his references to Geller, I know that he now believes that Geller is probably a fraud. But back in 1976 when this interview was published, the issue was far more clouded and most public information on Geller that Hynek had seen was probably positive. (The interview was probably in 1975.)

The only comment I really found upsetting in the interview was Hynek's single sentence about other planes of existence on p. 47. I would have to ask him about what he meant by that.

I will say that Hynek is probably more receptive to the reality of psi than I am, but I see no harm in this as long as he does not believe that the case for psi (as also not with the case of the extraterrestrial hypothesis for UFO origins) is scientifically convincing at this time.

As usual, you seem to think that reasonable men can differ about *some* things but UFOs and areas like Bigfoot, Velikovsky, etc., are not among those things.

This brings me to a minor point: my reference to Jerry Clark changing his mind concerned not the fairy photos but his book on UFOs where he propounded a psychic projection theory. This also brings up the interesting issue of the fairy photos. You seem to suggest that the only alternatives in the paranormal view on fairies were to believe in either a hoax or real little fairies. Though Conan Doyle believed in the little fairies, the dominant proponents of the fairy photos did not argue for real fairies but for psychic projections of such images on the film (as with Serios' alleged photos). Most proponents I have read about the fairy photos saw it as some sort of strange PK phenomenon.

Now I turn to the matter of Rogo. First of all, I consider Rogo an immoderate proponent because (a) I think he shows extreme credulity about some matters and does not seem to really understand the degree (quality) of evidence science demands for such extraordinary claims as he makes, and (b) I think his immoderation shows up mainly in the conversations I have had with him. In other words, his books are not put forward in nearly so immoderate a manner as his viewpoints when stated in conversations and in forums where I have heard him. Now, the criterion I use for moderate versus immoderate largely revolves around the question of whether he is open to arguments. I think that Rogo (like yourself) may not be as open to argument as he probably thinks he is. But this is an immoderation that mainly shows up in his dialogues and exchanges with others. In terms of the evidence he presents in his books, I think he is far less immoderate. I do not find him presenting totally unreasonable sources of evidence like revelation or authority.

This brings me to the matters of my brief reviews. To a point, you are probably right about my imbalance in reviewing Rogo and, say, Kammann. However, I think this is largely due to the fact that Rogo is not a professional scientist writing science books and is not perceived as such by those I know (except some critics like yourself). Rogo is mainly a professional writer who churns out books (much like Dan Cohen on the other side, and whom I treat about the same way) for laymen. There is little in the way of imprimatur of science about Rogo or his books. But Kammann is a professional scientist, and praised for his debunking by other scientists who should know better. He is expected to know how to do science and needs to be held accountable. I have frequently told you and others that I simply don't get as angry at occultists (who make no science claims) as I do towards parapsychologists-scientists who do bad work. Nonetheless, even given all this, I may sometimes act somewhat inconsistently—I'm only human. And another factor may be the depth of reading I have done in the book briefly noted in these short reviews. In the case of Kammann (as with Randi's book), I read it most carefully and felt ready to do battle in a full review elsewhere. I am supposed to write a full review of Marks & Kammann's book for the *Journal of the American Society for Psychical Research* (you can judge me more fully when you see that). I did not read Rogo's book with the same care for details. Someone writing me commented that he liked my reviews better after he realized "the code" in which I wrote them. I really do not intend to have any such "code," but I guess there is a pattern. I do get most upset about bad parapsychology books (like *Mind-Reach*) which present themselves as *the answer* just as I get pissed at a book like Marks & Kammann's for doing the same on the other side. But I try to be pretty balanced on books coming from the same sorts of parties on either side of the issues. Perhaps I am assuming tacit knowledge in my readers—who are supposed to mainly be serious people interested in *Zetetic Scholar* bibliographies and such and not simply a mass audience such as you and some others want *Skeptical Inquirer* to center in on—and perhaps that is a mistake on my part. I presume that a book like Rogo's one on phone calls from the dead is pretty outrageous right from the start and does not need a serious and heavy-handed critique. So, I just slap him on the wrist and presume everyone knows it is another potboiler to be read more far fun than real enlightenment. And I think Rogo views that book the same way. But I guess I should have taken into account that people like you might not approach it with the tacit ideas I expect. Otherwise you would not already be using cannons to slay fleas in *Skeptical Inquirer*. I am now thinking that perhaps I need to spell some of this out in a future issue of *ZS*.

I will drop a line to Hynek to see if I can get some clarifications about his views. (I am told that he largely corrected some of his earlier psi-oriented ideas at the Smithsonian "debate" on UFOs where he presented his most clear and newest position on the controversy.)

Best
Marcello Truzzi

[Woods End]   17 July 1981
Dear Marcello,

I have read your defense of Hynek very carefully, and find it unprofitable
to comment on it. I rest my case. I assume you have read his latest
pronouncements in *Technology Review*, July issue, and the article about it
in *Discover*, August 1981. No doubt you will consider the *Discover* page
highly biased.

Let me ask you a question about a more serious matter, not so much in
a spirit of criticism as of honest inquiry. As you know, the occult revolution
has been dwarfed by the Protestant fundamentalist revival. This has produced a large body of books and periodicals on creationism. It takes only
a glance at the better books to see that the literature is fully as impressive as any book by Velikovsky. V was a one-man phenomenon, but there
are dozens of creationists, some with science degrees, who are producing
well-reasoned serious books which purport to give strong scientific evidence
against evolution. They are playing the science game much better than V
because they are sticking to one hypothesis, and have no quarrel with well
established areas of orthodox physics, chemistry and astronomy. As you
know, the movement is having a serious impact on American religion and
politics.

If I were a sociologist of science I would find this trend far more significant than the books about sea monsters and children who bend metal or
affect strain gauges. From your own viewpoint, it seems to me a dialog on
creationism would be a topic of far greater cultural import than a dialog
on V who will soon be forgotten. Are you planning to devote an issue to a
dialog on this trend, which is the outstanding trend at the moment involving a conflict between orthodox and non-orthodox science? I am sincerely
puzzled by your seeming disinterest in this growing conflict. I heard a Phil
Donahue program last week devoted entirely to this (it is unthinkable Donahue would consider a program on V to have any current interest). The
creationist was quite sophisticated in his arguments; at one point he even
quoted from Karl Popper on how evolution was a "research program," not
a fact! Could you give me your reasons for evading this topic?

                                        Best
                                        Martin

CC: Ray

[Eastern Michigan University]   20 July 1981
Dear Martin,

Your questions about the creationist controversy and me and *ZS* really concerns two somewhat distinct elements. The first concerns my own personal views re the controversy and its sociological value to the sociology of science. The second concerns *ZS* policy towards discussion of those issues. There is some interrelationship, of course, but I see them as somewhat separate.

As for myself, I quite agree with you that the controversy is a most significant one for sociologists of science. I do not see myself as very central to it, however, because several sociologists, including especially Dorothy Nelkin have written extensively about it. As for myself, then, I see myself as in the background and not the forefront of sociological discussion of that particular controversy. Therefore I give it a personal (though not an intellectual) low priority for my own activities. But it goes somewhat further than that, for I see the creationist controversy as essentially quite different. I will try to spell this out a bit for you now.

I make a strong distinction between creationism and anti-evolutionism and hate to see the two equated. People like Popper and Norman Macbeth represent anti-evolutionary perspectives. Many people within orthodox science present revisionist and critical perspectives towards evolutionary doctrine as presented by Darwin. These are essentially scientific critics who are not so much offering a defense for an alternative (such as the Biblical explanation) as they are calling our attention to the incompleteness and tautological components in current evolutionary doctrine (especially as taught in dogmatic fashion in many schools). I would classify these anti-evolutionists as essentially scientific in spirit and essentially revisionists within the fold of orthodoxy (though some, such as some Lamarckians, get near the borders at times). But there are also anti-evolutionists who are not merely negative and who also are not creationists. An example might include the members of the Ancient Astronaut Society. Now this group does not do a very good job of presenting evidence for their case of extraterrestrial origin of mankind, but their doctrine is at least in principle amenable to empirical evidence and evaluation and that would appear to be how they want their case judged. If they were better organized and had some serious scientific membership, I might go so far as to call theirs a proto-scientific research program, but I recognize that theirs is a rather weak case, and their "spiritual leader," von Daniken, seems to be an outright charlatan. Nonetheless, I see theirs as at least presenting testable alternative hypotheses about origins and more than merely revisionist or anti-evolutionary theory. But when we come to Creationism, I find that by all the criteria I can muster, theirs is simply not a scientific enterprise at base. Creationism quotes the revisionist anti-evolutionist critics (often properly) but without really giving an alternative scientific set of conjectures to account for origins. Instead, they rely on revelation as evidence for their alternative to evolution. In this sense, unlike the revisionists and even the followers of von Daniken (many of whom seem to be sincere and impressed by exobiological work), I truly see Creationism as anti-scientific rather than protoscientific. The Creationists act as though, the demolition of current evolutionary theory (which is admittedly incomplete and changing) must result in their revelationary account being all that is left for any serious person to consider. In the final analysis, the Creationists are not trying to improve evolutionary theory by making it more accurate (even by turning it topsy-turvy), they simply want to deny the whole business and turn to scripture.

As a sociologist, I am interested in the question of how protosciences and not anti-sciences gain entry into and exit from normal science. The Creationist case study is certainly an important one, but I find the work by Nelkin and others pretty thorough. The only part of this debate I find of particular sociological concern is the way that matters have become so polarized that criticism of evolutionary theory is now most unfortunately seen as lending comfort to the Creationists. Thus, revisionist and protoscientific efforts in this area have a hard time emerging while being misviewed as anti-scientific. I find the pattern similar in other areas as well. You obviously dislike Velikovsky's theories in part because of the aid and comfort they might give to some creationists. You dislike Gauquelin's work for the aid and comfort it may give to some anti-scientific astrologers. The pattern is rather consistent. Many dislike psi research because it lends aid and comfort to those who believe in superstitious ideas like the soul and spirit. This is especially true if the protoscientific attempt is made by an exoheretic. Endoheretics seem much better tolerated; they are members of the club. Gould or Sagan can say critical things; we know which side they are really on! But it quite another for someone like Macbeth to criticize evolutionists for being tautological at times; he may be a closet creationist . . . .

In short, it is because I think there is a most important difference between anti-sciences and proto-sciences (a distinction that I constantly accuse you of not fully appreciating in your own practice of criticism), which I have not personally taken up the challenge of the creationists. I see them as based on metaphysical and other non-scientific ideas. Unlike the flat-earthers (who could, at least in theory, come up with an empirical argument for their perspective, and thus offer a protoscientific research program), the Creationists offer no empirical research program except a negative one.

Their positive program simply consists of reading the Bible more closely. Perhaps I do them an injustice, but that's the way I read what I have seen.

I turn now the matter of *ZS* policy. I do not object to taking up Creationism in *ZS*. I assume the reason I have not is largely the same as why *Skeptical Inquirer* has not devoted much space to it: I have received no contributions from authors about that subject. Perhaps you would be interested in doing a piece. I would welcome that from you. I think, like myself, most *ZS* readers perceive Creationism as a political rather than a truly scientific quarrel. Scientifically, I can see many problems with evolutionary theory. But I see no sensible scientific case *for* Creationism. Every debate arid interview I have seen with a creationist seems to ultimately boil down to their view being based on faith. Science to them is important only insofar as it can help support that faith, in this case mainly by pointing out weaknesses in the doctrine of evolutionism.

I most certainly do agree with you about the political importance of Creationism. I am very much opposed to them for their misuse of science in a negative fashion with no positive empirical research program while claiming to be a scientific endeavor. In so far as Creationism has made many evolutionary theorists have to start responding and questioning their own usually dogmatic acceptance of evolutionism, there may be some good that will come from this battle. Macbeth and others have certainly been correct in pointing out that much evolutionary doctrine is taught based on authority rather than evidence. But when it comes to Creationism claiming to be a scientific research program, suggesting a positive scientific alternative, that is (and I presume we here would agree) *real* pseudo-science. It is religion masquerading as science. Faith pretending to be based on empiricism when it is not. And, as I have said, it is really more than just false science (pseudoscience), it is ultimately anti-scientific and, given my own values, to be condemned intellectually. (Please note that I consider some pseudosciences to be a-scientific rather than anti-scientific; mere meta- physics posing as science is false, but it is not necessarily antagonistic to the rest of science. But Creationism seeks to *use* science as an instrument to produce faith through a totally negative research orientation that is ultimately against the very goals of science; it seeks to use science's anomalies to destroy rather than revise science.)

Perhaps it would be useful to make these ideas clearer within *ZS*. I can see several possible routes: (1) I could, with your permission, and if you think it would be at all useful, publish your inquiry letter and this response, in *ZS* in the letters column. It might be useful for readers to know more about where I stand on this, and they may even start writing in and develop a proper dialogue accordingly. Or (2) Perhaps you will accept my invitation to write a stimulus piece for a dialogue for *ZS*. I would send it off for comments to people both you and I chose, and you could respond in the same or a later issue. What do you say?

I might add, by the way, that it is not really right for you to characterize me as evading the issue. I have invited many people to get involved with *ZS* that would be likely to write on Creationism (e.g., Gould, Sagan, Macbeth, etc.) And I know Dorothy Nelkin knows of *ZS* and should feel welcome, too. I really have not actively avoided the issue at all, but I admit that I have not eagerly sought out coverage of it, and the reasons are pretty much as I have spelled out above.

I hope the above makes my position reasonably clear; more important, I hope that you find it not only consistent on my part but perhaps something you might end up agreeing with me about. I must stress that I do not underestimate the political importance of the Creationists and the New Fundamentalists. I can only hope that Gore Vidal is right in thinking that they are a passing fashion; and, like occultism, perhaps it is like the death pangs of a dying system we are hearing. But my strong political concern with the New Yahoos is not something I feel must necessarily be brought into *ZS*. On the other hand, it would be perfectly consistent with the militant rationalist spirit of *Skeptical Inquirer* to become more engaged in that battle than it has so far.

Cordially,
Marcello Truzzi

CC: Ray

P.S. I was just looking over my letter and yours again and was struck by certain differences in our outlook that may not be apparent to you immediately either. You want me and *ZS* to be concerned with nonfrivolous and serious issues. You see Creationism as one such issue and would see other issues like Velikovsky and UFOs as otherwise. Now I would completely agree with you that Creationism is a most serious and important controversy. But I do not see its resolution—one way or another—having much relevance for true scientific progress. Of course, I do see Creationism's attacks as impeding some good work, and if it were to become Reagan's version of Lysenkoism, it would be a severe set-back for progress. But (a) I don't think the latter would really happen, and (b) I really have great faith in the ultimate self-correction of science to the point of rejecting nonsense ultimately (just as Lysenkoism was ultimately rejected even in a totalitarian state). And perhaps I am not quite as fearful of the threat from the Fundamentalists as you are. Here I may be wrong, but I really trust the Gallup polls on American opinion more than I trust the electronic evangelists' description of it. I am outraged, but I am not yet that fearful of a return to something more like the Middle Ages (though I recognize abortion laws which I favor are very much in trouble over the next four years). But the criteria I use for my sociological, concerns are simply not such political measures of practical importance which are usually short-run concerns for science. My concern is with bringing controversies between dominant science and protosciences into full dialogue. This is not a matter of mass communications. It is a matter of discussion between elites with importance defined by non-political criteria. I recognize that *most* of what goes on about the paranormal does not constitute protoscience, and most may even truly be pseudoscience. But I am interested in preserving the baby, not the bathwater. I don't object to you and Randi taking on the charlatans and I applaud you for that. I only object when the same brush paints all. In some areas, I think you would agree. I think you would dislike seeing Bob McConnell put in the same pot with Hans Holzer, as parapsychologists. But you don't feel the same way about other protoscientists in ufology, etc. I also recognize that there are real protoscientists (like Gertrude Schmeidler) who sometimes say credulous things (as about Uri Geller). And certainly some (like Hynek) are not critical enough about other fields (as he is not of parapsychology). But, I am interested in the best work, the elite work of the marginal scientists. It too may be without merit, but only history and serious and legitimate analysis can demonstrate that. I think you are asking me to answer your questions and not the questions I am really asking. You may also question my questions, but you really have not done that analytically, only by shifting context into political or other such areas. I think my questions can be justified as reasonable ones given the current dominant questions in the sociology and philosophy of science. Your questions are really outside or tangential to those boundary questions, as I see it. In any case, I have always insisted with you that every scientist has the complete right to set his own priorities on a subjective basis and properly ignore some questions. I do not criticize you for disengagement. I have only criticized you for weak or superficial engagements once made but then left behind. You don't seem to want to let me have same latitude in picking my subjects for engagement. You see my neglect of Creationism as evasion. You seem to feel I should feel a duty to be engaged with that issue. I may eventually agree if the issues become redefined for me. Until then, I probably will steer clear more than you probably think I should. But it will not be evasion so much as preoccupation with so many other things with higher priority to me. My point in this postscript has perhaps been obscured by my vérbosity. It is: You would exclude Flat Earthers and such because of the outrageousness of the claims based on their contradiction to established theory which says they empirically should not have been possible. I insist that what empirically exists must force conformity from our theories in science. I would exclude something even so politically hot as Creationism because I see no profit from its claims in terms of helping us alter our theories constructively, I do not exclude even Flat Earthers (should they seriously apply) because they might at least potentially offer a positive research program which might push science forward (though I think we would both agree that the probability of that is very low indeed). But I see not even a low probability in the case of Creationism since it seems uninterested in a positive empirical program and turns instead to scripture.

[Woods End]　　　　　　　　　　　　　　　　　　　　　　24 July 1981
Dear Marcello,

Thanks for your letter of July 20. I suspect, from what you say, that you have not yet actually read any of the better creationist books. They offer arguments and evidence that is as strong as any in V, and just as capable of confirmation or refutation. I have often urged you to check Price's New Geology, the classic reference (V often cited Price, and the two were friendly correspondents), but since this book is hard to find, let me suggest the readily available The Genesis Flood, by Whitcomb and Morris. The foreword is by McCampbell, Ph.D., head of the geology department of a University in Louisiana. You'll find in this book a well-worked out, systematic and fairly subtle theory about the fossil record. It is much more impressive as a work of "protoscience" than any book by V.

It simply isn't true that the creationists do nothing but attack Darwin, then say "turn to the Bible" for the truth. That is the whole point behind their recent challenges. Over and over again they stress that they are *not* asking schools to teach Genesis, or any religious doctrine; they are asking the schools to teach creationist geology and biology as a rival *hypothesis*. Popper, by way, has repudiated his former careless statement about evolution (see Letters, Discover, Aug/81, p.6). I believe Sir Karl would be furious to know anyone called him an "antievolutionist" (another item for my Truzzi dictionary!).

Please don't be offended if I put forth what I think is the main reason for your lack of interest in the controversy. I think you find it all less amusing than spoon bending, UFOs, Bigfoot, and similar marvels. That is to say, less Fortean. Now there is no reason why a popularly-written debunking rag like the SI should raise its sights, but you are supposed to be engaged in higher things. The level of controversy between the creationists and evolutionists may be no higher than V, but it is certainly not lower, and it is kilometers above flat-earthery, and the topics Rogo writes books about.

I'd send you a copy of my review in the August 13 issue of *NY Review*, but my machine doesn't accommodate their large pages. You may not agree with my mathematical "realism," but you'll find my remarks about Popper and Lakatos at least provocative.

Best,
Martin

CC: Ray

[Eastern Michigan University]   28 July 1981
Dear Martin,

Re your letter of July 24th, I am most surprised to see you apparently saying that the "better creationist" books represent a positive scientific research program. Or is that what you really mean? If all you mean is that they present strong arguments that are capable, of confirmation or refutation that may well apply to the negative aspects of their criticisms of evolutionary doctrine, but it says nothing about their *positive* program. If you are arguing that the best evolutionists in fact state positive programs which are empirical and not at all based on biblical revelation, then I can only say that these would then be protoscientific research programs and not anti-scientific ones, and I would welcome their discussion in *Zetetic Scholar*. George M. Price, I presume, is deceased. I will check into *The Genesis Flood*. Can you suggest other positive creationists?

In my letter of July 20th, I distinguished anti-evolutionists from creationists. I have examined much literature from the Creationists in both California and locally (there is a center here in Ann Arbor), and none of that literature reveals a positive research program such as you seem to suggest exists. I heard numerous special programs and debates dealing with the recent California case. I watched Sagan and Gould on the Donahue Show. Everything I have seen except your comments indicates an absence of visible proponents such as you describe. This does not mean they do not exist, of course, but it certainly is not the position of the Creationists bringing the recent suit. They do indeed pretend to espouse an alternative hypothesis, but their alternative hypothesis is really a negative one in any scientific sense and only becomes positive when they turn to scripture. (By the way, I hope you note the irony in the fact that you seem to be arguing that the best Creationists represent a protoscientific position while I am arguing they are pseudoscientists—not our usual positions relative to such matters.) But you may be correct in stating that I am overlooking the "best" exponents, and I certainly don't wish to do that. So, I welcome any you can persuade to write for *Zetetic Scholar*, and I again extend my offer to let you have space in *ZS* to present your views.

In regard to Popper and evolution, I have seen the *Discover* letter (and the ones in *New Scientist* and in *Science*) and stick to my description of Popper as an ally of the anti-evolutionists. It is not merely a matter of his early comments about evolution being unfalsifiable (a position I know he now rejects); Popper endorsed Macbeth's book and its antievolutionist arguments and his last writings in his book with Eccles say many critical things about evolutionary ideas. He certainly is not a Creationist, and he is not an *Anti*evolutionist in their sense, but he is certainly a sharp critic of evolutionary dogmatism and fits the description I gave of those "antievolutionists" I described in my complete paragraph as revisionists within the fold of orthodoxy. I think my portrait of Popper is amply borne out by his writings.

I never said that the level of controversy between the Creationists and evolutionists was lower than that between Velikovsky and his critics. I only argue that it is qualitatively different in the analytic terms I specified as being my criterion for what I included in *ZS*. If that qualitative distinction is false, as you suggest it is, then by all means bring on the "best" of the Creationists.

Finally, I should mention something I forgot about. When I was editing *The Zetetic*, I got someone to review Wysong's book *The Creation-Evolution Controversy* and also obtained a long reply, which I intended to publish by Wysong—in which I think he made many good points. After I left *The Zetetic*, Frazier published a much edited version of Wysong's reply (about which I privately complained to Frazier, by the way). So, as a matter of fact, I *have* tried to bring Creationism into discussion in my journals (and I view *ZS* as really a continuation of my early newsletter and journal which I edited for CSICOP since I have changed my own views relatively little along the way). (By the way, Wysong apparently did not consider Price a Creationist since he does not include him in his very extensive bibliographies.)

So, in short, I welcome any Creationists who wish to argue for a positive scientific research program (in practice and not merely in rhetoric). I welcome your suggestions as to whom I should invite to participate in a dialogue for *ZS* and again invite you to be a prime participant.

As for myself, you are certainly largely correct in your view that I am personally little attracted to the Creationist controversy given my tastes for the bizarre. But I certainly recognize the relevance of the topic, especially if you are correct about the "best" proponents being truly protoscientific. The main reason, then, that I do not take more active part is because I really do not consider myself an expert in such matters; and those that are either have not seen fit to write for *ZS* or have ample opportunity to make their arguments in other journals.

I get *NY Review of Books* so will see your August 13th review. I look forward to reading it. (I think I got the issue already—is it predated?—but haven't gotten to it yet. I seem to recall your name on the cover.)

Thanks for the copy of *Origins*. Glancing through it, it appears to be most negative on standard evolutionism. I see little sign of any positive research program so it appears to be antievolutionary in at least the revisionist sense, but I get the impression that it may actually be a creationist journal since I see little indication that it seeks to reconstruct evolutionary doctrine. Its thrust may (on the whole) be more anti- than protoscientific. Individual articles, however, may simply be critical without reflecting a creationist attitude on the part of the author. I really know little about this journal and have not had time to carefully read through it. This particular issue strikes me as mainly revisionist rather than truly anti-scientific since I see no sign of overt reliance on revelation. A protoscience needs to be working with new variables or new relations between variables. This journal seems mainly to criticize the standard relations and variables without positing a new scientific theory of origins. So, I see it as more revisionist than either protoscientific or anti-scientific. I would have to examine a fuller set of the issues to reach any real conclusions about the matter.

Best,
Marcello Truzzi

CC: Ray

P.S. In reading over this letter and my last one plus yours, I wonder if perhaps I have not been clear about what I mean when I say a positive scientific research program (which I think Creationists lack). Your letter emphasizes that they have good or sophisticated arguments. I grant that, but contend these are negative arguments citing problems with evolutionary ideas. What explanatory framework (other than the will of God) do the creationists present to account for the sudden origin of man? What scientific theory of origin is put forward as their alternative hypothesis? It might be possible to show evidence of a sudden creation, but what non-theological explanatory ideas do they put forward? I would see a positive scientific research program as one that would offer such explanations. In this sense, at least the Ancient Astronaut promoters present an explanatory framework that I think is lacking for the Creationists I have read. (But I have not read all of them and am anxious to read those "best creationists" you make reference to whom I presume do present such a positive program.)

Dear Marcello,

It's hard for me to recall your various terminological distinctions. I certainly regard all the creationists as "pseudoscientists," and think that Price is almost as funny a crank as V. But Price did put forth a theory to explain the fossil record that is certainly capable of empirical testing, and in fact his followers (who often do not refer to Price) claim to be doing just that. I will cite only two testable aspects of his theory:

1. The contention that there are many areas where such a large number of prehistoric animal bones are found in one spot that the most reasonable explanation is sudden death by drowning. V cites one such example, (which he takes from Price) in support of *his* catastrophe theories. Incidentally, for Price the Flood was a tidal wave that swept around the earth, much like the floods caused by V's Venus. V, however, based his Noachian Flood on a cloud of water vapor ejected from Saturn. So far as I know, neither of *these* hypotheses have been subjected to any research program.

2. The Priceans contend, positively, that a study of the fossil-bearing strata show a random distribution of fossils, not an evolutionary progression. They keep locating spots where fossils are in an "upside down" order. They argue that there is no physical evidence to support the thrusts and folds that paleontologists invoke to explain these anomalies. Now this is clearly a testable hypothesis. There are various physical ways of deciding if there has been an overturned fold. The search for upside-down strata is certainly a low-level research program. If the distribution of fossils is indeed random in time, then it is strong evidence in support of the theory that they were all buried in sediment at one time, not over millions of years.

I am not arguing that the creationists have *much* of a research program, but only that it is as good a program as any proposed by followers of V. If V is worthy of your attention, so are the creationists. At least they have a number of centers where they *claim* to be doing positive research, whereas I know of *no* ongoing research whatever on V. Add to this the enormous social and political repercussions of creationist propaganda, I feel they deserve your attention more than V's followers who are no more than journalists writing about their hero.

Let me add this thought. The belief in a creation no more makes the fossil/flood theory untestable than V's belief in the creation of the universe makes *his* theories untestable, or a Catholic astronomer's belief that God caused the Big Bang makes his theory of quasars untestable.

At the moment, however, I am more interested in pursuing your strange tendency to speak of Popper as an "anti-evolutionist." Is this your own private term for him, or is there some critic of Popper I don't know about who refers to him this way? I was puzzled by your reference to Norman Macbeth. Who is he? Suddenly I recalled that he was a lawyer who once wrote a little book attacking Darwin's way of explaining evolution, and who provided grist for an incredibly stupid article in *Harper's* about five years ago called "Darwin's Mistake." Is that where you encountered him, or do you have his book?

In any case, we have to make a clear distinction between revisionists of Darwin, and anti-evolutionists. Darwin was a Lamarckian (he even had a well-worked-out theory of how acquired characteristics were carried by particles in the blood stream!). Since Mendel, all "Darwinians" are revisionists in the sense that they support the mutation approach as versus Lamarck. Today, the leading revisionists are the supporters of "punctuated equilibrium." This includes Gould himself. But this is a technical debate *within* evolutionary theory. No one who supports the "jump" theory of evolution would dream of calling himself an "anti-evolutionist."

So what can you mean when you apply this term to Popper? I have read all of Popper's main books, own them, and have half-a-dozen books about Popper. He has always called himself an evolutionist. In his autobiography (in *The Philosophy of KP*) he dubs himself a neo-Darwinian, and explains at length how he has applied Darwin's trial-and-error approach to the growth of science, and even to the history of culture, which he prefers to call World 3. Chapter 4 in Bryan Magee's little book, *Karl Popper* (1973) is titled "Popper's Evolutionism and His Theory of World 3." And now I come upon Popper's own book, *Objective Knowledge: An Evolutionary Approach*.

Could you kindly spell out what you mean when say Popper has an "anti-evolutionary" approach? If you mean nothing more than that he is critical of dogmatism about the *process* of evolution and how it works, then that makes Gould an anti-evolutionist, and me, and just about everybody who is doing current work on evolution theory.

Unless you have some very recent Popper pronouncements up your sleeve, I suspect this is another example of what I call Truzzese, comparable to your attempt a few years ago to persuade me that Ray and I should call ourselves "para-psychologists" because we study parapsychology. You say in your last letter that Popper's "anti-evolutionary" views are "amply borne out by his writings." What writings? Unless we can agree on simple terms like this, there is little hope we can agree on more complicated terms like "research program."

Incidentally, your remarks on Price not being cited by Wysong as a "creationist" makes me wonder if you know who Price was. As an introduction to the man, see the chapter about him in my *Fads and Fallacies*. I find him more interesting a personality than V, and much less egotistical. I will have more to say about him in my *Science: Good, Bad and Bogus*. My friend Ronald Numbers, at U of Wisconsin, is working on a book about Price. As I have said before, V was one of his admirers, and made good use of his *New Geology*. I am hoping that the new books on V, edited by Rose, will let us in on what V thought about the fossil record, and its origin.

Best,
Martin

CC: Ray

[Eastern Michigan University]   26 August 1981
Dear Martin,

I have been forced to delay answering your last letter of August 1 because (1) I have been swamped with higher priority items this month, and (2) I have been trying to run down my missing copy of Norman Macbeth's *Darwin Retried* so I could exactly quote you Karl Popper's very complimentary blurb written on the front of the paperback edition. Unfortunately, I don't seem to be able to find who borrowed the book from me (though the book may simply be misplaced within my rather large and disorganized library).

Let's take up the first issue first: my use of the term antievolutionist and its relation to Popper. I believe my first reference to "antievolutionists" was in my letter to you of July 20th. I specifically used the term in the context of trying to differentiate those who were antagonistic to standard evolutionary theory from those who were biblical fundamentalists or creationists. I specifically spoke of revisionists as being among these. Obviously, people like Gould (and presumably yourself) are such revisionists. Obviously, Gould and you would not want to be called antievolutionists in a general sense. I would never call you that in a public context for that reason. The term is normally full of political content. I was using it in a loose sense as defined in my letter and that mainly to indicate that there were those perceived by some as being "against" evolution and Darwinian ideas (Macbeth is a prime example) who were not on the side of the fundamentalist creationists either but who were frequently quoted by the creationists as support for their notions. In formal discourse, many of these would better be referred to as evolutionary revisionists. But please note that I spoke of these revisionists in the same paragraph of my July 20th letter, so I think essential meaning should have been clear.

Let me turn now to the matter of Popper. First of all, I am well aware that Popper does not consider himself an anti-evolutionist. However, his criticisms of evolutionary theory are in a very different and more fundamental category than those of simple revisionists like Gould, etc. His complaints concerned the fundamental scientific character of the enterprise. Until very recently, Popper clearly considered evolutionary theory a metaphysical research program. This was not simply a minor phrase tossed out somewhere as some of his apologists now make it seem. Nor is his new position on all this completely clear either. First. let me comment on the early Popper statements. The clearest negative statement that I know of (and I am no expert on Popper) is in *Unended Quest*, which also appears in the first section of the Schlipp volume on Popper. Section 37 titled "Darwinism as a Metaphysical Research Programme" is hardly a short glib comment. (This is not to imply that *you* ever termed it a short glib statement. I refer here to some of the apologists for Popper that have written on this controversy in *New Scientist* this last year. Some Popper defenders have made it sound like it was carelessness on Popper's part.) He says he speaks of not merely Darwinism but also of "neo-Darwinism" (p. 170). He says "It is metaphysical because it is not testable" (p. 171). But I assume you already know all this and would agree about the early Popper statements. Popper seems to have since changed his mind (as quoted by Hans Zeisel in Zeisel's letter to SCIENCE (May 22, 1981). Popper now says that evolutionism now *may be* formulated so as to be testable. He also says it "turns out to be not strictly universal. There seem to be exceptions...." (p. 873 in SCIENCE). Obviously, Popper has changed his mind. Even now please note that he does so quite cautiously. He says "may be formulated" and not that it usually is or has been so formulated. Also he cites that exceptions exist anyway.

Now as I see it the key difference between us re Popper concerns not the most recent Popper as found in this letter in SCIENCE. This represents Popper$_2$ and the issue concerns whether Popper$_1$ was reasonably termed an antievolutionist by me even in the loose sense I used in my letter to you of June 20th. It seems apparent to me that a good case can be made based on Popper's major writings. It is also indirectly clear since Zeisel found it necessary to cite a personal recent letter to him from Popper in which Popper himself refers to his new view as "a recantation."

In *Objective Knowledge*, Popper says (p. 69) "In other words, a considerable part of Darwinism is not of the nature of an empirical theory, but is a *logical truism*." Somewhat tangentially but not without relevance, Popper has repeatedly argued against any law of evolution (as opposed to a theory of evolution) which has also resulted in some terming him an anti-evolutionist within the broader debate that includes social evolutionism (see his *Conjectures And Refutations*, p. 340: "There exists no law of evolution, only the historical fact that plants and animals change, or more precisely, that they have changed.").

[You cite Bryan Magee's excellent little book on Popper for its title of a chapter "Popper's Evolutionism and World 3." That really is not relevant since this mainly discusses Popper's view of what others have called the "superorganic" or culture. It does not discuss the scientific status of natural selection theory in biology.]

In *The Self and Its Brain*, Popper closes the book with "I am somewhat critical of evolutionary theory and of its explanatory power, and especially of the explanatory power of natural selection." (p. 566).

All of this does not of course make Popper an anti-evolutionist in the full sense. But until his recent recantation, Popper's criticisms of natural selection and evolutionary theory were not simply revisionist criticisms. They were fundamental. Given his criteria of nonfalsifiability as the definition of a pseudoscience, the nontestability and tautological character of evolutionary theory (as he consistently called to our attention about it), would make this "metaphysical research programme" termed a "pseudoscientific" effort by strong antievolutionists who read Popper. Popper never terms evolutionary theory pseudoscientific, as he does Marxism and psychoanalysis, but the logical implication was clearly seen by Creationists and others who read him. And those who read Popper now can rightly call attention to the fact that astrology, Marxism and psychoanalysis (all pseudosciences and nonfalsifiable for Popper) "may be so formulated" as to also be testable. In fact, many neo-astrologers, neo-Marxists, and neo-Freudians do now present falsifiable and testable versions of their one-time pseudoscientific originals.

Unfortunately, I can not cite you the exact quote from Popper on Macbeth 's book. But I assure you that on that book's jacket (paperback edition) he has a very positive quotation about Macbeth's book and Macbeth's book is perhaps the best attack on evolutionary arguments by a non-religionist that I have seen. I urge you to get hold of it. (I just happened to speak to Donald T. Campbell, the psychologist and Popperian—he is in the Schlipp volume—and Don thinks highly of the Macbeth book too, though he is certainly also critical of it.)

In sum: Popper has been a sharp critic of evolutionary theory on not only empirical but on logical grounds. He has allowed himself to be connected very positively with Macbeth's clearly anti-evolutionary book (*Darwin Retried*). He is consistently cited by antievolutionists as presenting arguments that favor their position. (See the letter from the Creationists in the same issue of SCIENCE as Zeisel's letter.) All this does not make him an antievolutionist in the strong sense. But it certainly makes him more than a mere revisionist like Gould and many others who call themselves evolutionists.

In the final analysis, I think Popper is actually a bit "schizophrenic" in his writings about evolution. His early writings should have forced him to forthrightly conclude that evolutionary theory was pseudoscientific. But he never went that far. He obviously found that distasteful. It was one thing to describe astrology as such, but he could not go that far with evolutionary theory. At the same time, he did not feel it necessary to speak of astrology as merely a "metaphysical research programme." There he could be blunt.

I also think that Popper's inconsistency shows up in his anti-materialistic sections of *The Self and Its Brain*. Popper wants consciousness to play an important role in evolution and this also shows up in his World 3 discussions. I think there is much ambivalence in Popper about evolutionary theory and its proper place and role. He obviously is in many ways a neo-Spencerian of sorts, but at other times he is a sharp critic of all evolutionary thought on the most fundamental level.

I turn now to some of your lesser points in your letter of August 1.

You are quite incorrect about your characterization of the pro-Velikovskians, as not currently involved in a research program. I can only urge you to read *SIS Review* and *Kronos* to see this is the case.

I have been trying to find out more about the alleged "positive research program" of the Creationists. So far I can only find rhetoric and not concrete studies. I will write to the San Diego group to follow this up.

I admit to knowing little about Price but I certainly did read your section on him in *Fads and Fallacies*. I confess to knowing no one who currently is working with his ideas as a research program. Please tell me who is and I certainly check into that. On the other hand, you seem to be quite unfamiliar with Wysong which is all the more remarkable since Wysong's book is currently (and for the last several years) been a central document among the Creationists attacks on evolutionism. (You might also look at *The Catastrophist Geologist* for more current work stemming from Velikovsky. You can also get an up-date on Velikovsky's work and influence from writing to Mrs. Velikovsky who has a woman in charge of a kind of bulletin about new developments.)

You may be correct that the Creationists have as good a research program as the Velikovskians. I look for the evidence you might send me of their positive research program efforts and will try to get such myself. But the bottom line is that no Creationists have submitted anything to *ZS* whereas Velikovskians have. If I get such a submission from the Creationists, I will judge it on its merits. *And most important*, I again invite you to write something on the Creationists for *ZS* if you would like to see *ZS* get into this area. If the burden is on the proponent, you seem to be the proponent in this case and I would welcome your views in *ZS*.

Best,
Marcello

CC: Ray

P.S. You refer to my Truzzese in wanting to have you and other negative researchers and analysts of psi studies also be called parapsychologists as though this were original to me. I remind you that (1) the term as defined by Rhine did not include belief in psi and explicitly did so, (2) I have met many parapsychologists in the PA who are skeptics like Ray and me, (3) Hansel was a member of the Parapsychological Association. See also my article on Kurtz's views for more arguments on this point.
P.P.S. See, Rawlins isn't the only one who writes in all the margins.

[Woods End]   30 August 1981
Dear Marcello,

Here we are again, arguing not over anything substantive, but only over how to use words. In this case, the two words at issue are "research" and "anti-evolution."

First, research. To me, research involves scientists who are trained in a certain discipline, engaged in active work to advance knowledge in that area. I know all about *SIS Review* and *Kronos*. There is not an astronomer on the staff of either magazine. (I assume that V's major contribution was in astronomy, and that all his work rises or falls with his central astronomical scenario.)

I am glancing now through copy I saved of *SIS Review*, Spring 1980. It features a tribute to V by 17 people. Not one is an astronomer. Books are reviewed. Letters are published. Thornhill, a computer man, writes about Venus. Not a shred of personal research; just an examination of reports by other scientists in which he finds little tidbits that he thinks supports V. I skip over the archeological stuff because it is peripheral. On p. 109 at last we encounter a paper by an astronomer, Thomas Gold. But of course he did not write it for *SIS Review*. They picked it up from *Science* and reprint it because it deals with electrical currents on Io (as though this somehow supports V's wild views about the electrical fields that surround planets and act like gravitational fields). Ditto for the paper by astrophysicist Drobyshevski. It is picked up from *Nature*. Not a line in the issue indicates that anything has been done by anyone connected with the staff that remotely could be called research on V's astronomy.

Nor is there any evidence that anyone connected with *Kronos* is doing any astronomical research. It reprints articles by V. It prints pieces by such nonscientists as Rose. Most of it concerns archeological and historical speculations. Whenever you encounter a piece by an astronomer it is reprinted from an establishment science journal. Same for geology. In brief both magazines are edited by a group of V buffs who comb the current science scene from anything that they think supports V or damns his critics. I do not call these "research programs."

Now the creationists are not much better, but they are enough better to justify saying they are doing "research." Of a sort. For one thing, professional geologists and biologists are involved. In the second place, they have established what, they consider research centers, e.g.:

- Creation-Science Research Center,

- The Institute of Creation Research,

- The Creation Research Society (Ann Arbor). They publish the *Creation Science Quarterly*,

- Geoscience Research Institute,

and some 20 or 30 smaller groups with similar names. If you will obtain a copy of the 500-page work, *The Genesis Flood*, which I recommended to you before, you will capture the flavor of their efforts.

You understand, of course, that I consider this stupid, pseudo-scientific research. But it does involve professionally-trained geologists who do occasionally go out into the field and do what they believe to be research. It is singularly unimpressive. "Research program" is obviously an extremely fuzzy word. Personally, I would prefer not to apply it either to the creationist movement or the V cult. But my main point is that if you are going to maintain, as you do, that V's followers have a "research program," then I maintain that you have to say that the creationists have a "research program" that is *more* impressive, not less.

Now to "anti-evolution." Let's start with some kindergarten basics:

1. Evolution is the theory, now confirmed to an extremely high degree, that millions of years ago an extremely simple form of life appeared on earth—no one knows how—that was capable of self-replication. Over the millennia it evolved, descendant by descendant, until it produced all the present living forms, including humans. Fossils are the records of life of past ages, when this process was underway.

2. A person who believes the above is an "evolutionist." A person who disbelieves it is an "anti-evolutionist." In common discourse, an anti-evolutionist is a creationist. (There are some minority opinions that do not fit this dichotomy, but they are so minor that we can ignore them.)

3. Evolutionists disagree about the mechanisms by which evolution took place.

4. Darwin was a Lamarckian. Since the rise of genetics (about which Darwin knew nothing), all evolutionists today are Mendelians in the sense

that they believe that evolution is based on random mutations. They disagree over what causes a mutation, the exact nature, of a mutation, and how mutations interact with environmental factors. At the moment, the chief dispute is between the gradualists and the punctuationists over, whether evolution is best regarded as slow and gradual, or whether it is punctuated by jumps (a jump being some 50,000 years).

5. No scientist, philosopher, or any informed person known to me uses the term "anti-evolutionist" for any scientist who accepts evolution but holds a minority opinion about the mechanisms of evolution. In particular, I have never heard anyone call Popper an anti-evolutionist.

6. Macbeth's book is not a book attacking evolution. When you write that his book is "the best attack on evolutionary arguments by a non-religionist that I have ever seen," I could hardly believe what I read. Macbeth does *not* attack evolution. He attacks a certain Darwinian theory as to the *mechanisms* of evolution. On the jacket Popper is quoted as saying: "I regard the book as a really important contribution to the debate." I am quoting from Bethell's *Harper's* article. I have no quarrel with this quote, nor even with Macbeth's book, but you are making the same verbal obfuscatory mistake that Bethell made. You are not making a distinction between "evolution" and "Darwinian." The latter itself is a fuzzy word, because it can be taken in so many ways as to be useless. But "evolution" is sufficiently sharp to be used in the way I defined it above. Macbeth is critical of certain views about how evolution *operates*. Popper has been similarly critical. All the revisionists are similarly critical. But all of them are "evolutionists". Not one of them is an "anti-evolutionist."

And now for the heart of the matter. Like Humpty Dumpty you enjoy making words mean whatever you want them to mean, with little respect for common usage. What do you gain by this? Surely it spreads only confusion. Everyone, assumes you mean what everybody else means by a word, and this produces vast misinterpretations, which then necessitate elaborate explanations on your part. You may enjoy this sort of verbal fencing, but what a waste of time for everybody else!

There is a section in Mill's classic book on logic that is headed: "Evil consequences of casting off any portion of the customary connotation of words." Peirce was another philosopher who stressed what he called the "ethics" of terminology. Everything that you wrote in your letter about Popper's views on the *mechanism* of evolution, and the extent to which the theory is falsifiable, and so on, is simply irrelevant. Popper was an evolutionist. He believed in the theory. He called himself a neo-Darwinian. There is no way that you can justify calling him an "anti-evolutionist" unless you insist that everybody adopt your private vocabulary.

And it is this same curious propensity of yours to redefine words that was behind our early conflict over "parapsychology." Of course if one wishes one can define a parapsychologist as anyone who studies the claims of parapsychology. But that is not how the word is commonly used. I recall Ray, in my presence, saying to you in no uncertain terms that he did not want to be called a parapsychologist. Hansel would never dream of calling himself that. In a trivial sense, of course both Ray and Hansel are parapsychologists. But why try to twist common usage so flagrantly? Ray has recently been studying the history of phrenology. Should we therefore call him a phrenologist? Come, Marcello, let's stop playing these childish word games. In common usage a parapsychologist is a person who believes in, and does research in, the field of psi phenomena, An anti-evolutionist, in common usage, is a person who does not believe that life evolved.

I have no desire to write about the creationist movement. If I did, it would be a debunking article of the sort you deplore. But if you wish to continue in the tradition you have been following, and deal with current conflicts between orthodox science and off-beat science, I should think you would wish to devote an issue to a dialog between the Creationists and the evolutionists. If you ask Henry Morris, for example, to do an article on the scientific evidence for creationism (as distinct from Biblical revelation), I imagine he would be happy to do it. (He is the director of the Inst. for Creation Research, and author of many books on "scientific creationism", including the book I cited above.) You could have someone do a piece on the other side, and allow each the right of rebuttal. You would thus be grappling with a conflict of enormous current sociological import, far more significant than the bizarre little pockets of Forteanism that appeal to you personally.

It is your reluctance to get into these broader issues that continues to keep your periodical on a frivolous level; to which must be added that it has also become an organ of revenge. To use the magazine to fight personal vendettas against members of the committee you dislike, and to keep focusing on such trivialities as V, Gauquelin, Bigfoot, and so on, may be fun for you, but in my opinion it is unworthy of your intelligence and talents.

You express surprise that I do not know about Wysong. You are right. You asked me to tell you of someone working with Price's ideas. All the work at the Institute of Creation Research is based on Price. I enclose a page of an old catalog (several years old, that is) and have circled a documentary film that is based squarely on Price's central contention. Now in return please tell me who Wysong is, and who published his book, and I will get a copy.

Best,
Martin

[Page from a creationists book catalog]
_____

[Eastern Michigan University]   2 September 1981
Dear Martin,

    Just received your substantial letter of August 30th.

    The first of your letter dealing with evolutionism-creationism deserves
a proper response which time now constrains me from giving it, but I was
greatly disturbed by your next to last paragraph and wanted to immediately
get my reactions off to you. So, here goes.

    I can understand your remark that my lack of concern in *ZS* with "these
broader issues" keeps *ZS* on a "frivolous level." To a degree, I might agree,
but that is largely a matter of quite relative opinion. Others may agree or
disagree. But your next comment really disturbs me for what I view as its
incorrectness and unfairness in light of what should be the obvious evidence.
You say that *ZS* "has also become an organ of revenge," and say that I "use
the magazine to fight personal vendettas against members of the committee
you dislike." I ask you for the concrete evidence for that statement. I
think the charge is ridiculous in light of certain clear facts: (1) I have very
purposely avoided mentioning Kurtz, the Committee, and my reasons for
leaving CSICOP in the pages of *ZS*; (2) I have refrained from attacking
CSICOP, *Skeptical Inquirer*, etc. in *ZS*; (3) I have sought to involve as
many Fellows of CSICOP in *ZS* and its dialogues as I could, and I have
even invited Klass (whom I definitely do not like) to become a Consulting
Editor to *ZS*; (4) You left *ZS* because of the Velikovsky coverage—so far
as I know—rather than because of any so-called revengeful actions on my
part in *ZS*; (5) I have consistently invited Kurtz and other CSICOPers
to submit materials, join in the dialogues, etc. though they have pretty
consistently failed to accept my offers; (6) I have never refused to publish
anything sent to me by a CSICOP member for publication in *ZS*, i.e., letters
and such; (7) I have been much aware that since several of my consulting
editors were CSICOP Fellows, I should not offend them gratuitously; (8)
My nasty comments on Hansel's letter complaining about his review were
pretty much endorsed by Ray Hyman (and since reinforced by Ray's own
comments on Hansel in *Skeptical Inquirer*); (9) My negative comments in
my short book notes have been commented upon to you before and certainly
in no way are the outcome of any vendetta; (9) My only negative comments
in print about Kurtz have been in (a) my article solicited by the *Journal
of Parapsychology* (and not in *ZS* anyway), and (b) my interview in *Fate*
(solicited by Clark and Melton and really quite mild and balanced overall,
but in any case not in *ZS* and ignored by Kurtz and commented upon
positively by some CSICOP members at the time).

It seems to me that you are accusing me of being revengeful and fighting a personal vendetta in *ZS* when I am clearly trying to be fair-minded and responsible towards both sides. I don't claim I always succeed in being completely fair. But I assure you that I get more complaints about my not being harsher on Kurtz & Co. from readers than I get the other way around. In fact, so far you are the only person to suggest that I have been using *ZS* for revenge or a vendetta. If you really mean this, I don't see it at all and insist that you document your allegation.

Now, Martin, I am only human, and I am no saint. My feelings were hurt by what happened between me and Kurtz. I admit to some joy in seeing Kurtz having to defend himself now. But none of this is the result of *ZS*. I have scrupulously tried to keep my personal feelings pretty much out of *ZS*. I have not even gotten into any of the dialogues. Perhaps you see such revenge in my editorials somehow, yet I think I have kept these on a pretty high level. Certainly, Ray and other CSICOPers on my editorial board have never complained to me about them even if some of my statements might indirectly have implications for CSICOP.

Perhaps you see *ZS* in general as a revenge act. I remind you that (a) *The Zetetic* existed before the CSICOP, and (b) I told CSICOP while still on the Executive Council that I was going to start up something like *ZS* when I left the editorship of *The Zetetic* taken over by Frazier. In fact, the Council then wished me good luck with my new effort. It was only after Kurtz denied me a copy of the mailing list that the troubles between Kurtz and me really emerged.

Rawlins does indeed seem after revenge. And I enjoy watching him get it. But I have never really actively sought revenge myself in any direct way. I did, of course, hope to eventually show that the *ZS* way was the right scientific way to go in dealing with paranormal claims. But does that constitute revenge or a vendetta? Surely if I had been after revenge I would have done things very differently. I would have started a rival organization up, immediately after leaving the committee. I would have myself spoken of my departure as a schism (as did the *Science* writer). I would have openly attacked Paul and announced my departure at one of the CSICOP press conferences, I did none of these things. In fact, I pretty much went off quietly and licked my wounds.

I am absolutely convinced that good science can come only of bringing together proponents and critics in public dialogues. I want even Kurtz and those I think most ill of to come forward into such dialogues where we all can see how they treat and present evidence. I have great faith that, good science publicly done will reveal the shoddy critics as well as the incompetent proponents. Is such a desire to show incompetence on both sides your idea of my revenge?

I guess what bothers me is that I have indeed felt the desire for revenge but feel that I have rather scrupulously tried to keep such negative elements from diminishing my constructive orientation in *ZS*. I have cheered myself by believing that the truth about Kurtz and Co. would simply emerge through the due process of scientific investigation. By placing my faith in the scientific system, I have hoped for a final just outcome. Is that a vendetta? Can't you see that if I had really been out after Kurtz's blood I would have been able to do so much more in the way of revenge behavior than I have?

I have made my views known to you and others in my correspondence, but that again has nothing to do with the pages of *ZS*. I have come to pretty openly dislike Klass and Kurtz, but who else am I supposed to be seeking revenge against? I think I am on pretty good terms with Frazier, and I certainly hope I remain so with you and Randi.

I really think you confuse my search for fair-play with my limited desire for revenge. Or do you think I earlier had no desire for fair-play (was an active debunker) and moved into the middle position because of need for revenge? If that is what you somehow mean, if the desire for revenge has increased my openness to the side of the proponents so that I will now give them a fair hearing which I would not do earlier, then I can only pray that such a need for revenge strikes all my fellow critics including yourself.

I will write you later as regards the evolutionism matters.

Zetetically,
Marcello Truzzi

CC: Ray Hyman

P.S. Doesn't it strike you as odd that Rawlins (at one time my sharpest critic on CSICOP and second only to Kurtz as the likely target of any revenge I might seek) is also so critical of Kurtz? It seems clear to me that even though Rawlins and I are on opposite sides when it comes to our view of the need to mock and debunk the paranormalists (Dennis is actually, very close to your and Randi's position on this issue), we both seem greatly concerned about due process being followed by any scientific committee such as Kurtz claims CSICOP is. Both Rawlins and I are very idealistic and optimistic about science being a self-correcting system if only given a good chance to operate properly. Kurtz seems to see science as a political battle to be won by rhetoric rather than evidence. Your apparent cynicism can contribute towards people like Kurtz gaining power in the scientific community and thereby producing ultimately anti-scientific results such as we seem to have gotten in the Kurtz tests of Gauquelin's conjectures.

P.P.S. From what I have been able to learn, Sidney Hook did not support

Reagan this last election, but Kurtz seems to have done so quite openly. I still await tangible proof of this latter statement from Bette Chambers or someone in Buffalo, but I have heard this from several disconnected sources now.

P.P.P.S. Totally aside from any of the above, let me throw some particularly interesting references (for different reasons) your way:

- Andrew Pickering, "The Hunting of the Quark," *Isis*, 72 (1981), 216-236.

- Nicholas S. Thompson, "Toward a Falsifiable Theory of Evolution," *Perspectives in Ethology*, vol. 4 (1981), Chapter 3, (Edited by P. P. G. Bateson and P.H. Klepfer for Plenum.) Pp. 51-73.

- Irving Kirsch, "Demonology and the Rise of Science: An Example of the Misperception of Historical Data," *Journal of the History of the Behavioral Sciences*, 14 (1978), 149-157. (Very relevant to Kurtz's silly historical arguments about science versus the occult and irrational.).

- Eugen Weber, "Fairies and Hard Facts: The Reality of Folktales," *Journal Of The History Of Ideas*, 42 (1980), 91-113. (A very nice article on the positive functions of belief in fairies in an earlier age.)

[Woods End]   8 September 1981
Dear Marcello,

I hope you will soon respond to my claim that you are not using words responsibly when you speak of Popper as an "anti-evolutionist" (biological sense) and in contending that followers of V have anything that deserves to be called a "research program."

I was wrong to say that *ZS* has become an organ for a vendetta against CSICOP. This was careless language. What I should have said, and really meant, was that *you* have been conducting a vendetta against the Committee, and I see *ZS* as one part of this total behavior pattern.

Webster's Collegiate defines vendetta as: "a prolonged feud marked by bitter hostility."

Let me count some ways:

1. Your two-part interview in *Fate*. You checked and approved the text, and distributed copies to friends. Part 1 was headed: " ...a group of Humanist scholars marshal forces to save civilization—and, woe to those who get in their way." Granted you didn't write that, but you certainly could have guessed how *Fate* would handle it. The heading of Part 2 calls

the Committee's efforts "a bust." You say: "...the Committee's main impact has been in publicizing itself." Clark calls the Committee a "thing of the past," and says its "days as a media event have passed." It is. he adds, "little more than the committed talking only to the committed." "Nobody is paying much attention to it anymore."

Truzzi: "I think, you're quite right—a dissipation is taking place." You go on. The media, you say, is not only recognizing the Committee as made up of dogmatic advocates—they now see them as "not even very interesting advocates." (I assume you will review my new book as "not very interesting.")

The entire thrust of your interview is a strong damning of the Committee as consisting of bad guys in contrast to the open-minded, good-guy approach of you and your two interviewers. You even include a defense of *Fate*'s advertising on the grounds that even fortune-tellers do more good than harm!

2. In the *Chicago Tribune* interview you told Lyons that the Committee was a knee-jerk mouthpiece for orthodox science, acting as judge and jury. You spoke of its inflexibility as choking off legitimate research, You accused Kurtz of not understanding that pseudoscience is on a spectrum, and you cited (of all people!) Hynek as an example of someone to be admired for his "seriousness" and "caution."

3. In your lecture at the Clever Hans Conference (the book is now published) you pick up Hyman's reference to the "hit men" who discredit unfairly but effectively the proponents of the paranormal. You say that magicians who do mental acts hold the debunking magicians (e.g., Houdini and Randi) in "great disdain." It is true you mention no names here, but your correspondence leaves no doubt that you consider Randi, Klass and myself the principal hit men of the Committee.

4. In lectures and papers you seldom fail to mention your break with the Committee, citing the *Fate* articles as the basic reference, and portraying us as enemies of true science.

5. You have since made friends with a large number of proponents of the paranormal, and with the journalists who defend then. In your wide correspondence with these friends you constantly snipe at the Committee, and especially at Kurtz and Klass. You are not so naive as not to know that this is widely publicized in occult journals and books on the occult. Fuller never ceases to talk about your break, always portraying you as the good guy against the bad guys.

Need I cite more instances? Well, let me cite one. Last year Malcolm Dean published an unusually stupid book called *The Astrology Game*, with its 9-page bibliography of Gauquelin's great "scientific" papers. He retells the story of your break, picturing you and Rawlins as fine, honorable fellows, and lambasting Kurtz, the SI, and the Committee in general as the heavies.

6. You are planning an issue of *ZS* on the G controversy. You know full well that it will be a vigorous punch at Kurtz and Abell, and serve to fan the flames of Rawlins' vitriolic attack. I think it is this planned issue, more than any other one thing, that I consider springing from personal motives of revenge.

7. It is true, as you say, that your attacks on the Committee in the *ZS* have been between the lines. But they are clearly there, in your editorials, and above all in your constant sniping at any book by a Committee member. You have printed unfavorable remarks about books by Kammann and Marks, Hansel, Klass. This would not be so bad if they were balanced by similar blasts at books by occult journalists whom you yourself have declared, in a letter to me, to be "irresponsible." But the books by these hacks, some quite horrendous, are uniformly recommended as worth reading.

Let me sum up, I don't think Rawlins' intemperate attack, with its mixture of truths, half-truths, and lies, will do us much harm, even with your support in *ZS*. But you have the prestige of being the head of a department at a reputable university. I firmly believe that you have done the Committee more harm than any other single person, and that you will continue to do it as much future harm as you can. I think this is a reasonable interpretation of "vendetta."

You said at the meeting at which you resigned as editor that you were not the sort of person who "picks up his marbles and goes home." You not only picked up your marbles, you have been firing them at us ever since. As you surely must know, your letters to your friends are widely copied and circulated. Your ad hominem attacks on Klass, for example, who you called (in a letter to me) the "greatest" of all living "cranks," fully equal any remarks he has made about you. Do not suppose that Randi and I are unaware of what you have said to others about us. And please do not respond by defending what you consider to be the truth about what you have said; truth and falsity do not enter into the definition of "vendetta," as the word is commonly understood. As usual, the debate here is over the meaning of a word.

I am pleased to learn that you have backed down on your former statement that Sidney Hook supported Reagan vs. Jimmy Carter. It is a rare thing for you to admit a mistake. Now if I could only extract from you an admission that you have a compulsion to use words, such as "evolution" and "research program," in curious, private ways! As for Kurtz having favored Reagan over Carter, I still have no information and couldn't care less. If I remember it, I'll ask him the next time we speak.

Best
Martin

CC: Ray, Frazier, Randi

[Eastern Michigan University]   15 September 1981
Dear Martin,

I just received your letter of Sept. 6th and, especially as you have sent copies to both Randi and Ken Frazier (I expected you to send a copy to Ray since he has been involved in getting copies of our correspondence all along), feel I should respond immediately.

First off, let me comment that I will respond to your first paragraph—re Popper and evolution and Velikovsky and the "research program" related to his works—in a separate letter. I assure you I have *not* forgotten our discussion of that matter; I just need to find a bit of time for that relatively low priority item between us. I want here to comment on the rest of your letter.

Since you indicate that you were wrong in saying that *ZS* has become an organ for a vendetta against CSICOP, we can largely let that matter drop (though you quite incorrectly later seem to retract that somewhat when you bring up the matter of Curry's article and my short book reviews, both of which I will comment on later in this letter).

I think the thing that bothers me most about your letter is the clear implication that I am somehow being two-faced or hypocritical. You make reference to my letters which have been copied and sent around as though these were confidential missives now exposed and revealing of dirty-dealings somehow. That is nonsense and almost all of my letters get copied and sent to several people *by myself.* My letters are no secrets (unless—and this is most rare indeed—they are marked confidential, and I doubt that you have ever seen one of those), and I have actually gone way out of my way to circulate them. For example, I think Ray Hyman has gotten copies of almost all my letters that in any way concern CSICOP. I did this partly as self-protection because I feared that Klass or Kurtz (as they indeed have) might accuse me of positions that I do not in fact hold or express.

I indeed do have a prolonged "feud" going with Kurtz. But I would very much dispute your characterization of it as one of "bitter hostility." If Ray Hyman would call my attitude one of "bitter hostility," then I will accept that description—since he has known my views more fully than just about anyone else (Ray and I probably talk on the phone around once a month or more, usually for an hour). I reject the term "bitter" in your characterization because I think I have managed to retain quite a bit of humor throughout the whole thing. I am resentful, yes, but I think bitter goes *way* too far. I'm just not *that* angry about it all. I might add that some people have even expressed surprise to me that I am not angrier than

I am. I think Ray will recall even saying to me that if he had been in my place, he probably would have quit CSICOP earlier than I did. (Since Ray is getting a copy of this, he can correct me if I remember incorrectly.)

But at heart, I think you continue to miss the point I have been trying to make about CSICOP for a very long time. I do *not* now nor have I *ever* sought the demise of CSICOP. I seek its reform. And I see the reform as potentially going in one of two different directions. *Either* CSICOP should be openly the advocate for what I call orthodoxy (but which you and Frazier might prefer to call the general majority opinion or something like that), *or* it should truly take the kind of neutral and objective stance that its original goal statement proclaimed it was taking. Either route would be legitimate in my view. You, Martin, seem to clearly favor the first direction. In your introduction to your new book you speak of the Committee (at least indirectly named) as a "pressure group." You explicitly say that the extremes should not be reacted to on strictly rational basis, that a horselaugh is worth many syllogisms. I do not object to that position from an *advocate* body. We need that kind of balance in the spectrum of opinion that exists. But Kurtz and others have tried to consistently speak of the Committee as a *scientific* body. Kurtz has even expressed interest in becoming part of AAAS (an idea rejected by others among you, I know). If you are going to be a scientific body, you can not deal in nonrational argumentation and horselaughs. Your interest must then be in rational argument and evidence. When I was on the Committee, I wanted to take the second alternative. I knew that you preferred the first, but that was a matter of open difference and I always found you consistent. I disagreed with you but respected your open commitment to that position. But Kurtz always seemed interested in the second route as well. He was the one I disagreed with then and now. He wants to have his cake and eat it too. And I consistently have insisted that you can not do both. After leaving the CSICOP, I felt that my route was probably impossible for CSICOP, so I urged that you go the route of open advocacy, an honorable position if done openly. I *say* this in my *Fate* interview. I have consistently said that there is a need and a place for an advocate body such as I think CSICOP essentially has become. I do not condemn such advocacy. I only condemn it when it is done in the name of objectivity and impartiality that simply does not really exist. I merely want you to be lawyers *openly*, nothing else. What the hell is so wrong with that? How does that contradict *your* goals? You certainly say as much about your views in your book.

Now I do not deny a personal element is involved in my feelings about Kurtz. I only deny that these have seriously intruded in the way I conduct my efforts at doing science. As for Klass, I did not even really have anything against him when I first left the Committee. My later negative reactions to him were the result of his accusations about me after I left (I think I sent some of those letters from him to me—via xerox—to Ray). It was also well after I left that I became aware of (1) Klass's exchanges with various Ufologists whom he attacked ad hominem regularly through his so-called white papers; (2) Klass's vigilante attack on McDonald as documented in McCarthy's doctoral dissertation; (3) the crank plasma theory in Klass's first book (a position he *still* insists on; and I am told that your reaction to his first book was also quite negative); and (4) the critical literature responding to Phil's second book (which I at first thought was pretty good). The point in all this is that whatever vendetta I am supposed to have towards Phil is quite recent and is rather clearly documented in our exchanges as having been initiated by him, not me. (I think Ray might corroborate what I say here for you.)

Now let me turn to the particulars you mention in your letter.

1) I did not, in fact, see or approve the heading matter on my *Fate* interview. However, I was not really upset by it. I think you simply did not read my article carefully, I certainly am critical of the Committee. I remain so. But my criticism was of its pretended objectivity and neutrality instead of its open advocacy. I specifically (in *Fate*) say that there is a need for CSICOP as a balancing force if its advocacy will only be explicit.

I also remind you that quite a few CSICOPers were and are involved with *ZS* and I think you were around then, too (though you may already have left *ZS* at that point—it's not worth my checking the dates at this point), and not a single one of them raised any objection to my interview. In fact, several of them wrote me nice notes about it. This does not make what I said correct; but it does indicate that others did not apparently view it as full of hostile bitterness, as you seem to have.

2) The *Chicago Tribune* interview seems to have somewhat misquoted both of us. But in basic substance it was correct in its interpretation of my remarks about Kurtz. I stick by those. As for Hynek, I still view Hynek as essentially an agnostic about UFOs, so I think in that regard he is both "serious" and "cautious." I do not imply that I fully agree with Hynek. I do not agree with much in his *Fate* interview and I have told him so at a recent lunch meeting with him. But that is not relevant to the Trib article.

3) I did indeed characterize you and Randi as "hit men" in Ray's sense. I think *Ray* would agree with that characterization. I am surprised you would not. I do not think it is especially pejorative if you are *openly* that. I think that your comments in your intro to your new book openly say just that. I might add that I think you have sometimes transcended that role and done good scientific work (as with your nitinol article), but I think you usually go for the "hit" (translate as "horselaugh") rather than for the appeal to rational argument. I thought you *admitted* that. You seem to

suggest that I somehow misused Ray's term "hit man." Ask *him* if I did so.

4) Yes, I do indeed often mention my break with CSICOP. My zetetic position is in fact most easily defined by contrasting it with both advocate bodies, on either side. Besides, CSICOP is misperceived by most as a skeptical rather than denying body. I think this needs to be made clear especially in relation to the 2-sided skepticism I represent with *ZS*.

5) Sure I criticize CSICOP, but I also criticize the paranormalists. Sure Fuller praises me. He also praises Ray Hyman (and on occasion Fuller has praised *you* to me). What the hell does that prove? Guilt by association? Am I to be held responsible for those who praise me even if I publicly disagree with them about matters they choose to ignore?

You mention my praise from Malcolm Dean. Again the same thing holds. This upcoming issue of *ZS* has a most negative review of Dean's book. Does that show my bias? Dean knows I disagree with him. Can I help it if he thinks I am an honest critic and Kurtz is not.

Did you ever consider that just maybe Kurtz *is* the heavy and perhaps I *am* honest? You yourself have repeatedly pointed out that Kurtz runs the committee and that things like parliamentary procedure are irrelevant if he holds the purse strings. Yet you wince when I attack the Committee for Kurtz's actions. Dammit, Martin, we have been corresponding regularly for over 10 years. I thought we probably had more respect for one another than you seem to have for me. I have usually tried to attack Kurtz rather than the CSICOP when I could easily differentiate, but that just isn't possible most of the time. (E.g., some of you have recently claimed the Gauquelin study was not by the Committee, yet I have a press release from Kurtz stating it is a CSICOP study!)

I think what bothers me most about your letter is the implication that I am being hypocritical. I have never hidden the fact that I think you are learned and brilliant but that you are also—in my view—dead wrong in your strategy towards pseudosciences. I have told Klass *directly* that I think he is vicious and most recently wrote him that I thought him crazy. I have written Randi saying I thought he was a zealot. Yet you seem to be saying that these are things that have subtly leaked out to you.

6) You refer to my having a special *ZS* issue on the Gauquelin controversy. That is not exactly true. There will be a dialogue in *ZS* #9. I expect Abell, Kurtz, Zelen and any others (you included though you declined to participate and sent your card back saying no) to participate. The result could completely exonerate Kurtz and damn Rawlins. I don't expect that to happen, but it is possible. You refer to this as a planned issue. Nonsense. Curry wrote me months ago volunteering such a stimulus paper for a *ZS* dialogue. Should I have said "no"? I would think you might view this as a

chance for Kurtz etc. to set the record straight!

7) Your comment on my book notes was answered earlier but seems ignored by you. I condemn many books by paranormalists, too. My god, you'd think that ZS ran nothing but favorable stuff to the paranormal. I recommend you look at the issues. The new issue is almost all debunking stuff. (I might add that there is a very favorable review of Randi's book by Milbourne Christopher, too.)

(Interrupted in mid-paragraph at the office, I now finish up this letter on my better typewriter at home.)

I find your comment about my future review of your book particularly odd in light of the fact that (a) I really liked so much of it, and (b) have been recommending it to many people lately. But you seem to want to think the worst likely to emerge from me . . . In any case your remark that I "uniformly" recommend the books by hacks simply shows that you are not reading my book notes very carefully. I condemn most of them. However, where you probably would like me to say "book X is a bunch of crap," I usually just say something like "unscientific" or "unconvincing." As I have said before, the reason I have made stronger negative comments about some critics' books (who, by the way, don't seem as offended as you seem to be, e.g., I continue to exchange cordial letters with Kammann, and Randi has said nothing about the matter of my note on his book to me) is that they are supposed to be scientific works whereas most of the occult books don't even claim to be science works in the first place; so I try not to whip a dead horse. I might also add that I have by no means showed "constant sniping at every book by a Committee member." I have said positive things about books by Dan Cohen, Milbourne Christopher, L. Sprague de Camp, Eric Dingwall, Lawrence Kusche, Carl Sagan, and other CSICOP associates like William Nolen, and probably some others that don't come readily to mind. I think you are being most selective in your recall.

You say that I may "have done the Committee more harm than any other single person." If my efforts to get the Committee to either become clear advocates or start acting truly objectively towards the claims of the paranormal can be defined as doing "harm," then so be it. But I think my attitude is constructive rather than destructive. It is destructive to the power Kurtz has over the Committee. That is true. But in terms of the role the Committee can play in advancing science, I think my actions are positive (and I assure you that there are many who would commend them while not supporting the paranormal claims at all; most ZS readers seem to agree with my stance on this and see it as positive, too). Surely many might agree with me (including CSICOP Fellows) that a person who has done the *most* harm to CSICOP is Paul Kurtz. After all, it is his antics that have been criticized most by people within and without the Committee. I remain amazed that you continue to lend your reputation to Kurtz. I sometimes wonder, given your more-or-less admissions about his faults, just what it would take for you to finally disown him. Obviously, the scandals around him at the AHA and now in CSICOP cause you to defend him the more!

Re Hook and Reagan, I didn't say what you state I said (see my letter to you of Dec. 17, 1980 on this matter). I said that I had heard a rumor which I was trying to substantiate that Kurtz supported Reagan last election. I said this was ironic given Reagan's views on Creationism and astrology. I said that it might be explicable if he were simply following his mentor Sidney Hook and Hook's connection with the Hoover Institute meant that Hook was now pro-Reagan. I then recently found out that Hook was anti-Reagan, which I mentioned to you. In the meanwhile I have found that Kurtz definitely did support Reagan (according to an article-interview on him). All this really means nothing, but I only mentioned it because I thought it ironic and amusing as hell. This leads to my final item. You allege my infrequent willingness to admit I am wrong. I hope that is not true, for I think of myself as quite willing to admit errors, especially to my friends. Since my views were once much more in line with your own and now no longer are, I would think that would be a *clear* case of my saying I was wrong earlier. It seems to me that your complaint is that I *am* willing to admit I was wrong, not the other way around.

Frankly, Martin, I don't think you are cool and rational towards your differences with me, and I don't think I will change your mind about the contents of your last letter (though I really wish I could). But since you have sent copies to others, I feel I must reply to you fully here.

I continue to have great respect for your learning and intelligence, but I really don't think I can get through to you about our differences. I recently put our letters into a large notebook and was amazed to see that there must be 600 pages of correspondence between us over these last 10 years. Obviously, neither of us thinks it is a waste of time to write to the other. Obviously, we each have taken a lot of time in communicating with one another. I hope we will continue to do so, I don't claim to be absolutely right on all matters upon which we differ. I think that, within the mass of our past correspondence I must have said you were right and I was wrong about at least some point. I can only do what I am doing now: go over every point in your letter and try to respond to each as best I can. I have certainly tried not to neglect anything you have said to me. (Just as I will soon reply to our differences about the evolutionary postures by those like Popper, etc.)

The silly part of all this to me is that I get equally complaining letters from proponents of the paranormal who think I am stubborn in the other direction. (Thank god there are people in the middle who lend some support to my zeteticism!) Either I am doing something right or I am doing everything wrong, because I seem to have people who view each other as the enemy both attacking me for being soft on the other side. (This may be part of the problem, too. In general, I have not sent you copies of my critical letters to parapsychologists and proponents but have only sent you the letters criticizing the critics. So, in a way, you may have somewhat a lopsided view of my activities.)

Zetetically,
Marcello Truzzi

CC: Ray, Ken, and Randi

P.S. One last minor point that strikes me: In our correspondence on Creationism, you chided me for excluding it from *ZS*. I replied that I thought Creationism did not have a legitimate "positive scientific research program." You then came to the defense of Creationism to the degree that you claimed they did have such a program, and more of one than Velikovsky followers. I said that you may very well be right about Creationists having such a research program (I pleaded ignorance based on my admittedly limited reading), and now the debate has somehow shifted to the matter of Velikovsky's program and its status. I will reply on V in detail, but don't you find it ironic that I was more willing to dismiss Creationism as pure pseudoscience than you were?

[Eastern Michigan University]   17 September 1981
Dear Martin,

    A question about what I see as a separate matter from the issues in my last letter to you.
    You referred to Rawlins's "intemperate attack, with his mixture of truths, half-truths, and lies." I wonder if you would mind being specific as to what are the lies and half-truths. I followed the exchanges between Gauquelin and Kurtz quite closely (speaking occasionally with Abell and Ray Hyman about the controversies). My own general impression has been that intemperate as Rawlins's article is (and it is indeed a most angry piece), the basic facts are correct. If they are not so, I want to know where they are not so. I found two minor errors but neither affect his arguments. So far, the only alleged corrections of his article that I have seen consist of one matter raised by Frazier in a letter to me and the letter Klass sent to Rawlins (which for the most part is easily answered and presumably will be in print if Klass publishes his criticisms). In case you do not know it, Rawlins's article was edited from a 173 page document which I have not

seen but which I am told, by those who have, is meticulously documented. I understand that Kurtz is replying, and I look forward to reading that reply. But I still would like to know your bill of particulars when you speak of lies and half-truths.

Zetetically,
Marcello Truzzi

CC: Ray

P.S. I will seek to reply to you on the evolution matter this weekend when time should permit.

[Eastern Michigan University]   20 September 1981
Dear Martin,

This has been an extremely busy weekend. I had expected to get around to taking up our exchange re the definitions I have been using speaking of evolutionism/anti-evolutionism, etc. I need a block of time because to do it right I need to review all our correspondence on the issues. But I have just had too many higher priority items to deal with. (I think you would agree with me that these matters are not of great moment and urgency.) Anyway, I will try to reply this coming week. I have not forgotten the matter.

The reason for this note, however, concerns a matter that has begun to trouble me after I last wrote you. In your last letter you spoke of allegedly bad things I had said about you and Randi in letters you had seen circulated. Since I usually send around copies of my letters to people and generally write letters so they can be seen by any interested party, I simply assumed that you had projected something into my letters that was not there. But recent events have reminded me that in the past some clear lies were told about me by some members of the Executive Council to Rawlins (he told me of these and said they played a serious part in his conspiratorial attempts with Kurtz to scuttle me out of CSICOP, an action he says he now regrets having participated in). This suggests at least the possibility that you have been told I said something I said that I never have said. Therefore, I would appreciate it if you would tell me exactly what things I am supposed to have said about you and/or Randi that so strikes you as a sign of my hypocrisy. Since I hold you and Randi in high esteem even though we clearly differ on some important matters, I am anxious to clear this up if it is possible to do so. Perhaps you have already answered my last letter with a bill of particulars. If so, I will respond to that. The important thing to me is that you realize that I am not aware of any inconsistencies in my characterizations of either you or Randi to anyone. Why in the world would I do otherwise? I don't see that I have anything to gain from inconsistencies. I am sure that some facets of my impressions come out more in some situations than others. Thus, I find myself tending to defend critics and paranormalists to each other. In similar fashion, I tend to be more critical when talking to persons friendly to the person being criticized. This is because I want to be constructive and facilitate communication and internal criticism. I hope it is only this sort of contextual difference in my discussions that makes you think I have been two-faced. But since the real possibility does exist that I have been lied about, I would like to know exactly what you are referring to. I really should have asked you that in my last letter.

Zetetically
Marcello Truzzi

CC: Randi, Frazier, Hyman

[Woods End]   21 September 1981
Dear Marcello,

Two comments on yours of 15 Sep. Your major theme seems to be that a group must be either scientific or an advocacy group, but not both. You want us to "reform" and be one or the other, which apparently you cannot conceive of occurring unless Kurtz is booted out.

What's wrong with being both? I consider *Scientific American* "scientific," yet it has published a strong attack on creationist, it has taken strong advocacy stands on the proliferation of nuclear weapons, my December column will be a humorous, satirical attack on supply-side economics, etc. The AAAS is a "scientific" group, yet it sponsored a conference on V at which V was strongly attacked, and V allowed only a token reply.

In spite of what Paul may have said in the past, don't think any of us, including Paul, thinks of the Committee as neutral with respect to science vs. pseudo-science. And in spite of what Paul may have said, the Gauquelin business was not Committee business, but just a mistake on the part of Paul, who is only slowly learning how to deal with cranks. Whatever Paul says, we obviously are not set up to investigate and perform experiments. We have no lab. Members are spread over the US and have a hard time getting together even for an annual meeting. Your insistence that we be either an investigative, neutral group, or an open advocacy group is clouded by your inability to perceive that there are areas of pseudoscience so far outside the pale that no "scientific" body is going to treat them with anything except disrespect. To me, your "respect" for the views of V and G are only a cut or two above showing respect for hollow-earth theories. No scientific journal is going to publish serious dialogue between advocates and critics of hollow-earth theories. I really think that your attempt to treat V with respect, as a serious challenge to science, springs from the fact that you are not really trained in any of the physical sciences, and this leads you to imagine that V is somehow to be compared with, say, Wegener. The result is that *ZS* remains on a trivial Fortean level when it could be dealing with genuine challenges to orthodoxy.

You end by saying you are amused by the fact that I seem to consider the creationists as less crazy than V. What gave you that notion? I only said that the creationists are making stronger efforts to find scientific support for their theory than the V crowd, and therefore they have what more deserves to be called a "research program" (and hence *more* deserving of your attention, quite apart from the fact they outnumber V buffs by about a million to one). I consider creationism crazier than Velikovskianism, though I might consider them equally crazy if it turned out that V was a creationist. It was his admiration for Price that makes me wonder. It could be, however, that he took from Price only arguments that he thought would bolster his own crazy flood theory.

Best
Martin

CC: Ray

You mention an interview in which PK stated his support of Reagan. Do you have a copy of this? If so, I would appreciate a copy.

[Woods End] 6 October 1981
Dear Marcello,

I have delayed replying to your last two letters, hoping to hear from you about our argument over "anti-evolution," but since you may want more time on that, I will make some further comments about Rawlins and me.

In your letter to Paul (Oct 1) you state that R's motives and mental state are "quite irrelevant" to his arguments. On the contrary, it is the main reason he was booted out, and his mental state is the primary source of his radical distortions of everything he writes about. Surely this is evident from the material Ken sent you concerning his charge of "censorship."

Consider a parallel situation. The head of a law firm hires a new man. The man is a good, competent lawyer so far as his knowledge of law goes, but he has the unfortunate personality of being unable to get along with anybody in the firm. He speaks to them abusively and behind their backs.

He thinks they are all plotting against him. He gets the staff into such a constant uproar with his charges that, in desperation, the firm "fire's" him. He then goes about shouting that the firm is unfair, that is covering up its incompetence, persecuting him, and so on. Now *this* is the very heart of the Rawlins case. He cannot perceive that he antagonizes everyone who tries to work with him. *This* was the reason for the boot. It overshadows all other reasons.

You ask me to specify what I consider "lies" in Rawlins' outbursts. I am not going to list them for two reasons: (1) it might lead to dreary exchanges of letters with you over the meaning of the word "lie"; (2) if the list got to Rawlins, he would start writing; and (what is worse) telephoning, to argue endlessly over every item on the list. R. is friendly with you now because he finds you an ally in his neurotic desire for revenge. And you in turn, aware of R's personality problem, find him an ally in your vendetta.

As for your sniping at me? I did not imply you were hypocritical in saying things to others you do not say to me,. I think we have both been totally frank in speaking our minds in direct correspondence. The worst I have said about you are: You are naive with respect to the philosophy of science, relatively uninformed about the physical sciences, overly fond of bizarre, Fortean-type anomalies, uninterested in the kind of eccentric science that has the best chance of providing a new Kuhnian paradigm, and fond of sitting on the fence with respect to outlandish claims.

I consider all these charges mild compared to calling someone a "hit man" for the establishment. The term "hit man" cones from the world of organized crime. He is a hired gun who is paid to kill someone he may not even know, and who he doesn't care whether the victim deserves to be murdered or not. The implication is that I am a hit man paid by the establishment to murder the reputations of innocent men.

First of all, I am not paid by the establishment. Second, I can list (as I never tire of saying) hundreds of nonestablishment scientists (in astronomy: Fred Hoyle, Torn Gold, Roger Penrose, to name three) whom I hold in high esteem even though I think many of their theories are outlandish. My attacks are confined to those who in my opinion are dishonest (Puthoff), crackpot (Velikovsky), irresponsible hack writers (Rogo, Lyall Watson, Spraggett, etc). The people I attack are so beyond the pale that, with few exceptions, the establishment doesn't even know they exist, and if they did, would not lift a finger to attack them. Nobody hires me for these "shots." Finally, I lack the power to kill anybody's reputation. In fact, cranks generally thrive on attacks. The more the Beverly Hills diet is called harmful on TV by establishment nutritionists, the more books it sells. As Safire said in a recent column, all the Reagan "gold bugs" need, to put over the gold standard (which I consider quack economics) is to have Galbraith vigorously denounce it. In sum: I am personally offended by being called a hit man. Incidentally, Ray used the term in reference to Carpenter. So far as I know, he has never applied it to me. It is you who have done so.

But there is a more important point. You are, in your far-flung correspondence, doing far more than conduct a private war against Paul K and Phil K. It is a vendetta against just about everyone in a position of leadership on the committee except Ray.

Please don't forget to let me know your final thoughts on the advisability of calling Popper an anti-evolutionist,

Best,
Martin

CC: Ray

[Eastern Michigan University]   8 October 1981
Dear Martin,

I see that much time has elapsed since I received your last letter regarding the evolution-Popper business. I had really wanted to carefully go over our letters and write a full response, but things have been so hectic lately with my administrative duties plus much *ZS* and Rawlins nonsense plus my trip to China next week, that I just don't seem to have had the time. But rather than put it off longer, I will try to respond now even if inadequately; so this letter will probably seem anticlimactic after such a long wait for you.

Your main question is in two parts (a) am I using the term antievolutionist to describe Popper in irresponsible fashion, and (b) am I correct in terming what the followers of Velikovsky are doing a "research program"? A third question (c) refers together or not the Creationists do or do not have a positive (versus negative only) research program. Correct me if I misperceive the issues, of course.

In my original letter of July 20, I referred to Popper as an "antievolutionist." This seems to have started the matter between us. Looking over my July 20th letter now, I see clearly that I should have put quotation marks around the term "antievolutionist" because I did indeed (as you point out) not mean the label in the usual sense in which it is used. On this I am wrong and mea culpa. However, two factors, I should think, would mitigate my error. (1) The context of the paragraph in which I make this error should have made my special use more clear. And (2) my later letter of July 28 should have reinforced my real meaning. But, I admit to having been quite unclear and that is my fault. Let me try to clarify matters now. I fully recognize that Popper is not in fact an antievolutionist and does not perceive himself as such. He is clearly what I think both of us could term some sort of neo-evolutionist or revisionist-evolutionist. To properly use the terms involved—and again I admit that I did not earlier—I would suggest three categories (perhaps along a continuum): (1) Creationists (e.g., fundamentalists re the Bible); (2) Anti-Evolutionists (those like the Ancient Astronaut Society people and others, who deny evolutionary theory entirely—in anything like the Darwinian sense—but propose some non-biblical alternative origins), and (3). Revisionist Evolutionists (these obviously are of many types and might include neo-Lamarckians, people like Gould, Popper, etc. In my original letter of July 20th, my concern was separating (1) from (2&3). My paragraph spoke of the fact that I feared that the attacks on (1) might force people to stop taking (2) and (3) seriously as well. I did think (mistakenly) that you might be among those who might do that. I see now that you are actually yourself probably in category (3) and so would not make that error of failure to differentiate. However, I did have some reason to make this error. In particular, your seeming condemnation of Macbeth (who is clearly in category #3 along with Popper—who endorsed Macbeth's book) led me to think that you were lumping Macbeth into category #1. If not, we have no disagreement on this matter. It should be noted, however, that the Creationists in category #1 will and have used the names and arguments of those in categories #2&3—especially invoking Popper's early statements and even now pointing to some ambiguities in his current "retraction"—to support their attacks against evolutionary theory. That is, from the point of view of the Creationists (and even of some pro-evolutionists), Popper has been perceived as an enemy of evolutionism and thus might be called by them an "antievolutionist." I would not agree with such a label for him by them for the same reason I raised in my original argument. I don't want people to lump these categories together. I don't think you do, so there is—I hope—now peace between us on this point. Since I was so unclear, I must take the blame for the error produced. I don't think I was wrong in my conceptualization of matters, but I was clearly wrong in the way I expressed myself. I hope I have at least now expressed myself clearly. You may, of course, still disagree with me at this point, but I think probably not. OK on that matter?

A side matter might be mentioned. I did use the term in a special way that was not clearly characterized as special by me. So I plead guilty on this matter. But I do *not* at all think I am generally guilty of doing this in regard to our earlier disputes about the use of such terms as crank, crackpot, etc., terms which are in fact in most vague common usage and which I was trying to clarify through discussion and more careful definition (not the other way around, as in this case).

I turn now to issue (b): Velikovsky followers and their alleged research program. The degree to which the Velikovsky followers represent a research program is obviously a question of degree. The existence of (1) the many articles in *Kronos*, only about half of which attack V's opponents and seek to defend V, (2) the similarly non-defensive articles in *SIS Review* and *Catastrophist Geology*, (3) the on-going posthumous publications of Velikovsky elaborating his views, and (4) some articles stemming from his views in regular journals (admittedly few), all—to my mind—indicate that there is at least a *minimal* research program going here. It also seems clear to me that what research program does exist is not a negative one but a positive one. Obviously, it is "negative" in the sense that so much of it is defensive and attacks its critics, but that is not negative in the sense that it primarily seeks to destroy some existing scientific theory, (as with the Creationists' attacks on evolution). If the V-followers are negative about anything, it is in their being anti-uniformitarianists. This was a major issue when they started, but today's astronomers are not the uniformitarians they were when *Worlds In Collision* was first published. I do not claim that the Velikovskians constitute an important protoscientific effort. Unlike yourself, I do not put them in the same category as the Flat-Earthers. I have pointed out earlier (my editorial in *Zetetic* #2) that the Velikovskians represent what I would term a crypto-parascientific set of claims and these make it very very difficult for them to gain scientific respectability (since both their crypto-historical claims and the para-astronomical relationships they claim are *both* controversial; that is both their variables and the relationships between them are extraordinary). I find their claims unconvincing, but unlike you, I do not find them preposterous and so implausible that they are like the Flat-Earthers' ideas. I also think that you may be terribly wrong in associating V's ideas with the anti-scientific notions of the Fundamentalists. (As V supporters repeatedly point out, V claims that the biblical miracles were in fact not supernatural miracles at all but quite normal extraordinary astronomical events; thus V is very much opposed to the idea of biblical miracles endorsed by the Fundamentalists. For V, it was an act of Venus and not a special act of God.) I believe that your opposition to V stems in part from your (mistaken) belief that V's ideas give aid and comfort to the Fundamentalists whom you see as the important challengers of science. Since I do not share this perception, I see V as a most radical protoscientist (frequently an incompetent one) with a fledgling positive research program developing from his efforts. I do not see his as a fully developed research program such as that found in the protoscience of parapsychology.

I assume it is not necessary for me to actually list the articles in *Kronos* and elsewhere that I consider to be the non-defensive efforts of a positive research program underway. I assume you can examine the issues and have.

In looking over your letter of August 30, I see that you seem to define a research program by the Velikovskians as having to be strictly astronomical. You refer to other articles being *merely* archaeological or historical speculations. First of all, I would definitely consider these archaeological and historical writings as scientific efforts and part of the interdisciplinary scientific research program that Velikovsky is developing. We are, after all, talking about history in Velikovsky's writings. His case is based on his allegation that certain historical events actually took place in certain ways. He uses these events to argue for the accompanying necessary (to him) astronomical events, themselves part of the history of the solar system, not something currently observable (directly) by astronomers. You want to simply write off the historical and archaeological efforts as "speculations." The strategy of the pro-V group is to try to first establish the alleged crypto-historical events and then move on to the para-astronomical claims. Obviously, you give the astronomical difficulties of V priority. You see geology and astronomy as the more fundamental sciences, I think. I am personally inclined to agree with you and that is why I do not find V convincing. *But* I remind you that Lord Kelvin said that the age of the sun precluded the possibility of an earth old enough to allow biological evolution as Darwin claimed. Luckily, the evolutionists did not stop work because of these objections by the physicists. Biology was right and physics later changed its mind. It may be a long shot for the pro-V followers to hope that the astronomers will come around to their point of view re the historical and archaeological record, but it is not as unreasonable as I think you find it.

I also think you are incorrect in saying that there is no astronomical or geological part of the pro-V research program. Of course they reprint stuff supportive to them from other journals. Why not? Obviously, you want to discount any work in geology or astronomy done by V himself or by his "nonscientist" followers like Rose. Much of the work of such amateurs is, of course, geological or astronomical. (I note that when you speak of the people in V's camp, they are "nonscientists," but when people like Kurtz or Oberg, etc., write on your side, they are not referred to as "nonscientists" or even as talented amateurs.) Let me put it differently. There are so far as I can tell, only about 50 people entirely who are regular contributors to the pro-V literature of this tiny research program. Only a couple of astronomers are among them. Suppose you were to criticize parapsychology for having so few professional, degreed psychologists among them? Even Rhine was not really a trained psychologist, after all. Psychologists are the main opponents of parapsychologists, just as astronomers as the main opponents of V and his followers. This surely is not the way to argue. It is the quality of the work done, whether by amateurs (Rose, etc.) or professionals, that needs to be evaluated. I do believe that *some* of the defenses raised by these amateurs against the professionals like Sagan are quite impressive. Even Jastrow has written that on one point at least, Velikovsky was the better astronomer. (I think my views of V, by the way, are probably quite similar to Robert Jastrow's, from what I have read of his views so far.)

Let me turn now to issue (c): whether or not Creationism has a *positive* and not just negative research program. I am not sure what I can add here. I have tried to look at their materials. I agree that they *claim* to have a positive research program. I do not see evidence for one. I see only attacks on evolutionism and no constructive program of their own (in any scientific-empirical sense). This may be due to my ignorance, but you have yet to enlighten me with citing some positive research studies produced by the Creationists to demonstrate (corroborate) a single creation event as biblically stated. I have seen only negative (anti-evolutionary) stuff and nothing constructive. If it exists, make me specifically aware of it and I will try to read it. I live in Ann Arbor and have meant to go by the center here but just have not had a chance during office hours there. But since you are making the claim that they have a positive program, the burden falls on you to demonstrate that to me. If you show me you are right, I will certainly welcome such dialogues in *ZS*.

Obviously, if you don't wish to write on the subject, that is your prerogative. I merely invited you to do so since you seemed to be the one hot on my covering such matters in *ZS*. As you surely remember, I have always argued that any scientist has the right to set his own priorities. Creationism is not one of my priorities because I see it as negative rather than positive in its research program. If I am wrong, it should be among my priorities; but I can not yet say how heavily weighted it should be among my priorities.

Re the ad you sent me from the Institute of Creation Research, all the stuff advertised therein seems to be negative in character. Re Wysong, here are the facts you requested: The book is R. L. Wysong's *The Creation-Evolution Controversy*. I enclose a flyer on it for you. I think the book is also available through William Corliss's book department of *The Sourcebook Project*. Wysong's book was reviewed in *The Zetetic*, vol. 1, #2 (1977) and there were some letters following that. I am surprised you missed it and/or the exchange therein.

Looking over your letter of Aug. 30, I see that you think I am playing Humpty Dumpty in my word redefinitions when I spoke of parapsychology as also including disbelievers but active researchers in parapsychology. First of all, I think you are ignoring my earlier remarks on all that. The term was one coined by Rhine, after all, and he clearly meant it to include those who might disbelieve in or express nonbelief in psi. More important, this explicit stipulation was made by the Parapsychological Association and was one of the reasons the AAAS granted affiliation (also explicitly noted by the AAAS statement on all this). Clearly, the parapsychologists use the label to include nonbelievers in psi. Also, there are nonbelievers (me and others) and disbelievers (Hansel) who have been in the Parapsychological Association. I do indeed recall that Ray did not like the term used for himself. But this is because the critics like yourself have taken the term to mean believers in psi. This is a clear distortion by the critics. Also, Ray is primarily a general psychologist and thus prefers that label for himself, just as I prefer the term sociologist for myself. I might add that there is a subsection (specialty area), of the American Psychological Association called the Parapsychology Section. People belonging to it (and not all are in the APA) would also use the general term psychologist to describe themselves. Obviously this is a matter of definition, but I would argue that the critics are the ones playing Humpty Dumpty and have distorted the terminology. Similarly, the critics have taken the term "skeptic" and its original and well-established historical meaning of doubter or nonbeliever (an agnostic position), and given it the meaning of disbeliever. I am not the offender of meaning in this case; the critics have been the ones creating semantic confusions. In both, these cases, the literature backs me up thoroughly.

If I have missed some, points in your earlier letters, let me know. I turn now to your most recent two letters still unanswered.

Re your letter of Sept. 21: You misread me when you say I think a group must be either a scientific or an advocacy group. I do not say that. Advocates can be scientific in the same sense that lawyers can be proper officers of the court playing by the proper rules of evidence, etc. The distinction I make is between advocates and judges/juries. CSICOP has publicly presented itself as a non-advocate body. It claims that nothing will be prejudged prior to inquiry. It claims that it is a purely objective and fair-minded body of judges. It is obvious to all from its activities that it represents an advocacy position. The position of the CSICOP (and its journal) is not doubt (true skepticism) but denial. The main function of the journal is not to promote inquiry but to dismiss and discredit nonsense. You and others see the CSICOP as an arm of Science being used to combat Pseudoscience. For the most part, you already know what the "pseudosciences" are. An inquiry body would normally seek to obtain the very best evidence from claimants of the paranormal and honestly seek to evaluate that evidence. Rather than going after the best evidence, you have gone after the most widely believed (i.e., media disseminated) claims. Thus, instead of examining the claims of people like Honorton, Sargent, Palmer, Rao, etc., you attack parapsychology via *The National Enquirer* (which the Institute of Parapsychology and the Parapsychology

Association refuse to communicate with at all).

CSICOP is a valuable part of the total scientific effort. I have always maintained that CSICOP can and sometimes does perform useful functions. I keep having to remind you that I have never called for its dissolution. Unlike Rawlins, I never called upon other Fellows to resign. Ask Ray if you don't believe me. Even in Ray's case, I remain surprised that he can feel comfortable, in CSCIOP and wonder how long he will remain aboard, but I have never insisted that he was unreasonable in staying on. The only person I have even come close to trying to persuade to leave CSICOP has been Ray. And that is because I feel that Ray's efforts are about the only thing that has saved CSICOP from mistakes that Kurtz would otherwise have made. Ray, to me, is the token-liberal on CSICOP's Executive Council. I believe Ray has been responsible for CSICOP not being clearly perceived as the advocate body it is. The rest of you are so hard-line, it would soon become apparent to everyone how biased the CSICOP really is were it not for Ray periodically restraining Kurtz and Randi from some efforts that have been considered by the Councilors.

I am delighted to see that you confirm much about the "mistakes" of Kurtz. You see these as his failure to know how to deal with cranks. Since I do not view Gauquelin as the crank you do, we disagree on this. Besides, I also think that Kurtz's "mistakes" may not in fact be honest ones. On this matter, I can only urge you to check with the AHA leadership about Kurtz's past history and the view there on his being an honest bumbler rather than a most Machiavellian and dishonest advocate. I recently got some letters from Bette Chambers on this matter. She has asked me not to xerox them, but I do have her permission to read them to you on the phone (which I did to Ray), which I can do if you want me to. I did not call you about this because I perceived you as largely trying to stay out of all this Rawlins stuff. But if character is relevant to Rawlins, I think the same can be said of Kurtz given the strong views of him at the AHA. Let me be clear: I do not know that the views of Kurtz at the AHA are true. I only know that they are strong and most damaging *if true*.

I presume you got the article on Kurtz's views on Reagan which I sent you earlier. I turn now, finally, to your letter of Oct. 6th. I do not think Rawlins's personality is irrelevant to his motives or to the reasons for his expulsion from the Executive Council of CSICOP. I grant you that completely. But I do think that Rawlins's personality and motives are quite irrelevant to the issues he centrally claims about CSICOP (i.e., Kurtz's) alleged cover-up re the Gauquelin studies. I see these allegations as the only serious matters that should concern those of us outside CSICOP. I do personally believe that Kurtz may have in part wanted him off the Council because of his troublemaking re Gauquelin. But I am sure the rest of you "failed to renominate" him because of these other matters. You may recall that I was the one most interested in seeing you dump Rawlins before all this mess emerged because of Rawlins's attacks on me and because of his nonsense re the U. of Toronto astrology conference.

But I do not believe that Rawlins's ejection as a Fellow of CSICOP—independent of his ejection as a Councilor—was independent of his blowing the whistle on Kurtz & Co. It seems quite clear that Kurtz did not send out any ballot re ejecting Rawlins (you wrote me you did not get any ballot and would not have approved Rawlins's total ejection), and yet Kurtz clearly told the other Councilors at the last Executive Council meeting that he had done a balloting. When Rawlins polled you all about any balloting, not a single reply came back to him saying there had been a balloting as Kurtz claimed (according to what Ray told me and surely a balloting was necessary for any such ejection). You, Randi and Ray all three told me you had not ever received a ballot (Ray said only that he did not remember any such ballot and is trying to check further in case he was wrong about this). So, it sure as hell looks like Kurtz unilaterally bounced Rawlins off the list of Fellows. I can see no reason for this other than Rawlins's complaints about the Gauquelin matter and Kurtz's handling of it. But even this is a matter of internal CSICOP business which is not really central to my or others' concerns. If Kurtz violated your own CSICOP by-laws, that is largely your own CSICOP business about which the Fellows and not I should complain (though there is the matter of your being a non-profit corporation with tax exemption so we general citizens do have some interest in such matters). But I am willing to pretty much ignore all this.

The fact is that Rawlins is making some serious charges about improperly done science in the name of (if not directly by) CSICOP. This includes claims of cover-up (only a small part of which involves Rawlins's personal ejection from CSICOP). Until now I had not been including you in my sharing of matters re the Gauquelin matter. I thought you considered him such a crank that he never should have been taken up by the CSICOP at all. So, I have just left you out of the many exchanges re all this Rawlins stuff. However, since you now bring it up to me in detail, I think it best that I share some of the more interesting stuff with you that I have already sent Ray and others. So, I enclose two items: (1) my tentative outline of the central controversy, which I am already getting some corrective feedback on from both sides, but which is generally still correct and might be helpful to you in understanding my views of the matter, (2) a copy of the Curry article which you earlier indicated you did not want to comment upon but which I think you might now want to look over (and which you remain invited to comment on should you change your mind), and (3) some letters between myself and Ken Frazier re McConnell's letter (which I presume you got from McConnell).

Naturally, I am delighted to know that you merely consider me un-informed and naive rather than hypocritical. That should settle that. I remain in disagreement with you about my purported vendetta. But I see nothing I can add to my earlier comments to you on this in my letter of 9/15/81.

Re the matter of your and Randi being "hit men," I do not mean the term in any way other than Ray defined it. I do not mean to imply that you are in any sense in the employ of the establishment. I don't think Ray meant that about Carpenter, either. I also mentioned that I see Randi as the true "hit man" and that you function that way only at times. I can see why that term might be obnoxious to you whereas I saw it mainly as amusing. There was no intention to be offensive to you or Randi. I believe you are both perfectly honest and sincere. What makes you function as "hit men" in my view is your willingness to violate evidence and argument in favor of ridicule and the "hard-hitting" approach versus the more reasonable one. Thus, Randi consistently attacks all of psi research by pretending that much of the nonsense he debunks is truly representative of the best stuff. In your case, you feel that some things are so far beyond the pale of reasonable science that you go for the horselaugh or irrelevant issues (like attacking Hynek's work on a film rather than what he says in his book you are reviewing). You are correct in pointing out that I, not Ray, have referred to you as a "hit man." I will not speak for what Ray thinks. But I must mention that I have frequently referred to you and Randi as hit men in conversations with Ray and he never corrected me once. So, I have always presumed he put you in the same category as Carpenter. If I am right about Ray, I am sure he did not mean the term in an offending way either. Obviously, you and Randi are not in the pay of some nefarious organization. You may be out to assassinate nonsense, but you do so as self-appointed knights against the dragon or irrationality. I just think a lot of these dragons are myths. But most important, I really do not see anything wrong with people functioning as "hit men" in the way I characterized you. I think you and Randi add important balance to the controversies. I remind you that originally, you, Ray, Randi and I were going to start up something very much like CSICOP just because I believe in the need for balance. But— back to the point at the beginning of this long letter—CSICOP claims not only to bring balance into the picture; it claims to bring objective judgment. A good hit man is merely a hard-hitting lawyer/advocate. Nothing wrong with that so long as the lawyer is not speaking for the scientific community as a judge. (You may think I exaggerate this role for CSICOP, but I remind you that the journal *Science* had one negative reviewer of a parapsychology article submitted refuse to consider it until it had passed the approval of

CSICOP; this makes CSICOP act as an inappropriate gate-keeper rather than as an advocate). I assume that by "hit man" Ray mainly had in mind that someone from outside the regular scientific community (a science writer like yourself or a magician like Randi) is functionally "used" by the science establishment to try to eliminate the "pseudoscientific" opposition. I think that is a fair picture. (I don't want to make much of this last point, but I note that you say that you are not paid by the establishment. Obviously, you write what you believe in, but you certainly are paid by what most of us would consider the Establishment, e.g., *Scientific American, NY Review of Books*, etc. It is obvious that you and Randi are lent support by the forces of orthodoxy, which cheer your efforts. No one ever suggested that you or Randi or Carpenter were prostituting yourselves. A "hit man" can agree with the goals of his employer.)

Finally, you refer to my "private war" against Kurtz and Klass. I have already said that I think that is too strong for my admittedly negative feelings about Kurtz. But in no way am I even that concerned about Klass. I pretty much ignore him except when he writes me a letter. I have already told you I consider him a crank; he is not really worth my time. I should mention however that CSICOP will eventually be most embarrassed for not realizing what a mistake it is to let him be your principal (so far) defender of Kurtz. For example, Klass just sent Rawlins an absurd legal challenge (for $10,000 of course) if Rawlins can prove he was ever an "Associate Editor" of CSICOP. (a) Rawlins admits this is literally a mistake, anyway, and (b) it is almost completely irrelevant to all the charges that are important.

My god, I can not believe how long this letter has become. But here it is. . .

Next week I go to China to see the kids for whom demonstrable psi-powers are being claimed. It sounds so far like standard deception, but I am keeping an open mind. In any case, I have never been to the Orient before and am really looking forward to it. We will visit Peking, Xian and Shanghai. I also hope to contact people there about the "wild man" allegedly in China and will try to visit with the growing ufology movement there.

Best,
Marcello Truzzi

CC: Ray

P.S. I hope Ray (who is getting a copy of this letter) will write you directly about his views on who is or is not a hit-man, with a copy to me so I can be enlightened also. Actually, I have hoped all along that at some point Ray would write us both commenting about our exchanges, but as far as I know, he has not yet done so. (What do you think, Oh Silent One?)

[Woods End]   8 October 1981
Dear Marcello,

Although I regard the question of who Paul K voted for as irrelevant to your dispute with him, I finally got around to asking him when we last talked on the phone. He told me he voted for the libertarian candidate, but later regretted it because he decided the libertarian party was too extreme. He favored, he said, the Reagan tax-cut program, and the Soviet containment policy, but could not vote for him. Nor could he vote for Carter. Let me add that as far as pseudoscience and/or the paranormal is concerned, I see the two candidates as about even. Reagan seems to believe in astrology, but I consider him a phony fundamentalist, and believe that he made his statement about evolution solely to capture the fundamentalist vote (which he did). Carter, on the other hand, I believe to be a sincere evangelical, very close to fundamentalist, and who seems to believe that UFOs have other than conventional explanations.

I might add that I have little respect for the libertarian party. My forthcoming philosophical book has a chapter devoted mainly to attacking the extreme libertarianism of Harvard philosopher (Harvard no less—the citadel of liberalism!) Robert Nozick, in his book *Anarchy, State, and Utopia.*

Just out of curiosity, who did *you* vote for president? I think I mentioned in a previous letter that I did not vote for any presidential candidate.

I am having a very funny exchange with Puthoff over the number of days involved in the dice-box test with Uri. I enclose his latest letter. I take it that if he doesn't know whether it was three or two days, there must not be even a *written* record on each trial!

                                        Best,
                                        Martin
CC: Ray

[SRI International]   5 October 1981
Dear Martin,

To answer your latest question: The dice-box trials were carried out over a two or three day period, a few trials per day, sandwiched in among other experiments, until a total of ten trials were collected. As I have stated elsewhere, of the ten opportunities to make a call, Geller called eight times and passed twice, one of the passes being filmed. Each of the eight calls was correct.

The length of time per trial, from when I began the shake to when I opened the box, was relatively short—30, 40, 50 seconds. The one you see in the film is quite typical and it is well under a minute.

With best regards,
H. E. Puthoff

[Zetetic Scholar]   14 October 1981
Dear Martin,

Just a most quick note. Off to China tomorrow but got your letter of 8 October today.

Apparently, Hal Puthoff will be on my China trip among our group. I will try to find out more about the 2 vs. 3 day business and the apparent absence of a written record.

Thanks for the information re Kurtz's voting preference (at least so he tells it; I have come to suspect Kurtz, perhaps more than I should, of saying what sounds best). It is, of course, irrelevant, but interesting. Particularly in that about the only place where I am truly libertarian is in regard to science (where I think it works because the system is self-correcting and self-critical). Like Feyerabend, I am quite libertarian in my view of how science should operate, rather like allowing 100 flowers to bloom. I find it interesting that science is the one place Kurtz seems to be nonlibertarian. As for my vote last election, I was so scared of Reagan that I choose what I felt was the lesser of evils most likely to overtake us: I reluctantly voted for Carter. My wife abstained. I almost did. I was tempted to vote for a socialist candidate but it seemed a waste to me. I think you are right about Reagan really believing in astrology but not in creationism-fundamentalism. However, I think you underestimate the sophistication of Carter (I refer you to his better (unofficial) biographers on this point). The private Carter seems far more sharp than the public one.

Best,
Marcello

[Woods End]   18 October 1981
Dear Marcello,

To avoid another overlong letter, I'll confine my remarks to two points:

1. Yes, we are now in complete agreement on Popper and the word anti-evolution. I hasten to say that not for an instant did I misunderstand what you actually meant. I was concerned over your way of stating things which I believed caused only confusion. I am willing to call myself a "revisionist" in my views on evolution, but even this term is useless because, as I pointed out, all modern evolutionists are revisionists in the sense that they have gone far beyond Darwin. There is so much contemporary ferment among the evolution theorists that it is not easy to define an "orthodox" view, and for this reason the word "revisionist" becomes extremely fuzzy.

Your tendency to use words in strange ways crops up again in your last letter. Eager to defend your view that I am, in a sense, a hit man for the establishment, you mention that, after all, I am paid by such "establishment" journals as *Scientific American* and *NY Review of Books*. I know, of course, what you mean. *Scientific American* often ventures into fringe areas—its article on the physiological effects of TM (now known to be nonexistent—I mean the particular effects cited in that article), Premack's article on talking apes (he himself has now repudiated a large part of it), Weber's article on his discovery of gravity waves (now totally falsified), and a cover story on the minority view that dinosaurs were warm-bloodied and evolved into birds (most geologists reject this). What you mean, of course, is that *Scientific American* doesn't run pro-parapsychology articles, and doesn't accept advertising for crank V literature. As for NYR, you are on much shakier ground. You're the first person I ever heard call this anti-establishment journal an establishment one. (Although it now regrets it, the NYR at one time gave lots of space to a wild conspiracy theory about the Kennedy-Oswald killing.) It has a strong leftist, political slant (featuring such socialist writers as Heilbroner and Izzy Stone), and is noted for its savage attacks on establishment writers (e.g., its recent hatchet job on *The New Yorker*'s Pauline Kael). What you mean, however, is clear, it occasionally allows me to attack books on pseudoscience.

2. I am well aware that V thinks some OT miracles have natural explanations (see page 30 of my *Fads and Fallacies* where I discuss this). But so does George M. Price. If you will ever look into his masterpiece, *The New Geology*, you will find that he has more convincing *natural* explanation for Noah's flood than V. Like Price, and many other fundamentalists (though a minority), V believed that God works some of his miracles by using natural laws, though the laws may not yet be fully understood by science. So far as I can recall, you have never admitted that V was a devout orthodox Jew. For him the OT was a revelation from Jehovah. The essence of his approach is that the OT miracles were *simultaneously* the work of God and the work of nature. When V explains the parting of the Red Sea, he means that it has *both*, a natural explanation and a divine one. (Not being a

Christian fundamentalist, he naturally does not take this approach toward NT miracles.)

Exactly how V thought life got started on earth, and how it evolved, is not yet clear. That he had considerable sympathy for Price (whose flood theory is the basis for most present-day creationist "research programs") is evident from his quotations. As to where he disagreed with Price, we will have to wait and see if he left any papers that spell it out.

I am happy to hear that you are keeping an "open mind" on the possibility that the Chinese children can read print with their ears. You and Krippner are certainly an "odd couple" to be investigating this, and I shall be most eager to know your respective opinions on Chinese dermo-optical perception.

Best,
Martin

CC: Ray

(Ray: I don't expect you to waste time writing to me about any of this.)

[Woods End]   18 October 1981
Dear Hal [Puthoff]:

You probably won't read this until your return from China. Truzzi tells me that you are expected to be on this trip. Of course I will be anxious to hear from Marcello what you and the others think of the Chinese children who claim to read print with their ears.

I write because your brief reply to my letter of 13 October contains some big surprises. I did not know that the dice test for *Psychic* magazine, made in Geller's motel room, had been videotaped. Nothing on this test is reported. in the *Psychic* interview. May I ask what the *results* were of this test? I think it is important that the results be known, especially if the test was considered of sufficient importance to be videotaped.

But my main reason for writing is your last sentence which is; "I continue to entertain Randi's hypotheses, but he has to do better than saying that Geller peeked in the box. Go back and view the film—that's what we have to deal with."

You refer, of course, to the filmed single trial. But if no other trials had been filmed (according to Pressman this was the final trial), is it not obvious that Geller (assuming he used a peek method before) would not employ it, knowing the trial was being filmed and therefore could be viewed by a knowledgeable magician? And not having been free to use the method, of course he would pass, since he would have obtained no information about the die. With one chance in six of getting a hit, the hit is not in any way surprising.

Surely we can agree that with respect to Randi's theory the film of the pass trial is irrelevant. The only data available now, aside from your memory (which as all magicians know is totally unreliable with respect, to deception by a magician who is as skilled in misdirection as Geller), are the written records. Would you be willing to Xerox these for me, at my expense of course?

In the interest of seeking the truth about this historic test (in which the results were so unambiguous and so overwhelmingly against chance), it would be enormously helpful to see these records. I want to be completely open. I know a great deal about dice-cheating techniques, and it is my belief that Geller did indeed peek by a method similar to the one Randi conjectured. The written records may cast no light on the matter, but at least they could be of help in pinning down the exact protocols.

All Best,
Martin

CC: Truzzi

[Eastern Michigan University]　　　　　　　　　　　12 November 1981
Dear Martin,

I enclose two papers which you might enjoy and find worth reading. The first is by Jerry Clark and will appear in *Frontiers Of Science*. I think it demonstrates my earlier argument that Jerry is a responsible proponent of the paranormal and amenable to reasonable arguments such as those I think I have had with him. It should, in any case, set matters straight as to just what his views now are.

The second paper by Henry [L. D. Henry?] makes frequent references to you in it, and I presume you probably do not get the *Journal Of Popular Culture*. So, I thought you might like to see it.

The trip to China was fascinating. We saw about nine of the children and they did nothing paranormal in their demonstrations for us. However, the work being done by the Chinese scientists, if valid, is quite extraordinary in its claims, and we expect to learn much more about that as we get the various papers translated and our communication with them increases over the coming months. I perhaps should mention that I got to know Hal Puthoff reasonably well over the 14 days of our trip. I am now still of the opinion that the SRI work is probably based on errors, but I no longer believe—as I once did—that Puthoff is simply a charlatan. I think he

is quite sincere, and I also now believe that much of the criticism about
SRI work (especially Randi's) is in error. There remain many unanswered
questions, but I must say that Hal made a good case about many matters.
He persuaded me that I should conduct a rather simple remote viewing
experiment, which I now plan to do with my class later this term. I intend
to have him approve my protocols prior to the run so that he can not later
post hoc away my failure if the experiment results in chance results.

Have you read Jim Alcock's new book on parapsychology? Some very
nice things about the book even though I have some serious points of difference with it.

I trust you saw the *New Scientist* review of your book.

Cordially,
Marcello Truzzi

P.S. I sent you a comp copy of *Zetetic Scholar* #8 last week. Starting with
issue #9, I will no longer be sending out comp copies as I have been, due
to increasing postal and production costs.

[Woods End]   16 November 1981
Dear Marcello:

Many thanks for the Henry paper, which I enjoyed, and for Clark's
article. I am pleased to see how Clark is changing, and also pleased to
learn that you do not consider Puthoff "simply a charlatan." Not once
have I ever doubted Puthoff's sincerity. Where you and I have differed
with respect to Puthoff, Gauquelin, and others is that you tend to think
that sincerity rules out charlatanry, whereas I am impressed by how often
the two go together. I have no doubt that Dr. Levy is a true believer,
and totally sincere in his beliefs; and the same for Soal. Of course Puthoff
is "sincere"! Has anyone doubted it, including Randi? But I think the
evidence is quite strong that Puthoff sometimes lies, and is often extremely
deceptive in the way he phrases his reports.

Clark's article prompts a few observations.

1. Finding the drawing of the fairies in a book has no effect whatever
on the thoughtography explanation of the photos, which is what I assume
Eisenbud still believes. The girls simply saw the drawings, and put them
on the film paranormally, just as Ted did with *National Geographic* pictures
that correspond even *more* accurately.

2. Clark's remark: "Don't assume that the experts are always fools,"
reminds of a sentence I once came upon in a book on psychosomatic disorders by an orthodox Freudian. He cautioned analysts not to assume

that certain visual disorders (such as near sightedness and astigmatism) were psychosomatic merely because orthodox ophthalmologists said they couldn't be!

3. I find your distinction between "doubting" and "denying" of little value because of the fuzziness of the continuum. For example, even you would not say "I doubt that the earth is flat." It would sound too funny. If you were to say "I doubt that Chinese children can read print with their ears," I would find it just as funny. There is a sense, of course, in which everything outside of logic and math must be doubted, but at extremes of the continuum there is nothing wrong in denying. I "doubt" that ESP and PK are valid phenomena. I "deny" that children can read with their ears. To carry Forteanism to such an extreme that one speaks of doubting *everything* is to destroy all of Fort's rich humor in the name of a skepticism so extreme that it makes only for confusion, not light. Look up sometime the fable that Carneades, the Greek skeptic, liked to tell about the man in the cave who thought he saw a snake.

Best,
Martin

[Woods End]   9 December 1981
Dear Marcello,

Your last letter called both *Scientific American* and *The New York Review* "establishment" journals, which I took to be a derogatory remark about my affiliations with both. Now I am told that ARPA not only paid for your trip to China, but that ARPA is paying you $5000 a year as a consultant. Is that true?

Please—I see nothing wrong with this, if true, and congratulate you on the arrangement. I bring it up only as part of our continuing battle over the funny way you talk. Are you willing to say: "I now work for the establishment"? All words are fuzzy, but "establishment" is unusually fuzzy. Do you use it in such a private way that it can apply to the *NY Review*, but not to ARPA?

I must add that I am overwhelmed by the sense of disproportion revealed in your recent letter to Paul. Here are you, howling with rage and righteous indignation over the distribution by mail of a few dozen copies of a demented letter in which Rawlins said awful things about *you*. But when your friends at *Fate* allow hundreds of thousands of readers to see Rawlins's lengthy and equally demented diatribe about Paul and others, then you chuckle with glee, and are doing all you can to keep publicizing Rawlins's attack.

I suppose you justify this in your own mind by assuming that Rawlins is, naturally, *wrong* in everything he said about *you*, but since then has had such a change in his personality and his perception of reality that he is *right* in everything he says about Paul, Abell, Klass, Randi, me, etc., etc., etc.

Rawlins is easy to understand. But you, a professor of sociology and a consultant to ARPA, are surely capable of more mature behavior.

Best
Martin

CC: here and there

[Eastern Michigan University]   11 December 1981
Dear Martin,

Received your letter of Dec. 9 today. You indicate that copies are being sent "here and there." I assume Ray is among those so am sending him a copy of this letter; I'd appreciate your sending further copies of this letter to the same people who get yours.

First, there seems some confusion about this "establishment" business. The original reference came from our discussion surrounding Ray's use of the term "hit man" for the scientific establishment. I never initiated that stuff, but you wrote me denying you were part of any establishment. I responded with the fact that most people would consider *Scientific American* and *NY Review of Books* "establishment" publications. You make it sound like I once said there was something wrong with being on the side of the establishment. That was not and is not my position.

Next, you are partially incorrect about your facts re DARPA. DARPA was contacted by me about possible funding for the China trip (not necessarily from them). They replied by giving me $5000 for the trip through a grant to CSAR (the trip was to cost most of this money). There was no other money involved and there are no promises of any more. So, I am not being paid $5000 a year via DARPA, as you apparently thought. I would certainly consider my trip to China, therefore, financed by what I would call an "establishment" source. Obviously, DARPA represents part of the military-political establishment and indirectly is related to the science establishment. *Scientific American* is part of the science establishment and *NY Review of Books* is part of the general intellectual establishment. All this business stemmed from the context of the "hit man" discussion and in no way denigrates any establishment per se.

Finally, re the Kurtz and Rawlins matters. You completely miss the points involved. Rawlins is a private individual. I do not particularly approve of his actions nor do I agree with all the charges he has made against CSICOP. I have told him about the points of disagreement. Kurtz, on the other hand, represents an organization operating under the pretense of being a scientific body. The sins of CSICOP reflect on more than merely Kurtz and vice-versa. Kurtz did not send the letters out to only a "few dozen" people. There are over 70 people associated with CSICOP and others in the press and outside CSICOP also have seen his mailings. Both Rawlins and Kurtz have apologized to me about these letters. But Kurtz is the one who continues to disseminate them despite his apology. Kurtz is the only one who can do anything to make his apology known to those to whom he sent his copies. Whatever wrongs Rawlins may have committed, these do not somehow excuse Kurtz.

Rawlins has said he was wrong about me, to both me and others. I do not forgive Rawlins for his past actions, but I certainly do accept his apology. Rawlins has since done nothing to suggest his apology was false. Kurtz, on the other hand, continues to act like a scoundrel about these matters.

You seem to imagine that I somehow have ignored Rawlins' letters as far as Rawlins is concerned. That is nonsense. I remind you about my complaints to Rawlins and all the CSICOP Councilors about Rawlins' first letter about me making similar charges. I would have replied to Rawlins about his second letter if he had ever sent it to me. But he did not. The first time I ever saw this letter was when I got it in the packet sent out by Kurtz. At that time, I complained to both Rawlins and Kurtz about it. But only Kurtz has ignored his apology and promise of correction of any misimpressions caused.

You seem to somehow think that I completely agree with Rawlins' criticisms in "sTarbaby," and that I do so merely from a superficial reading of that polemic. I remind you that I have been receiving correspondence relating to these matters over the last three years. I remind you that Patrick Curry's article on this whole matter was submitted to me before I ever read "sTarbaby" or knew the full character of its charges. And what have I done since then? I have gone way out of my way to (1) urge detailed replies to Rawlins by Kurtz, et al. (2) I have invited all the principals to give their side of matters on both Curry's and Rawlins' charges in *Zetetic Scholar*. (3) I have urged participation in the dialogue around all this by everyone concerned including the CSICOP fellows and "members." (4) I have made no public charges myself as to the validity of Rawlins' charges and have urged others, including the Parapsychological Association, to wait for the CSICOP replies, (5) I put together a synopsis of the principal charges which I circulated privately to the leading principals on both side for their critical comments to help me clarify my understanding of matters. I certainly do have my beliefs and opinions, but they remain tentative and subject to new information. I have done everything I could to obtain all information available. I welcome further suggestions on how to get more. Do you really consider all this demonstrates my bias? Most people would see it as my desire to bring all the evidence and incidents into the open. I do believe that all the evidence I have seen to date (and I have read a great deal more of it than I think you have) makes a most damning case against Kurtz and CSICOP. But I not only think my mind could be changed, I have done everything I could do to get any such evidence from Kurtz et al. that might change my mind. I will continue to do just that. On the other hand, I think you really know very well that the evidence is damaging to CSICOP and do not want it brought forward. I hope I am wrong in this assessment, and urge you to persuade me otherwise.

I don't claim to be emotionally indifferent to all these matters. I have come to view Kurtz as a devious scoundrel. I earlier hoped CSICOP would reform itself at the Toronto meetings, but all evidence indicates nothing to make matters better was done. Instead, I see stonewalling and vague counter-charges (particularly character assassination by Phil Klass) that hardly help CSICOP's credibility. But whatever my emotional likes and dislikes, my commitment to truth and fair-play (even for Kurtz) remains more important to me than such petty and personal matters. I have repeatedly stated that I think Curry's criticisms are far more serious than those so far brought up by Rawlins (Rawlins apparently has further charges waiting in the wings). As always, I invite you, Kurtz, Klass and anyone else interested or involved in these matters to enter into the public dialogue in *Zetetic Scholar* (or to publish your papers in other public forums including *Skeptical Inquirer*). I believe in open science. I once thought you did too.

Perhaps you are not aware of the vast number of documents involved in this whole affair. From my first attempts to get you involved with the Mars Effect controversy, you brushed it aside as crank science not worth your time. You write me now as though you know the full range of the evidence and arguments While all I have read is Rawlins' attack in *Fate*. I will let those who might eventually read our letters decide which of us is acting in a mature and scientifically responsible manner.

Zetetically,
Marcello

[Eastern Michigan University] 15 December 1981
Dear Martin,

I would appreciate it if you could throw some light on a matter.

On Dec. 17, 1980, I wrote you asking about whether or not you ever voted to expel Dennis Rawlins from the status of Fellow of CSICOP, as opposed to his removal from the Executive Council (where I gather he was not voted out per se, merely not renominated for a renewed term along with all the other Councilors). On December 24, 1980, you replied to me as follows: "As for the 'balloting,' I assume you mean a hand vote, because so far as I know all voting has been done that way, I can't recall voting on *anything*, but that may be because I missed the morning business session and arrived only for the afternoon one. If there was a vote to expel Dennis from the Council, I would have voted for it, because Dennis's tactics make it impossible for any group to function with him on it. If there had been a vote to remove him as a fellow, I would have voted against it. I honestly don't know if there was such a vote or not."

Later, when Rawlins circulated his questionnaire among the Councilors to learn about the alleged (by Kurtz) ballot on his expulsion, you returned the ballot to Rawlins with some personal comments on it but without filling in the questionnaire part about the ballot.

These acts seem quite consistent to me, and I presume the thrust of the matter is that you don't fully remember about any ballot but (a) did not vote on written ballot and (b) were disinclined to vote Rawlins out as a general Fellow (though you favored his expulsion earlier as a Councilor).

Now what I am hoping you can shed some light on is the following. Numerous inquiries have gone to Paul Kurtz about this purported ballot that expelled Rawlins from the CSICOP. Not one Councilors replied to Rawlins' questionnaire indicating that there had been any ballot. At the Toronto meeting on Oct. 23, I am told that Kurtz produced a different ballot in which new Councilors Fellows were elected. I am told this "new" ballot said something along the lines that one of the new proposed Fellows would replace Rawlins. Though most readers of such a ballot would infer that Rawlins had already been expelled and his vacancy was now being filled, Kurtz claimed that this ballot was actually the expulsion ballot for Rawlins. That is, Rawlins was indirectly being expelled via the statement that new proposed Fellow "X" would be replacing him. (I find this a far-fetched tale, but let us accept it at face value for the present.)

The problem is this: Kurtz just wrote at least one inquirer that (a) you voted on the ballot that expelled Rawlins, and (b) you voted in favor of expelling Rawlins from CSICOP fellowship.

I think this is contradictory on two counts: (1) a ballot is being claimed when you thought, there was never a written ballot you had voted on, and (2) you are described as favoring Rawlins ejection as a Fellow.

I am inclined to think your original description is completely correct and that Paul Kurtz's memory is inaccurate, to say the least. There maybe a way that both your and Kurtz's descriptions can be correct, but I can see no way to reconcile the two.

Any light you can shed on this would be appreciated.

Since Paul may be able to clear it up if you can not, perhaps you should ask him about it. I would write him myself, but he has not responded to the last couple of letters I have written him.

Finally, since the basis for my understanding about the Toronto meeting and the mysterious ballot comes from my conversation with Ray Hyman, and there is chance I misunderstood him somehow, I am sending him a copy of this letter, too. He can correct any misimpression I may have received from him if that is the problem.

Thanks for your help on this.

Cordially,
Marcello Truzzi

CC: Ray Hyman

[Woods End]   17 December 1981
Dear Marcello:

As you requested, I am sending a copy of your last letter (Dec 11) to those to whom I sent a copy of my previous letter. Copies have gone to: Kurtz, Frazier, Klass, Randi, and Diaconis. I did not send one to Hyman because you have already done so.

Since all our argumentation turns out to be so useless and time-consuming, I'll make no comments on your letter.

Best,
Martin

[Postcard]   18 December 1981
Dear Marcello,

In view of the fact that you are doing all you can to discredit the committee and its officials why should I collaborate in your vendetta? Please—don't write me any more letters.

Best,
Martin

[Eastern Michigan University]   22 December 1981
Dear Martin,

I write to make the record clear.

I yesterday received card from you in which you state: "In view of the fact that you are doing all you can to discredit the Committee and its officials, why should I collaborate in your vendetta? Please—don't write me any more letters."

I must presume that is in direct response to my letter to you of Dec. 15 in which I ask you to shed light upon what appears to be a direct contradiction between your version of events on Rawlins expulsion and that given by Paul Kurtz. I believe most neutral observers would view my letter to you as a request for information that might clear up rather than discredit anyone.

Your card indicates that your mind is made up about both my motives (though I have consistently denied that I have any vendetta against CSICOP or all its officials) and you apparently do not even want further information or possible new evidence. Since there is no reason to think you have more information about these matters than I do; and there is much reason to believe you have a great deal less, I must describe your attitude as dogmatic. That is a pity; for I have up until now respected you even though we disagreed.

Though you ask me to send you no more letter—and this will probably be my last one to you personally therefore—I will continue to send you whatever copies of letters I may send out to Fellows or Councilors in general. You are free to ignore them. I however, do not wish to be accused of withholding any information from you by others or have anyone think I in any sense speak behind your back. Unlike Paul Kurtz, who seems to want to respond to public charges through documents marked "confidential" and "not for publication, distribution or quotation"—a device which helps limit the opportunity for factual rebuttal by his critics—I will not indulge in such private tricks and gambits.

I am truly disappointed in your request that we stop corresponding. I have gained much from your letters over the years, and I had hoped you had also found my letters of equal interest. Since at one point, Ray even suggested publishing an edited form or our exchanges on pseudosciences. I don't think "our argumentation turns out to be so useless and time-consuming" as you stated in your Dec. 17 letter received a few days before I got your card. I have enjoyed our correspondence and regret your decision to discontinue it. Perhaps time will mellow your views and the reality of my motives and view of Kurtz will penetrate your dogmatism. I remind you that at one time Rawlins thought I was out to get CSICOP and you defended me against his charges. He now knows he was wrong and admits it. He also knows that I am now after reforming CSICOP and not its destruction. You may eventually also agree. I hope so.

Sincerely,
Marcello

CC: Ray Hyman, Persi and Randi

[Eastern Michigan University]   30 April 1982
Dear Martin,

I hesitate to write you since you asked me not to, but I do so because it is bad enough that we have philosophical differences without having factual errors get into our opinions of one another.

Dick Kammann mentioned to me that you believed that I had seen copies of the actual ballots which ejected Rawlins from CSICOP. This is absolutely incorrect. I have never seen these ballots, and even now do not know the votes therein. I did request that Kurtz send me a copy of a blank ballot when the note in *Skeptical Inquirer* mentioned that they were at CSICOP headquarters (presumably available to inquirers). Please note that I did not ever ask for copies of the actual ballots; I have no interest in knowing exactly how each of you voted (that is none of my business—though Kurtz has told others what the votes were). I received no reply from Kurtz to my inquiry, but I did receive a letter of indignation from Ken Frazier (with no copy of a ballot) for being audacious enough to ask to see one. The reason for my request for a copy was because Ray Hyman had told me that the wording was ambiguous and that he had not understood that it was a ballot to eject Rawlins when he had received it. The notion that the ballot was ambiguously worded was not my invention (as you suggested to Kammann); I derived that view from what Ray had told me. And even then, I did not simply assume it was ambiguous; I asked to see a copy.

Recently (about 2 months ago at most), Ray Hyman did send me a letter in which he quoted the actual wording of the ballot to me. It clearly is quite ambiguous. I am, of course, delighted to learn that Ken knew exactly what he was voting for. I understand that you told Dick Kammann that you clearly knew what you were voting for, also. Yet, I remind you that you wrote me to the contrary in your letter of Dec. 24, 1980. I do not accuse you of lying; but I certainly do point out that this is a contradiction. I make no accusations, and I am open to your clarification.

If you really believe that I am playing tricks and acting deceptively, I hope you will confront me with such charges directly, for I would be happy to answer them. I would have expected such direct confrontation from you rather than hearing about your allegations through third parties.

Sincerely,
Marcello Truzzi

CC: Dick Kammann, Ray Hyman

[Woods End]   5 May 1982
Dear Marcello:

I reply so I can send Dick a copy. When I told Dick I had signed and understood the vote, I said to myself: "As soon as Marcello hears this, he'll protest that I once told him otherwise." Sure enough, here comes your letter.

It is true that when you asked me about the vote I could not recall it. Had you asked me if I had voted Abell in, I would also have said I couldn't recall it. The vote was such a trivial part of my life that I had totally forgotten it. It was not until I received copies of all the voting sheets that I remembered it. I don't know about you, but the slipping of trivial events out of the memory is common with me, especially as I get older. Recently I bought a book on recreational math, took it home, and began reading and underlining. Halfway through, the pages began to seem familiar. I checked my shelves and found that I had bought the same book two years before. Worse still, I had earlier underlined the same passages I was now underlining.

Sorry I assumed you had seen the voting sheets. Only Ken, as you know took the trouble to write in his sheet that he favored Abell in but not Dennis out. I myself had, I now recall, similar misgivings. Ken and I no longer have such misgivings.

I am astounded that you are concerned about such nitpicks. Don't you have more important things to attend to? You are supposed to be publishing a periodical about anomalies and possible new paradigms. Yet here you are devoting the bulk of your last issue to CSICOP, with promises of more to come. To just about everyone except yourself, your behavior is becoming more and more obsessive. Can't you rise above the Fortean level, and personal animosities, to grapple with some of the marvelous new offbeat theories in a dozen sciences, minority viewpoints opposed by the establishment, which have a fair chance of becoming new paradigms in Kuhn's sense? Who gives a damn about the Mars effect, or Velikovsky, or bigfoot, or whether UFO's are extraterrestrial or some other sort of "unexplained" phenomena?

Please, please don't bother me with any more questions about Rawlins's battles against everybody except (at the moment) you. I have already wasted a half-hour of valuable time composing and typing this letter.

Best,
Martin

CC: Dick K

[Zetetic Scholar]   22 August 1983
Dear Martin Gardner,

I write you about 3 matters for the record. I don't expect a reply from you.

1) I am told that in your new book *Order and Surprise* you republish a review of the Clever Hans symposium in which you (a) refer to *ZS* as a Fortean publication, and (b) refer to my being a follower of Feyerabend re relativism in science. The word "Fortean" is a loose enough term that I would not argue that overmuch but would merely call your attention to the fact that I have both publicly and privately criticized the mystery-monger approach of most Forteans and disassociated myself from that. But more serious is your apparent mistaken idea that I share the view of Feyerabend that there is no such thing as scientific method. I thought I had repeatedly made my differences with Feyerabend clear, including in past letters to you. I do think you badly misunderstand Feyerabend—as does he—but I am not in agreement with his position.

2) At the recent panel of magicians I convened for the P.A. meetings, mention was made of the fact that although Randi seems to want to consult with and be a friend to parapsychologists, you recently wrote that magicians

were naturally the enemies of parapsychologists and that their bad vibes would alone eliminate any positive PK results. Commenting on this, Randi said to us all that he had spoken with you re the column in which you wrote thus and that you had indicated that you did not mean it after all. I hope, therefore that you will make some sort of correction/clarification about this in a future column in *Skeptical Inquirer*. Either that or correct Randi.

3) In your *The Whys of a Philosophical Scrivener*, in the chapter on why you are not a paranormalist, it is most interesting that you reject the common definition of the "paranormal", which is to be found clearly in the literature. This ignores the actual demonstrable etymology of the term and its common usage by authors who call themselves paranormalists. By that common definition, you are indeed a paranormalist, as you noted. So, welcome to the club. But you reject that definition (as would an offending Humpty Dumpty whom you cite later on in another chapter) for an idiosyncratic one by pointing to the use of the term by journalists. But the fact is that by the definition of "paranormalist" that you seem to accept. I simply don't know anyone in parapsychology who would then use that label to call themselves a paranormalist. You seem to me to be opposed to the views of an empty set in so far as your own definition would apply to those who call themselves paranormalists. The term as used by some journalists is clearly an abuse of language just as is the way some journalists would call some things both of us would call crap "science" (or the way some journalists abuse the sociological term charisma).

Zetetically,
Marcello

CC: Randi, Ray Hyman

P.S. I see I forgot to comment on your P.S. comments. I certainly agree that G's claims are extraordinary and require extraordinary proof. However, I do not agree with you about his claim being more extraordinary than remote viewing, and I do not think his evidence offered is less extraordinary than Targ and Puthoff's. Again, I think you misunderstand what G is claiming. He claims merely to have found a consistent correlation. He does not infer from this alleged data any form of causality such as in astrology. He does not claim that the planets cause occupational choices. He simply claims a correlation exists that needs explaining. He has seriously looked for alternative variables that link these two phenomena and has tried to control for such. But he claims the correlation persists despite whatever controls he has been able to come up with. He still thinks there may be some common factor linking the two things. He has also suggested that the causality may be the reverse of astrology's claimed direction; that is, there may be something genetically present in the fetus that gets triggered somehow by cosmic events and somehow signals that it is time for the fetus to emerge. He admits this is not very plausible either. He claims a consistently observed phenomenon. He would like to have others confirm it; but in the meantime will further investigate it himself in the hopes of finding the causes involved. If his data is correct, he may have found something important. Maybe not. Compare this to Targ and Puthoff. Their claim is far more extraordinary and their evidence far weaker. The claim is that the human mind can transcend space and time in ways completely alien to modern physics, physiology and psychology. When the subject remotely identifies a target, they consider this a direct causal relationship via clairvoyance/precognition. If they simply pointed to a curious correlation between a subject's guess and a target, said they didn't know what produced this remarkably consistent coincidence, and were willing to consider serious alternative explanations (alternative variables), their position would be more like Gauquelin's. But their claim is far more broad and extraordinary. They insist on a process that we would label magical. Gauquelin does not do this. He would accept an explanation, that was quite natural (e.g., athletes are mesomorphs, mesomorphs tend to come from mesomorph mothers, mesomorph mothers tend to be more erotically inclined when weather is colder so have more sex then, and this tends to produce mesomorph babies more frequently during months when Mars is more commonly in a certain position). As for evidence offered, Targ and Puthoff offer some dubious experimental data which really can not be checked (as can G's data) and the data they offer is not the evidence itself; the evidence is certain inferences they make from the data. Gauquelin's data is itself his evidence. Both Gauquelin and Targ and Puthoff cite non-chance patterns. But the non-chance pattern itself is Gauquelin's claimed anomaly. For Targ and Puthoff the inference they make as to the causes of this non-chance pattern (clairvoyance/precognition) is their anomaly, not just the non-chance pattern itself. So, you see, G's evidence is really much stronger than Targ and Puthoff's to my mind. But most important, his claim is itself not so extraordinary since I can concoct numerous possible explanations which might explain his data should his data be confirmed. And these explanations need not be very implausible at all.

[Eastern Michigan University]   10 January 1984
Dear Martin,

I was aghast to see your outrageous reply (*Skeptical Inquirer* Winter 1983-84) to my corrections of your *Skeptical Inquirer* column. I had thought you might simply not have realized that you were in error and give me a simple apology. Instead, you reply with a casual regard for the truth and with a sophistry that should not (but I am afraid will) mislead readers.

There seems little point in my replying to you further in the pages of *Skeptical Inquirer* since you would always have the final word there and I have no assurance of fair play from that quarter. But because I do not want my silence to imply consent to any of our mutual friends, and because I want the record set straight, I write you now.

You assert that my corrections are valid but you cleverly describe them as mere "quibbles" even though you must fully recognize that your original erroneous assertions would be damaging to my character and reputation. You then distort my points.

1) Randi did not demonstrate that the Mac Lab committed a "compromising statement or act" in the sense of actually doing anything pseudoscientific since their private belief in the authenticity of the boys did not result in their publicly, asserting any scientific validation for the boys "powers." Though you and I would agree that the Mac Lab personnel acted gullibly, they did not confuse their belief in the boys with scientific demonstration. Also, I used the term "entrapment" in my letter to Randi, but that clearly referred to Randi's then *attempt* to entrap the Mac Lab personnel. Randi may have engaged in entrapment in the sense of laying a trap, but the fact is that the trap was "not" successful. He did not actually entrap Phillips; he only apparently hoped to do so. I also use the word "sabotage" in that same letter (in the very same sentence) and that term (unlike "entrapment") refers to Randi's actions then and now. Finally, on this point, your original column suggested that I introduced the term "entrapment" in this dispute as though Randi's opposition picked up that term from me. That is what I was correcting, for I don't think I ever used that term outside this semi-private letter to Randi (which was sent to Randi's friends, not his opponents).

2) You initially said that I knew about the trap from the beginning because that was intended to indicate I was in some sense an accomplice via my silence. Readers would naturally infer that I initially approved of the scam. You also later imply (at least readers would naturally infer) that I was let in, on the scam by Randi, which you know from my letter to Randi was not at all the case. I objected from the first I heard about the matter from others than Randi.

3) You say "I had assumed he kept his word to Randi that he would not reveal the hoax." When am I supposed to have given my word to Randi that I would remain silent? I can only presume that you somehow got this from my letter to Randi of 7/29/81 a copy of which was sent you. There I stated: "I do not plan to do anything about all this until at least a week before the PA meetings—if I do anything at all." You must have read that statement.

How do you get a promise out of that? Readers of *Skeptical Inquirer* would likely get the impression from your writing that Randi told me about the scam and extracted a promise of silence from me. You must know that is nonsense. In addition, you ignore the facts which took place after my letter to Randi. Among other things, Randi's reply to my letter was a phone call in which he completely denied that any scam was taking place at all. Only when I confronted him with my source of confirmation (Ray Hyman) did he admit that Alpha existed. I also had other exchanges with Randi which convinced me that the outcome was likely to be destructive rather than constructive to good science.

4) You end with the assertion that by my telling a number of scientists and science writers about Alpha the tale would "obviously" get to Prof. Phillips. Not only was that not the case, for the science writers I communicated with were not friendly to parapsychology (in fact one of them was *you* via the letter to Randi of which I sent you a copy—for all I knew, you did not know about Alpha until you got my letter), but a prominent rumor at the PA was that Randi and Phillips were in collusion against others at the PA since Phillips' tapes of the boys were so similar to the tape Randi showed of trickery. It is interesting that you earlier criticized me for supposedly keeping silent, and now manage to criticize me for telling anyone.

I find it very hard to believe that these continued "errors" on your part are the result of your honestly trying to search for the truth rather than simply being rhetorical ploys to obfuscate matters and deceive the *Skeptical Inquirer* readership. All this is particularly regrettable to me because I once sincerely admired you. I used to think that we had our differences but that these were the result of different views about strategy in dealing with extraordinary claims. I did not then think you were intellectually dishonest. Alas your contradictory statements during the Mars Effect controversy and your recent misrepresentations of events about which you must know better leave me little choice but to conclude that the truth matters little to you. I think you know from my past correspondence with you that I have always sought to present arguments and evidence rather than horselaughs. I may likely give some response to your comments in a future issue of *Zetetic Scholar*. If so, you are always welcome to reply in the pages of my journal, and I shall try to give you more fair play than I think you have given me.

I am sending copies of this letter to those who received my original 7/29/81 letter to Randi, plus some other relevant persons, for whom I wish to see the record set straight.

Sincerely,

Marcello Truzzi

CC: Randi, Hyman, Diaconis, Frazier, Kurtz, Phillips, Hoebens, et al

Chapter 4

Return to Cordiality

After over a year the correspondence resumes with little rancor. The topics strayed away from the demarcation problem, but not entirely. This chapter provides a pleasant denouement to the stridency of the foregoing letters.

In a 1991 interview Marcello stated:

> My main concern is getting people together who agree with the basic scientific ground rules. The committee was set up to be a business. The name of the game was to get a publication out there on the newsstands, they actually had a calling, a mission. Originally, they were supposed to be in investigation, inquiry and research; what they did they botched up, and now they don't do that at all anymore. They're having a lot of trouble; major lawsuits, Randi has resigned and so on. Some people hear the death-knell. Some of it was foreseeable, actually I was surprised they lasted as long as they did ...

(This states that CSICOP was "supposed to be in investigation" which was never true.)

Clearly Marcello would be surprised that the *Skeptical Inquirer* has just celebrated 40 years of publication. The Committee (now called CSI) is going strong. While it still engenders complaints, it continues to attract many new people. By contrast, the *Zetetic Scholar* was only published 12 times until 1987, irregularly at the end. Both have their distinct legacies.

[Postcard]   26 March 1985
Dear Martin

Though I do recall your asking me not to write you, I feel I must do so out of courtesy in this case. I am currently writing a feature piece for *OMNI* on the relationships between parapsychology and quantum physics, for which I have been interviewing many. I do not want to misrepresent you, so I write you now to ascertain if you have changed your position in regard to these matters. I believe I have everything you have published on the subject, but there is always the small chance that you have reconsidered some aspects of these matters since then. If I don't hear from you, I will assume that you essentially stick with what you have published so far. As you must know, there have been some changes in this area of late which might have prompted some shift in your views—though I suspect not. Anyway, I don't want to misrepresent your current views if they have changed. So let me know if there has been any shift (in either direction). Thanks.

Sincerely,
Marcello Truzzi

[Woods End, Inc]   29 March 1985
Dear Marcello:

I assume you have my Discover article on "Quantum Weirdness." I accept the Copenhagen interpretation of QM as the best available, though I remain open to new evidence that might restore classical causality along the lines Einstein hoped. I don't know of any new developments in QM that have any bearing on ESP and PK, only new claims by parapsychologists that QM may explain their data. The recent confirmations of EPR merely confirm the formalism of QM. Einstein took for granted the EPR would be confirmed, and regarded this as evidence that QM was incomplete not wrong, but just not the full story. I know of no better books on the new physics than Pagels *Cosmic Code* and his soon-to-be-published *Perfect Symmetry* that covers the latest theoretical developments.

If you care to let me see a typescript, I'll let you know if I think I'm being misrepresented. As for the views of Schmidt and others, that a record of QM events (Geiger clicks or bubble-chamber photographs) are not fully determined until perceived by a person (or a mouse?), I assume you know that only an extremely tiny minority of physicists buy this extreme solipsistic point of view. I hope you don't try to elevate Walker into the ranks of QM experts!

The answer to your specific question, of course, is no.

Best,
Martin

P.S. You must know of Dr. Santilli's recent sad blast (in the book he privately published) vs. the physics "establishment," and his new magazine *Scientific Ethics*. If not, I can send you details.

[Eastern Michigan University]   8 April 1985
Dear Martin,

Thanks for your letter of March 29. Yes, I am familiar with Dr. Santilli's recent books (I trust you also saw the rather favorable review he got in the Harvard paper). I have not yet seen his new magazine but was surprised to see Mario Bunge's name associated with it.

Thanks for telling me about Pagel's new book. I liked his *Cosmic Code*. Have you perhaps reviewed Fred Wolf's *Taking The Quantum Leap* anywhere? (I recall your negative remarks about his earlier book with Sarfatti.) I wonder about your opinion of it since it won an award (I presume you dislike his *Starwaves*).

Although I would not elevate Walker to the rank of QM expert, the fact is that he has published on QM in regular physics journals as well as in the psi press, and I must admit that I found his response to your critique (in his rebuttal in the current *Journal of Parapsychology*) to be quite good. I hope you will respond to him at some point. The irony to me about Walker is that he is actually conservative rather than radical in trying to redeterminize the world via the introduction of consciousness as a hidden variable. Thus, he—like Wigner—is arguing for the incompleteness of QM.

I have been particularly impressed by Henry Stapp's efforts, and let me call your attention to the remarkable new article by A. J. Leggett ("Schrodinger's Cat and Her Laboratory Cousins" in *Contemporary Physics*, 25, 1984) which argues for QM paradoxes on the *macro* level! I am awaiting a copy of a tape of Bohm's talk about a week ago to the ASPR, and I just finished listening to a tape of Feynman's talk to the Parapsychological Association last summer. Feynman took a far less harsh line than is found in his "cargo cults" paper.

Don't know if you have kept up with Sarfatti in recent years. He seems to now be prepared to lay it all on the line with an experiment he has proposed to Aspect and others. He flatly told me that if it failed to bring confirmation for his predicted outcomes, he would admit he has wasted the last 20 or so years. I have to hand it to him for moving away from what physicists like to call "hand waving" to a falsifiable hypothesis that may soon be tested.

By the way, I wonder what you thought of Steve Shore's essay. If I read him right (and Mark Sher, an anti-psi physicist-critic who recently talked to the Bay Area Skeptics, seems to agree with me), he fundamentally disagrees with you about "quantum weirdness" and does not understand the difference between simply not knowing the eigenstates (indeterminableness) and having them be in an uncertain state of superposition (fundamentally undetermined). So, it seems, critics don't all agree with one another about QM when it comes to discussing paraphysics.

If time permits, I will send you a copy of the manuscript I am doing, but I will in any case check, with you if I go beyond direct quotation of you. Frankly. I think you'd be hard to accidentally misrepresent—you write so clearly—but I did want to make sure you had not shifted your views in light of so much recent stuff coming out that (I think) should perhaps moderate your views somewhat. If rumor be true, even John Wheeler may be shifting in his prejudices somewhat (I understand Bohm has spent the last week with Wheeler visiting him in Texas).

Re the matter of who is a proper QM expert, I am mainly sticking to the "founding fathers" and the clearly recognized folks, but I am impressed by the receptivity to consciousness intervention ideas among those like d'Espagnat and Shimony, and I do think Bob Jahn understands the field very well indeed. And I must say that I have become growingly impressed by the reputations and "straight" publications of some others I had previously thought of as wild, e.g., Elizabeth Rauscher. It's all complex.

Anyway, I am learning a lot researching the article, and I have certainly benefited from reading your own pieces in this area. Our own major difference seems to be that you insist that psi needs to be solidly established prior to using QM to explain it. I think the relationship between theory and data is more reciprocal than that. I remain quite unconvinced by psi claims, but whereas I doubt, you seem willing to dismiss (and even ridicule). Unlike yourself (apparently), I feel that since the burden of proof is on the claimant, and extraordinary claims require commensurate proof, that heavy burden on the claimant allows the scientist to be tolerant rather than dismissive toward unproved claims. Here I agree thoroughly with Huxley on the need for agnosticism even towards the wildest of claims (such as his example of a guy claiming $2 + 2 = 5$). In a way, if you go beyond that, it really shows that you don't sincerely believe the burden of proof is on the claimant at all; for if the burden is truly his, why go beyond merely saying his case is unconvincing and therefore unproved? Why block inquiry further?

Well, I see I have probably written you back more than you really want to read from me.

Best,
Marcello Truzzi

P.S. By the way, I was pleasantly surprised to learn that Fred Wolf is a long time conjuror and admirer of yours in that realm. Seems he uses two headed coins in some of his demonstrations about superposition.

Let me also call your attention to the amusing "The Classic Paradoxes of Quantum Theory" by Asher Peres in *Foundations of Physics*, 14 (1984). A series of dialogues you'd probably enjoy reading.

[Woods End, Inc]   15 March 1985
Dear Marcello:

Not suing Targ—just Random H, and *only* because they gave me their word, in unequivocal letters, that the offending paragraph would be removed from the paperback. This almost a year before it was published. They did not remove it. (I had promised to take no action if it were removed. They went back on their promise, hence...)

Can't reply to your Wheeler question without typing five or more pages, and I don't want to drift back into exchanging long, time-consuming letters.

Your father is mentioned on page 160 of Quine's just published autobiography.

Best,
Martin

[Woods End, Inc]   1 July 1985
Dear Marcello:

Dear Marcello:
I wish you'd try to check on what I write before writing to me about second and third hand reports. I know nothing about any stacked balls on the foot, or about a deflated ball. Ron Graham, the Bell Labs juggler, once told me that your father was the only juggler he knew who would spin three balls on a finger. Ron has an ironclad proof that this is dynamically impossible. He asked me to ask you about it, and I did. You told me that

one of the balls was weighted inside, and that was what I reported. If this is incorrect, or you think I misunderstood what you told me, please set me straight and I'll correct my review (of a book on lightning calculators) before it goes into a book.

Best,
Martin

[Eastern Michigan University]   9 July 1985
Dear Martin,

Re your note of July 1 in reply to my card comment on your piece and its reference to my father.

Sidney Gendin (you probably read his anti-psi article in *Skeptical Inquirer*) read your review of *The Great Mental Calculators* and told me about the supposed reference to my father and the three balls trick. He was supposed to bring the article to me but never did so. I sought to get a copy from our library (it was already off the newsstand here), but someone had ripped off the copy. So, I did indeed try to check on your report prior to writing you. It is no big deal, but I happened to think of it while writing the last card to you and had room on the card to mention it.

If you would like to send me a copy of the paragraph involved, I would certainly be glad to confirm or deny its account. But here is what I can say based on the comments in your letter to me. Ron Graham is absolutely incorrect if he asserts that my father ever demonstrated spinning three balls on a finger. My father never even did the two balls spinning trick (which Francis Brunn and some others have done). What my father did do (and which I remember once talking to you about when we met at your house with Ray Hyman) was stack up four balls on his foot and apparently balance them all. I quite agree with you and Graham that this is physically impossible without a gaff. The top ball of the four balls had an internal weight, which I did tell you about. The second ball up had an entirely different sort of gaff—which I don't mind describing to you privately but which I do not wish to do if you plan to expose the method to the general public in your article. As with conjuring effects, I think such tricks should be kept within the fraternity of jugglers. (Frankly, I am a bit surprised that you would blow the gaff on such a trick at all in your general article.) So, if you want to know about the gaff in the second ball, just assure me that you will not publicize it, and I will be glad to tell you. (I'd do so in this letter but I feel it improper to tell people secrets first and then try to obligate them to keeping them even if they did not first agree to do so.)

I think a far better example of this sort of thing—in terms of exposing to the public—is the common trick of seemingly stacking many rods one atop the other (apparently balanced but actually connected) with a basin spun on the top (which even some clowns do). This really would not be giving away a secret of any significance since anybody with brains should see that that is basically impossible.

Best,

Marcello Truzzi

[Woods End, Inc]   12 July 1985

Dear Marcello:

Glad to have cleared up the great spinning-ball mystery. Obviously I was either unclear in my question, or you misunderstood it, or both. I had recently been in Ron Graham's office where he demonstrated his skill in spinning two rubber balls on a finger. He explained how one had to move the hand in a manner opposite to that of spinning one ball, which made it extraordinarily difficult. I asked if three could be spun. He said no, then added that he had heard that the great Italian juggler Truzzi did it, but he must have had a gaff. I told him I knew you, and he suggested I ask you about it. *That* was my question. I recall your saying the middle ball had the weight. Naturally I took this to be the middle of three spinning balls, and naturally you meant one of the middle balls four balanced on a foot. I shall remove the offending passage when the review goes into a book, and henceforth will check all references to you or your father in anything I write in the future.

It would be fruitless to debate by mail what should or should not be exposed. On the continuum involved, along which all edges obviously are fuzzy, I distinguish roughly between magic which everyone In the audience knows to be trickery, and the use of trickery by magicians pretending to be psychic, by acrobats like Unus (though since his career depends on his one trick I wouldn't expose it), and by jugglers and lightning calculators who are believed by the audience to be doing everything legit. It's one of those many situations where your ethics and mine are not quite the same.

A personal note on my Quine review. Ed Haskell and I were friends at Chicago, both living in a two-dollar-a-week fleabag on Dorchester called The Homestead Hotel. I often wondered what happened to Ed, and almost fell out of my chair when I came across him in Quine's autobiography. You may have bumped into him sometime because he moved around a lot from campus to campus. I'd be interested in knowing if he is still living.

19 July 1985
Dear Martin,

Thanks for your letter (of 12 July) and the copy of your review. Thanks for your intention to make the correction in future.

Will try to learn of Ed Haskell for you. I presume you have checked with Jay Marshall (who seems to know everyone in magic).

Seen David Price's extraordinary new history of conjuring? Delighted with its section on the early mentalists. I am doing a piece on the relationship between conjurors and psi research for the next *Advances* volume for Krippner and this supplements stuff I have.

Have you seen anything negative written about Joseph Newman's machine and his fight with the Patent Office? Have you seen his book yet?

Did you know of Ken Frazier's brother's heavy involvement with UFO contact research? Funniest skeleton since news of Richard de Mille scientology connections!

Best,
Marcello

P.S. Random House settle out of court with you yet? I'd expect that.

[Postcard] 19 July 1985
Dear Martin,

In reading your *NY Review of Books* essay (which is otherwise quite excellent, I think), I discovered another mistake re my father. My father always worked under his full real (and not stage) name, Massimiliano Truzzi. Posters sometimes used only his last name, but I know of no occasion when only his first name was used. I don't know where you got the idea that he ever used just his first name as a "stage name." His real full name was Massimiliano Enricovich Truzzi. I wonder about another item: You seem to believe S. Devi cheated re her *Guinness Book* record. Do you have independent reason (aside from Smith's chapter on her) for thinking so? If so, I'd love to know the basis for your skepticism going beyond Smith's. Has anyone written the Guinness people for confirmation and exact conditions? They usually require documentation for the newer inclusions.

Best,
Marcello

[Woods End] 24 September 1985
Dear Marcello:

Herewith the *Foote Prints* piece. *Amazing* has asked me to write an article about Ray Palmer, as a forgotten UFO pioneer. Would you know whether Dr. Raymond Bernard, who wrote *The Hollow Earth*, was one of Palmer's many pseudonyms? *Fate* advertised the book heavily, and the book is filled with quotes from Palmer and praise of Palmer as the "world authority" on UFOs. The book was distributed by Fieldcrest Pub. Co, NYC. J. Clark might know, and surely C. Fuller would know, but I hesitate to write to either person for info.

Best,
Martin

[Eastern Michigan University] 27 September 1985
Dear Martin,

First, thank you for sending me your *Foote Prints* article which I am very glad to have. I eventually want to do something on the Joe Newman vs. the patent office issues and your piece should prove useful.

Regarding Ray Palmer's possibly being Raymond Bernard, I can understand your speculation, but so far as I know it was not one of Palmer's several pseudonyms. I am dropping a line to Jerry Clark and Curtis Fuller to see what I can further learn. You might drop an inquiry to Jim Moseley [address withheld] since he might be someone who would know more about Palmer. If you don't get Moseley's *Saucer Smear* you should ask him to send it to you (it's free). Strangely, Moseley is quite good friends with Phil Klass, of all people.

You probably know Eric Norman's *The Under-People* (NY: Award Books, 1969), but if not, you might find it relevant to your concern with the hollow earth claims. Indirectly related to such stuff is the marvelous hollow moon and other alleged NASA-cover-up stuff: Don Wilson's *Our Mysterious Space Ship Moon* and *Secrets Of Our Spaceship Moon*; William L. Brian's *Moongate: Suppressed Findings of the US Space Program*; Bill Kaysing's *We Never Went to the Moon*; and—perhaps the ultimate crazy book—Joseph H. Cater's *The Awesome Force: The Unifying Principles for All Physical and Occult Phenomena in the Universe* (the subtitle modestly tells it all!).

Another person who might be of help to you re whether Palmer was Bernard is Dan Cohen. Other possible sources could be Sprague de Camp or Asimov, but I suppose you already thought of them. Anyway, I will let you know what I turn up via Clark and Fuller.

Best,
Marcello Truzzi

P.S. You might not know that Guy Lyon Playfair is doing a new book on Geller and his critics (Uri now lives in London, I gather). Perhaps you can cast some light on a minor mystery. You will recall that Ray Hyman sent you a copy of his 13 pages of notes on his visit to SRI in 1972. I presume, based on what Ray has told me, that Jaroff probably got a copy via either you or Morrison, but do you know how Puthoff got a copy of it, so soon after you got it from Ray? Ray was very surprised at that and wondered if there might be a CIA connection in that somehow. Do you know where Hal got his copy? On another matter: Can you tell me anything about the new Institute for Advanced Studies at Austin to which Hal has moved from SRI (at reportedly double his old salary plus virtually unlimited research support)?

[Woods End]   8 October 1985
Dear Marcello:

Thanks for having Clark write to me about Bernard. What he says has been confirmed by friend Bob Shadewald.

I'm no help on any of your questions. Never heard of Einstein as a dowser, don't know who Hapgood is, and have no recent info on Newman's hearing.

Best,
Martin

[Woods End] 10 October 1985
Dear Marcello:

Just remembered that in my note I forgot to reply to questions in our earlier letter. Yes, Jaroff got his copy of Ray's letter from me, but I have no idea how it found its way to Puthoff. At the time, John Wilhelm was on Leon's staff, and since he interviewed Puthoff at length for his book, my guess is that Wilhelm gave it to Puthoff.

I know nothing about the Austin institute to which Puthoff is said to have moved.

Best,
Martin

[Postcard] 17 October 1985
Dear Martin,

Sarfatti (who now has a visiting professorship in Brazil) just sent me a most remarkable article from the abstracts of the recent Finland symposium on the EPR anniversary: T. Hellmuth, et al., "Realization of a Mach-Zehnder, 'Delayed Choice' Interferometer," which seems to be an empirical demonstration of one of Wheeler's wilder ideas about backward causality. Sarfatti seems to be getting some serious attention lately.

Also thought I might call your attention to the Aug. 6, 1984, House Committee on Science and Technology report on "Subliminal Communication Technology," (if you are not already aware of it).

If you hear anything about Newman's big press conference/demonstration of his machine in New Orleans last week, please let me know.

Best,
Marcello

[Eastern Michigan University] 12 August 1989
Dear Martin,

I presume you have seen "Response to Martin Gardner's Attack on Reich and Orgone Research in the *Skeptical Inquirer*" by James DeMeo. (If not, I can send you a xerox.) I am wondering if you may have responded

to this piece someplace, in which case I would like to see your reply. If you
have seen it and replied, perhaps you could share a copy of or the citation
for your reply.
    Thanks.

                          Curiously yours,
                          Marcello Truzzi

P.S. Art Lyons and I are currently working on a book on the use of alleged
psychics by law enforcement and government agencies. If you know of
anything you think I am likely to have missed seeing in that zone, I hope
you might let me know. We want to be as exhaustive as possible (so far
about 98 psychics and about 300+ bibliographic items). We are finding
some real surprises (both pro and con).
    The fastest address for me [address withheld].

[Glenbrook Drive]   14 October 1989
Dear Marcello:

    Many thanks for the Oz piece. I indeed would not have known of it
otherwise. I thought she did a fine job of summarizing Baum's Oz books,
though I didn't care much for her suggestions at the end for future Oz
books. Amen to her attack on *The Return to Oz* (last footnote).

                          Best,
                          Martin

[Eastern Michigan University]   9 January 1990
Dear Martin,

    A couple of items you might find of interest enclosed. A small chance
you can help me on two things I trying to run down. 1) Bert Reese is
alleged to have also had some fame as a dowser and was said to have found
some sites for Rockefeller. Know anything about this? 2) I am trying to
run down all I can on the psychic entertainer Eugenie Dennis (I have the
NY Times stories from 1924), in particular her or her mother's friendship
with Houdini. If you have no leads for me on either item, you don't need
to bother replying to this query.
    Happy Decade.

                          Best,
                          Marcello

[Culver Road]   20 February 1990
Dear Martin,

Mainly a note to say "thanks" for the article on carnival speech. I am glad to know about *Word Ways* which I have not previously run into.

I noticed that Prometheus's new paperback of your *Science: Good, Bad & Bogus* has a 1989 as well as the 1981 copyright notice. I can not see any differences. Did you add some new material? If not, you need not reply on this; but if you did, I would like to be aware of it to read it.

I just got the old Richard Himber "Mind Reading Is Bunk" series from *The American Weekly* (1944). Apparently there were four parts, but the legendary piece by Himber exposing Dunninger is not among them. Kreskin tells me that this fifth piece was never published. I had the impression from Stewart James and others that it had been. Do you know the facts on this?

Just finished the chapter on Peter Hurkos for our book on psychic detectives. I think it is pretty devastating to him. One small thing I wonder if you might know about. He was arrested for impersonating an FBI man back in 1964, but so far as I can ascertain, he was acquitted. Randi (in *Flim Flam*) says he was convicted and fined. Milbourne Christopher and others seem to think Hurkos' version of the events was bought by the judge. I have not been able to pin this matter down absolutely. Do you happen to know?

Re Randi, I gather that he and the lawyers had their oral deposition from Uri in NYC last week with the interrogation of Randi to soon follow. I suppose those transcripts could eventually become part of the public court record and would be fascinating reading (or viewing since on videotape). I still can not really believe Randi let himself get into this mess. He was always careless with facts, but his actions on all this strike me as just plain stupid. Randi may have some subtle strategy that I as an outsider don't perceive, but it sure looks from here like he has really hung himself with his interview comments on Byrd and Geller. Since Randi seems to have believed what he said, I keep wondering if he was somehow set up. I have occasionally thought of someday writing a book about both Geller and Randi. I may yet.

Anyway, thanks again for the carnival article. Hope I can reciprocate with something for you soon.

                              Best,
                              Marcello Truzzi

[Glenbrook Drive]   27 February 1990
Dear Marcello:

Nothing new in the new printing of *S:GBB*.

Wish I could help on the other questions, but I did not even know that Himber had written a series on mind-reading is bunk. I do know that Himber disliked Dunninger intensely. Nor do I know anything about Hurkos's arrest.

The Randi situation is a sad one, because Randi did speak carelessly about Byrd and Uri. It is hard to predict how the court cases will be resolved. Geller repeatedly (like Reagan) said "I don't know" in his deposition, and I would guess that his case vs. Randi will go nowhere. As for the Byrd case, "I don't know."

I have a contract with S&S for an anthology of parodies of "The Night Before Christmas." Do you know of any I might not know about? I have more than fifty, some pornographic, but many are too poor to use. I am particularly anxious to locate a parody by a deceased journalist named H. I. Phillips about Santa's visit to a poor working girl in NJ who yells "Sez you!" when Santa goes up the air shaft. Probably appeared in a NY newspaper. The book will be similar in format to my *Annotated Casey at the Bat*, recently reissued by the U of Chicago Press.

                              Best,
                              Martin

[Eastern Michigan University]   1 March 1990
Dear Martin,

I have no current leads for you re the H. I. Phillips parody, but I will make some inquiries. I tried to locate what I thought were some parodies of "The Night Before Christmas" in some college humor anthologies, but no luck. One such collection seems to be misplaced, but I will check it out when I locate it. I seem to recall seeing a parody (or maybe even more than one) in *Mad* magazine over the years. You probably already have checked there. I presume you are familiar with the old one liner: "Twas the night before Xmas and all through the house, not a creature was stirring—no spoons."

The Randi situation is indeed a very sad one. I think it likely that Byrd will win his suit (or at least get a fat out-of-court settlement). I suspect that the more it looks like Randi's private life will be brought into the case, the more likely it may be that the other defendants will look for a settlement. It is all really a very nasty business. As for Uri, I think he intends to continue legal actions for many years and probably will destroy Randi financially, whatever the legal outcome. Obviously, Geller's main money targets are the other defendants, and since he can easily show Randi had malice, Geller's case is a pretty strong one even for a celebrity suit.

What do you think of Uri's second suit already filed against Randi and others over comments in Japan where Randi is supposed to have told an interviewer that Uri was responsible for the suicide of Wilbur Franklin? Though there were some rumors that Franklin might have been murdered, I can not figure out where Randi got the idea that Franklin shot himself. But then I can't figure out where he got the nonsense about Byrd either. (Could someone be feeding Randi disinformation to set him up?) Meanwhile, the evangelists, who I expected to see sue Randi first given their extensive investigations into him, now lurk like vultures in a tree watching the animals fight it out below. I presume they have not yet dropped their plans for litigation. So, the current suits may just be the beginning of Randi's problems.

I found your comment about Uri's deposition particularly interesting since somewhat different from what I was told about it. However, my sources on this may be inaccurate. Still, if you have not actually read the documents, I urge caution should you be relying exclusively on Randi's description of their contents.

Best,
Marcello Truzzi

[Glenbrook Drive]   5 March 1990
Dear Marcello:

Please don't waste valuable time looking for parodies for me. I only meant for you to let me know if you happened to know of any. Yes, *Mad* ran four parodies, three of them by Frank Jacobs who is a friend by correspondence, and I have permission to use all of them. S&S phoned last week to say they couldn't do justice to a fall 1990 pub date, so they have extended it to fall 1991—a great relief because I have lots more research to do) not to mention the task of clearing copyrights, etc.

I'd better not comment any more on Randi's legal problems except to say that Byrd's past is pretty unsavory, and it will be fascinating to see how it all works out.

I subscribe to *Fate*, so I had already clipped the short piece about J. Z. Knight, but thanks for sending it. I am wasting far too much time trying to

run down the channeler of *The Urantia Book.* Mrs. Harold Sherman knows,
but she says she can't divulge the identity until her husband's papers are
opened at a future date set by his will. The channeler, she says, is one of
the adopted sons of Dr. J. H. Kellogg, the Adventist and cornflake king,
but he had *lots* of adopted children I would need a detective agency to trace
what happened to them.

Best,
Martin

[Eastern Michigan University]   9 March 1990
Dear Martin,

You probably saw this article re Einstein's first wife, but if not, I thought
you might be interested if only because of Evan Harris Walker's role in it.

Glad you have more time for your S&S collection. I will not search but
will keep alert to any new parodies that should some come my way.

A cautionary word: Randi a couple of years ago told me of Byrd's
allegedly unsavory past. I can only tell you that what he told me was
immensely distorted. You may, of course, have independent sources of information. But I thought I should say something to you on this because not
only was Randi's statement in his magazine interview wrong and libelous,
he added further nonsense in his talk to the NY Skeptic's gathering (which
you may not know about). I've talked to people who claim to know Byrd
very well over many years, and aside from his interest in porno pictures (of
adult women), which few today would call particularly unsavory, I don't
think Byrd has much to hide compared to Randi. But as you said, it will
be interesting to see the eventual outcome (which now may be as early as
June, I am told).

I also enclose an interesting cold fusion item, which you may not have
seen.

Did you hear about Geller's appearance on Soviet TV?

Best,
Marcello

[Eastern Michigan University] 21 April 1990
Dear Martin,

Thanks for sending me the *Physics Teacher* article. I hope you might send me the second part later.

Re the levitation by finger lifting, I am surprised you did not mention Hereward Carrington's doing it with people on an ice house scale purportedly showing a decrease in the total weight when the lifted man goes into the air. I once heard Uri claim that the trick was sometimes dangerous for he once continued floating on up to the ceiling!

I am told Randi says the lawsuits have so far cost him $70,000 out of pocket. Since Uri expects the suits to go on over several years, it looks bad indeed. I understand that the Byrd suit is now only going to trial around October due to a new judge.

I was very pleased to see that you have revised and updated your *Ambidextrous Universe*, which I always thought among your best books.

I can't remember if I earlier mentioned E. H. Walker's current interesting revelations about the first Mrs. Einstein's possible coauthorship of relativity theory. Been following it? Rather funny, too, since I hardly would classify Harris as a feminist.

Best,
Marcello

[Undated copy of the above returned with marginal comments by Martin]

[PSYCHIC ENTERTAINERS ASSOCIATION] 27 April 1990
Dear Martin,

Boy, do I feel stupid! When I got your PK article, I read over it very quickly and wanted to get a note off to you about it before I went to the post office that afternoon. Obviously, I either failed to remember or properly note when reading it hurriedly that you certainly did mention the Carrington weighing. I am, of course, properly embarrassed. I should not have written you about it before reading it again carefully (now done). In small mitigation: I did not intend my comment as a criticism, only as something to say I thought you might find interesting. I usually double-check before criticizing. Anyway, I clearly goofed, and it teaches me a lesson. Mea culpa.

I disagree about your view of Randi's chance for getting anything through any countersuits, but time will tell.

You may be right about Walker's feelings about Einstein. I hadn't thought of that as related to his motivation. However, since Walker claims consciousness is the hidden variable that would go against standard quantum explanation, he may see himself as supporting Einstein's position against Bohr.

Just finishing up our psychic detectives book. I think you will like parts (like the sections on fraud and pseudo-psi) but probably will not agree with the positive sections; I hope you might eventually decide to review the book (preferably somewhere other than *The Skeptical Inquirer* since I would prefer to answer you—if necessary—in a less biased forum).

Best,
Marcello

[Glenbrook Drive]   7 July 1990
Dear Marcello,

Thanks for that great Oz joke, which was indeed new to me. I doubt if we'll see it in *The Baum Bugle!*

C. Leroy had sent me a copy of his letter with the GK quote. (We have become pen pals since he became disenchanted with V.) GK is probably the most quotable of all recent British authors. Here's one I should have known about, but didn't come upon until recently: "When persons stop believing in God they don't believe in nothing; they believe in anything."

This will amuse you. John Grafton, Dover's top editor, made the mistake of asking me to write a foreword to their next printing of Crowley's *Magick*. I agreed provided Dover allowed me to blast the book as rubbish. So I turned in a foreword which essentially says that the book was not worth even its *first* printing, and that having read my foreword, the reader should throw the book away. Mr. Cirker, owner of Dover, is currently trying to decide whether to: reprint without my intro, print the book again *with* my intro or to drop the title from his list. He hesitates to drop the book because it has been one of Dover's best sellers. We shall see soon what he decides. If he uses my intro, which goes into detail about the life of Crowley, including his influence on some hard rock musicians (his picture is on the cover of Sergeant Pepper), it will be some sort of first in the history of publishing—a foreword that tells the reader he is a fool to buy the book if he thinks he will learn some black magic. (On page 95 of *Magick*, Crowley recommends the sacrifice of highly intelligent baby boys as the most effective way to gain energy.) A check of *Books in Print* discloses that almost all of Crowley's books are in print, and the chain stores are now carrying a mammoth volume giving all the rituals of the Golden Dawn! There must be lots of simple-minded people who are taking all this seriously.

Best,
Martin

[CSAR Center for Scientific Anomalies Research]   8 July 1990
Dear Martin,

I don't know if you get the *Journal of the American Society for Psychical Research*, but in case you don't, I thought you'd want to see the enclosed article on Stepanek and the review of your book on him, both in the same issue.

Is there a good existing essay on the criminological approach of Father Brown? I am seriously thinking of doing something on it but am new to the literature on him so may be thinking of reinventing the wheel. As his annotator, I assume you probably know the critical literature on him. I find it interesting to look at the differences between Brown and Holmes, which rather parallel (in some ways) the differences between Christie's Miss Marple and Poirot, and all of them have their methodological parallels among sociological theorists.

Cheers,
Marcello Truzzi

[Culver Road]   10 July 1990
Dear Martin,

I was much amused to hear from you about your foreword to *Magick*. I suspect you may not know that his biographer, John Symonds, feels about the same way you do about Crowley. I enclose a review of Symond's newest version by Wilson that you probably did not see when it came out.

I have always been amused by Crowley's book titled *Liber Oz*. It is his main book current Satanists claim as one of their own by The Beast. What would Baum followers say?

Actually, what most people underestimate about Crowley was his sense of humor. An old friend, Roger Staples, was writing his dissertation on Crowley's poetry (before Roger decided to chuck academia), and he convinced me Crowley was actually quite a good (if often pornographic) poet. Eric Dingwall knew Crowley pretty well and confirmed for me that much

Crowley did was strict tongue in cheek (perhaps too often somebody else's cheek).

Best,
Marcello

[Glenbrook Drive]   7 August 1990
Dear Marcello:

Thanks for Colin Wilson's review, which I had not seen. Dover is still undecided about whether to reissue their edition of *Magick*, with my introduction which tells the reader he is a fool to have bought the book, or to drop the book from their list. If they drop it, I'll place the foreword somewhere as an essay about Crowley.

Thanks, too, for the two pieces from the *Journal of the American Society for Psychical Research*. I did not have a copy of Kappers' report, but Beloff had sent me a copy of his review. I have returned the favor by reviewing Beloff's new book for the *London Times Literary Supplement*, and will try to remember to send you a copy when it appears.

The only essay I know that comes close to discussing Father Brown's methods in a way relating to sociology is the enclosed. All other commentators content themselves with saying that his methods are more intuitive than Holmes, based on FB's ability to get himself into the mind of the criminal, etc. Assume you know my little piece on Holmes and Brown meeting one another (it is reprinted in *Gardner's Whys and Wherefores*).

Just got back from a trip to Pittsburgh to check a private collection of material relating to Moore's poem about Santa, donated to the Carnegie Mellon library. The trip paid off, with half a dozen parodies I hadn't known about. I now have almost 100, including an amusing one in which St. Nick brings Holmes a Stradivarius, contradicting Watson's cover story about how Holmes bought it from a pawnbroker.

Best,
Martin

[Eastern Michigan University]   8 July 1990
Dear Martin,

Thanks for the interesting piece on Father Brown and historians. I have only recently become aware of what a great deal of writing already exists re Brown and Chesterton. At this point, I feel the task I had in mind might require far more reading on my part than expected, so I must assign a lower priority to the project given my other commitments. But I do hope to eventually do the piece when time permits my catching up with the Brown literature. Re your own Holmes/Brown piece, I think that must have been what subliminally sparked my enthusiasm, for though I had read it, I was not thinking of it when I first wrote you about Brown. Thanks for reminding me about your excellent piece.

Yes, please do remember to send me a copy of your review of Beloff's book (which I just got yesterday).

I am currently researching the late "Doo" R. C. Anderson, the Georgia psychic. Aside from two books on him, I have a batch of newspaper articles about him from the Chattanooga papers plus a video of a TV show on the paranormal that he appeared in. Do you perhaps have any stuff on Anderson? I am particularly interested in determining the truth about his alleged 16 hits (out of 16 tries in divining oil wells for John Shaw). Anderson was apparently quite a character since he earlier had been a circus strong man, a wrestler, and even a matador.

Though Randi sent out a press release indicating a "victory" in his court battles with Geller, Steve Shaw tells me that Randi has so far spent about $120,000 on his legal expenses. And I am told that he agreed not to counter-sue for his costs in exchange for Geller's agreement not to appeal. As far as I know, the suit by Byrd remains active and may go to trial in November depending upon some legal maneuvers happening this month.

In case you did not see it, I enclose a copy of John Maddox's letter in the current issue of the British & Irish skeptics' publication. I am told this was a result of Geller's threatening Maddox with a libel suit over Maddox's earlier statements.

Also enclose a copy of a review of Randi's *Nostradamus* book from the *Village Voice*, which you may already have seen.

Finally, I enclose a copy of the latest *News of the Weird*. If you don't already get it, you'd probably enjoy it.

Could you perhaps supply me with William Poundstone's address? I'd like to contact him.

Best,
Marcello

[Culver Road]   20 August 1990
Dear Martin,

Just a note to thank you for the Doc Anderson interview, which I am very happy to have. (Actually, I had the magazine but had not realized his interview was in them. One problem with all my damned periodicals is the absence of an index to them.) If you ever need anything from such mags, let me know since I have a great many. In this case, I simply did not think to look through *Psychic*.

You may have heard the sad news that D. Scott Rogo was stabbed to death last week by some sort of burglar. He was only about 40.

                              Best,
                              Marcello Truzzi

[Humorous stationery]   10 May 1991
Dear Martin,

Thought you would enjoy the enclosed recently sent to me (and which you may already have had sent to you).

Looks like Randi is finally getting called on his frequent claim that Geller's methods can be found on the back of a cereal box (I never expected Geller to bring suit over that matter, but I guess the issue of priority is important in pseudoscience as well as normal science).

If you have not yet read it, I urge you to take a look at Eugene F. Mallove's *Fire from Ice: Searching for the Truth Behind the Cold Fusion Furor* (Wiley, scheduled for July 15th release). Things seem to be taking a sharp new turn in this debate.

                              Cheers,
                              Marcello Truzzi

P.S. Have you seen Dennis Rawlins' new journal *DIO*? Fascinating.

[Willis Road]   30 November 1991
Dear Martin,

I am hoping you can correct or amplify my memory about something I now unclearly recall from over ten years ago when I heard about it from either you or Ray Hyman (at about the time that I visited you with Ray at your home). I am pretty sure I heard it from Ray, but since the story was about you, it seems best that I write you directly about it. (I may have something about it in our past correspondence, but it seems easier just to write you than to go through that mass of paper.) Here is what I recall, and please let me emphasize that I realize I well may have the story rather confused:

When Targ and Puthoff (T&P) were at SRI and you were interested in their work on their psi-teaching machine, you got some information from an inside (on-site) source (I think it may have been Persi Diaconis working for NSF if not the military) looking into things there "undercover." This source told you that his examination revealed that their results were due to some flaw, which I'll just call method X. You wished to expose their experimental incompetence but had the problem that describing method X would blow the cover of your informant (since T&P could then deduce where you learned of it). So (as I recall the story I was told), you carefully examined the published reports and saw that there was a possibility that another flaw (method Y) might also produce the seemingly incredible psi results. (That is, as a conjuror, you saw the possibility for another method to accomplish the same effect.) When you published your critical piece citing method Y (which you already knew was not the actual method or flaw in their work), T&P responded by citing some additional and probably sound evidence (not given in the original description of their experiments) demonstrating that your proposed method/flaw Y could not actually have accounted for their results. You then ignored their reply because (a) you already knew that it was method/flaw X that was the real reason, anyway, and (b) you still could not mention method/flaw X without embarrassing your original informant (probably Persi).

This is the way I recall what I was told. Do I have the basic story correct here? I'd appreciate whatever correction or amplification you might be willing to give me about it.

Best wishes (and Happy Holidays),
Marcello Truzzi

[Glenbrook Drive]   5 December 1991
Dear Marcello:

The only fact in your account is that Persi was my informant. He had been hired by the Jet Propulsion Lab, which had been monitoring the P and T experiment, to visit the SRI Lab under an assumed name. At the time, he didn't want his cover blown. Now he doesn't mind. It was Persi's negative report that had a role in the refusal of further funds to SRI (which P and T had requested for repeating the failed test).

Persi told me he had been shocked by the state of the tapes in the first phase of the experiment, which P and T themselves treated as relatively uncontrolled. The machines had been placed here and there in schools, the tests supervised by untrained teachers. When machines broke down, as they often did, Persi told me the students had continued keeping the records *by hand.* He said the tapes were in "bits and pieces," not in unbroken runs as claimed. After I reported this in my column, saying that it came from an observer I could not name, *Scientific American* received a letter from P and T, in which they accused me of saying that they had "torn" the tapes. I never said this. Nor did I ever say to anyone that I made up the "bits and pieces" just to make a case. *Scientific American* then offered, in a letter published in *Scientific American,* to pay for an independent statistician to go to the lab and verify whether the tapes were in good shape. P and T never answered this request. So far as I know, no outsider has yet been allowed to inspect the tapes. I assume the tapes are still at SRI, where no doubt they are classified.

The totally false rumor that I said I made all this up was passed on by you to Ron McCrae, so he told me, who in turn passed it on as an unverified rumor to Targ. Against the strong advice of Harary (who has recently written about this), Targ then put it in their book, even putting quote marks around my alleged statement as if he had a tape recording!

I wrote to the editor of Villard Books, a firm owned by Random House, and said I would take no legal action if he removed the libelous statement from their paperback. He assured me this would be done. It was not done. The editor was so unconcerned that he did not even check the paperback and write to me an apology. The book was on sale for months before I discovered that nothing had been removed. This was the *sole* basis of my legal action. I have you to thank for it all because Random House settled for ten thousand dollars, and I received an abject apology from the editor. At the time I was a Random House author (several books), and have since done another book for them.

The most amusing aspect of my deposition is worth telling because it shows how sneaky Targ can be. I had given the RH lawyer the date on which I first was in contact with Ron McCrae. He then staggered me with a copy of a letter I had written to "Dear Ron," about the SRI experiment— a letter a year or so in *advance* of the time I said we first were in contact. I worried about this all the way back from NY. Fortunately, I had preserved in my file all my letters about the SRI experiment. There I found that the letter had been sent to the SRI PR man, whose name also was Ron! I can only assume that the lawyer had been given the letter by Targ and told it was to McCrae when of course he knew otherwise.

I still believe the tapes are in bits and pieces, and that that is the main reason no one has been allowed to see them. I have no theory as to how the first phase came up with positive evidence beyond the possibility of careless recording and a fudged statistical analysis. As you may recall, when the experiment went to the second phase, in which a computer was used to eliminate all such errors, the results were negative.

I have a long article on William James and Mrs. Piper in the next issue of *Free Inquiry*, and I am currently doing a book about the Urantia cult. The movement interests me mainly because of the curious role played by Dr. William Sadler, the famous Chicago psychiatrist, and also because of the large chunks of science in the Urantia Book.

Best,
Martin

[Willis Road]   31 December 1991
Dear Martin,

Just a belated note to thank you for your detailed correction of the tale I dimly remembered. Obviously, I did not hear my warped version from you, so I am almost certain I got it via Ray. I definitely did not get it from Ron McRae. I will have to check with Ray to see if I can figure out just how I got it all so distorted. I do recall also talking to Persi once about the story, many years ago, too, but he did not know the details of it at your end. As for McRae's version of things (which he told me came from Puthoff when I first asked Ron about its inconsistency with what I thought happened), Ron confessed a few years ago to me that he made up several other things in his book, so I am now doubtful that he got his version as he then told me.

Did you see Geller's interview in Stan Allen's *Magic*? I look forward to reading Randi's reply next month. Meanwhile, as you may know, Randi (via his attorney) is threatening to sue his old friend Jim Moseley over Moseley's coverage about the Byrd & Geller suits in Jim's *Saucer Smear*. And, I think Geller is about to sue Kurtz and Prometheus, too. So all the litigation goes on and on (to the joy of the lawyers involved, anyway).

Martin, do you perhaps have a complete list of the Members of CSICOP? The listings published in The Skeptical Inquirer are stated to be partial. Thus, it is impossible to know if once listed but now deleted people are no longer Members or if there are never-listed people who are Members. Since

CSICOP is supposed to be a scientific organization and not a secret society, I expected Kurtz and/or Karr at least to reply to my repeated requests for a full membership list. If you have such a list, and assuming you agree with me that such things should be public, I wonder if you might share a copy with me. (Perhaps I should mention that George Hansen tells me that when he asked Kurtz for a copy of the membership list, Paul denied there were any "members" George reminded him that the CSICOP by-laws explicitly define the Fellows and Technical Consultants as the Members. George was not able to get a list from Kurtz, either.) This reticence, especially since Paul has been cooperative with me about some other matters, seems most odd to me, and it raises some obvious questions. Can you help me on this?

Again, thanks for your correction of my apparent misinformation on the SRI story.

Best wishes, and Happy New Year,
Marcello Truzzi

[Glenbrook Drive]   4 January 1992
Dear Marcello:

Thanks for yours of 31 Dec. I have a vague recollection of saying to Ray, on the phone, that quite apart from Persi's then secret report on the tapes of the P and T experiment, that my saying in *Scientific American* that I had it on "good authority" might be a way of forcing P and T to let an outsider examine the tapes. Ray may have misinterpreted this as my saying that I made up the remark about an informant.

I certainly have no list of CSICOP members. I assume Kurtz may not want the list circulated for fear it would be used as a mailing list.

Best,
Martin

[Consulting Anomalist]   11 January 1992
Dear Martin,

I haven't yet reached Ray as I intended (he seems to be out when I have called), but your reconstruction of things seems quite likely.

Regarding your conjecture that perhaps Kurtz has not sent me the list of CSICOP's members because he "may not want the list circulated for fear it would be used as a mailing list," there are several problems with that speculation. (1) I presume almost all the members' names and affiliations are already published in *Skeptical Inquirer* and all I want is the presumably very few additional names. (2) I don't seek addresses, only names. (3) He has ignored my requests rather than simply given me the reason you conjecture he has (and he has not ignored other letters of mine so presumably is not simply generally avoiding communication with me for some reason unknown). And (4) as I indicated to him my reason for wanting the list (in connection with an article I wish to write) and not to have a mailing list, and this is a legitimate reason in terms of scientific norms about disclosure, it certainly does appear that he wishes to keep something about this information from public view. The fact that you as an Executive Council member do not have a list yourself further seems to suggest that, I think. (You obviously have greater trust in the uprightness of Paul Kurtz than I am afraid I have learned not to have.) I must say that I am a bit surprised about all this since such lack of public disclosure is contrary to what most scientists might expect from a purely science-promoting rather than a quasi-religious (Humanist) organization (which many CSICOP critics have accused CSICOP of actually being).

I am told Randi's reply to Geller in *Magic* is now published, but I have not yet seen it. I am told Randi claims he has been misquoted and mistranslated from the Japanese. Perhaps so, but that should be pretty easy to demonstrate from tapes and formal translations when it comes to the courtroom. Odd, too, that Randi has not bothered to simply transmit his explanation to the Japanese court rather than ignore those proceedings as I am told he has done.

By the way, I hope you caught the interesting review of *Galileo's Revenge* in *Science* a few weeks ago.

Best,
Marcello Truzzi

[Glenbrook Drive]   2 May 1994
Dear Marcello:

I have nothing in my files on either Alexander Cannon or Manley P. Hall. I wasn't able to check indices of books because my books are all inside cartons to be moved to a new house we bought, in the same town. I won't be sending out change of address cards until we sell our two condos (one I use as an office) and move, which may not be for many months.

All best,
Martin

[Consulting Anomalist]   28 August 1994
Dear Martin,

    Other than Schwarz's book on him, and the few references to other stuff therein, Jacques Romano seems to have little about him published I can find. He was a pseudo-psychic, con man, and (according to Max Holden in 1929) one of the ten best card magicians in the New York area. Dai Vernon knew him, too. Anyway, I wonder if you ever knew Romano or might have something in your files about him. He seems to have been a fascinating scoundrel. Among other things, he told everyone he was 20 years older than he really was and even had phony passports made. He seems best known between around 1925-1960. So, I thought you might have run into him in the NYC area. Can you help me get information on him?

    If you are working on anything my files might have something on, let me know.

    If you have not yet seen Loyal Rue's *By the Grace of Guile: The Role of Deception in Natural History and Human Affairs* (Oxford U. Press), I think you would find it of interest. A strange theological call for a "noble lie."

Best,
Marcello Truzzi

[Chestnut Road]   22 January 1995
Dear Marcello:

    Thanks for recent clippings you have sent.

    I'm curious. Do you share Westrum's UFO views?

    Please—no ambiguous answer. On a probability scale of 0 to 10 (zero absolute disbelief, 10 absolute certainty) do you believe with Westrum that aliens are visiting earth in UFOs?

    Am checking page proofs on my 600-page in depth report on the Urantia cult. Charlotte thinks writing it was a waste of time, and she may be right.

Best,
Martin

["Low-Keyed Prof Believes in UFOs," *Tulsa World*, 17 January 1995, about Ronald Westrum.]

[Eastern Michigan University] 26 January 1995
Dear Martin,

Re your letter/note of 22 January 95, I will try to answer your question without ambiguity.

No, I do not share Ron Westrum's UFO views. On a scale of $0 =$ absolute disbelief to $10 =$ absolute certainty, I would place my own view toward the claim that aliens are visiting earth as about 1.

A small word, however, about Ron's belief. Though Ron's own belief is that aliens have made contact, he would, I think, distinguish between his own belief and what he would assert was a scientifically convincing case for others. In other words, he would (as do I) distinguish between belief and knowledge. Alien contact for Ron is an hypothesis which he thinks is very likely. But he does not assert that the case for the hypothesis is one he expects any reasonable scientist to also accept. (This strange typewriter suddenly started double-spacing on me, so make no inferences from this odd pattern. I am borrowing the secretary's exotic new IBM machine full of bells and whistles.) I would not characterize Ron as a True Believer. He admits he may be wrong. It is just that he finds no satisfactory explanation for the reports and finds Baker, Klass, et al., unconvincing in their debunking efforts. I share some of Ron's critical view of these would-be debunkers. Ron and I have debated our differences a lot, and I think we simply agree that reasonable people can differ on this matter. Despite our differences, I do think Ron is a reasonable person and though I think he is dead wrong, I do not think he is irrational.

In the final analysis, however, Ron and I part company on this issue. Ron seems largely influenced by the many contactees he has interviewed. Perhaps if I had his experiences with field investigations, I might be less disbelieving. But at this point, I definitely disagree with Ron. Hope this answers your question adequately. Since you might find it of interest and since you are mentioned in it, I enclose a copy of a paper I just finished on pseudoscience and its definition/demarcation.

I suspect I will be one of the few people who will read your Urantia book when it comes out. I am glad you undertook the task, but I doubt many people will be turned on to the subject unless your new revelations entice them into it.

Best,
Marcello

[Eastern Michigan University]   1 February 1995
Dear Mr. Gardner

Marcello Truzzi was kind enough to share his correspondence with you on the UFO article with me. I thought I would let you know that the *Detroit News* article severely misquoted me. Although the differences might seem small to you, I told their reporter not that I knew the government was withholding information (e.g.. about Roswell), but rather that I *thought* a big cover-up was taking place. I told him that while I thought the experiences with aliens are real, I have no idea where they come from or what they are. The second paragraph in the article is almost a pure fabrication on the part of the reporter. It turns hypotheses into certainties, speculations into claims.

The enclosed article is a far better representation of my thoughts than the one you sent Marcello. Since the reporter had it when he wrote his piece, obviously one must conclude that the school where he studied journalism did not teach him that truth is the first priority.

Best wishes,
Ron Westrum

[Marcello: I appreciate your sharing your correspondence with Gardner. Ron]

[Consulting Anomalist]   11 July 1995
Dear Martin,

I was just "amazed" to learn that Randi is soon going to be doing a column for *Scientific American.*

Because some changes to the text in Randi's recent *Encyclopedia* made by the lawyers at St. Martin's in the American edition were *not* made in the U.K. edition, he is about to be sued in England. If you have read even the U.S. edition of his book, you probably can see how full of errors it is (e.g., look at his entry for Lady Wonder whom Randi explains away as an example of ideomotor action despite the obvious explanation of quite intentional cues from Lady's owner). I hope that your friends at *Scientific American* provide Randi with a proper fact-checker. If not, I think it likely that they, too, will be seeing some lawsuits.

Have you had any reaction from the Urantia folks to your book? Do they see you as a minion of the devil or just as misguided? I suspect the reaction to you will be much as it has been in the past for other such groups (e.g., the plagiarisms of Blavatsky).

If you have not looked at Philip Kitcher's *The Advancement Of Science: Science Without Legend, Objectivity Without Illusions* (Oxford U. Press), I think you'd enjoy it.

This year's Parapsychological Association meetings will be in Durham, and it is the centenary of Rhine's birth. Lots of people will be there (Dan Cohen may even make it), and I presume this is near you. want to encourage you to come by if so. It would be good to see you. Despite our differences, I think you know that I hold you in some esteem and recognize that we probably agree on more than we disagree.

Best,
Marcello

[Chestnut Road]   13 July 95
Dear Marcello:

Thanks for letter. I'll check the new Kitcher book.

I do get occasionally to Durham, but I shy away from parapsychology gatherings, though it would be good to see you after so many years.

I'm getting some favorable mail from readers of my Urantia book, but as you can imagine it is being severely blasted on the Urantial internet, a group of several hundred Urantians who send postings back and forth.

I, too, was surprised that *Scientific American* hired Randi. You probably saw the spread about him in the current issue. I no longer know *anyone* on the staff. As you may know, last year the German conglomerate (it owns Macmillan, St. Martin's Press, *Nature*, and a few dozen other companies) fired everybody, including Jonathan Piel and Phil Morrison, book reviewer. When I was with the magazine, Dennis Flanagan's policy was not to dignify pseudoscience by publishing *anything* about it. Evidently the new staff is following the line laid down in the enclosed essay.

Best,
Martin

[Willis Road]   18 July 1995
Dear Martin,

Thanks for your letter of July 13. Sorry you are unlikely to attend the Parapsychology Association meetings, it would indeed have been good to see you. Pat and I will be driving and if Hendersonville is near, I might be able to go there for a chat with you. Would you be available?

Thanks for sending me the *NY Times* column by Park on "Voodoo Science" which I had missed seeing. I can certainly see why its stance would probably appeal to you, but I'm afraid that I still agree with Sagan that the best antidote to bad science (so-called pseudoscience) is good science, and I saw a lot of good (highly critical and even debunking) science going on when I attended this spring's conference on research methodology sponsored by the very NIH Office of Alternative Medicine which Park condemns in this column.

Have you seen Bob McConnell's recent book-length attack on the influence of New Age crap on proper parapsychology? In many ways, the parapsychologists are the last positivistic holdouts, insisting on things like double-blind controlled experiments. They are really now more methodologically conservative than many conventional psychologists. They, too, are being attacked (when not ignored) by the extreme relativists and postmodernists/deconstructionists that you (and Park) deplore. As I see it, though you may disagree with McConnell's conclusions about psi (and perhaps other substantive issues), you (and I) share his perspective on scientific method and that is what is sad about current matters.

In reference to my last letter to you, according to my latest information, it looks like Geller will be suing Randi's British publisher but not Randi himself. I suspect the suit may trickle down to the U.S. Publisher and Randi at some point. Anyway, I am looking forward to seeing Randi's new *Scientific American* columns. If they are as inaccurate as his *Encyclopedia* entries, *Scientific American* may regret its decision.

I wonder if you have seen the extensive documentation by Jim Lippard and some other skeptics of the seeming plagiarisms by Robert Baker (which go well beyond the materials presented by Hines in the current issue of *Skeptical Inquirer*). I hope CSICOP will squarely face this issue and not just ignore it or cover for Baker (as so far seems the case to many of those commenting on it). What makes this particularly interesting to me is that Baker has often cited cryptomnesia as an explanation for seemingly paranormal matters but steadfastly refuses to claim that as his own defense.

Best,
Marcello Truzzi

[Chestnut Road]   23 July 1995
Dear Marcello:

Thanks for letter and enclosures. Hendersonville is in the western mountains of NC (it's a long east/west state), about 300 miles from Durham. The route is 40 to Asheville, then south on 26 to Hendersonville. A visit would be fine if you and Pat are willing to endure such a long trek.

Hendersonville has dozens of motels, ranging from moderate to costly, and lots of good restaurants.

If you decide to come, I'll send you a map showing how to find our house.

Best,
Martin

[Chestnut Road]   14 March 1996
Dear Marcello:

Thanks for the audiotape. I have the text of Keillor's "Casey"—it ran in the *NYT Magazine*—but its good to have his recitation for my files. I quote from the parody in the revised edition of my *Annotated Casey*, recently issued by Dover. I'm now collecting parodies of *all* famous poems for a book to follow my Dover collections: *Best Remembered Poems*, and *Famous Poems of Bygone Days*. Not interested in parodies of *style*, but only funny parodies of *specific* well known verse. "Casey" leads the list in number of parodies, with "Night Before Xmas" a close second, followed by the "Old Oaken Bucket."

Best,
Martin

Dear Marcello:

An excellent article, with valuable quotes and data, and a good bibliography. Here are some random observations.

I would like to have seen a sharper distinction between laboratory evidence for ESP, based on statistics, and physical manifestations such as the rising tables and floating trumpets of the great mediums, materialization of objects (Sai Baba, and Uri), Uri's translocations of objects (including himself!), metal bendings, eyeless vision, thoughtography, Kulagina's PK work, and so on. My guess is that if you polled those magicians who believe in ESP, you would find the vast majority denying belief in such physical phenomena i.e., events that *look* like magic tricks.

I'm not surprised so many magicians believe in ESP. In my experience, magicians are a fair cross-section of the populace. I would expect them to conform to Gallup poles. Assuming that "magician" means, say, a member of the SAM or IBM, I would guess about half would believe in UFOs, astrology, and angels. I've known many magicians who were fundamentalists and believed in Satan, hell, and that Jesus walked on water and turned water to wine.

As you perceived, stage magicians are not very knowledgeable about magic. Of course "knowledgeable" is a fuzzy term, but there is a wide gulf between the average person who calls himself a magician, and extremely knowledgeable magicians. On page 7 you cite Krippner, Targ, and Hastings as "involved with magic in their youth." I suppose this is true, but it is hard for me to think of any parapsychologist more ignorant of magic than those three. Anyone who owned a magic set as a boy might be said to have been "involved with magic."

A knowledgeable magician clearly is no more competent to evaluate laboratory testing for psi than anyone else. (It took a statistician to catch Soal cheating.) But when it comes to the work of Geller, Kulagina, etc., the kind of magician most qualified to detect trickery is the close-up specialist, not a mentalist. When you (p. 27) say that mentalists are the best qualified to detect "pseudo-phenomena," you should qualify this by saying pseudo-phenomena of the sort that mentalists perform. A good mentalist need not know anything about close-up magic.

I think it is a misuse of words to suggest that the chap who identified phonograph records by studying their grooves would have been dubbed "paranormal" had he claimed psi powers. Like all words, "paranormal" is fuzzy, but Randi's simple tests showed there was nothing that could be called "paranormal" about what this chap did. There is a wide gulf between what he did and the physical phenomena of psi.

Did you know that "Doc" Tarbell wrote a hardcover book on physiognomy, and that the "Dr" was based on a degree in napropathy, a weird offshoot of chiropractic?

In your bibliography it would be good to substitute *Magic Numbers of Dr. Matrix* (Prometheus 1985) for the book you list. The 1985 book includes the earlier one, with many new chapters added.

I assume you noticed that in Stein's Encyclopedia I also wrote the entries on Piper, eyeless vision, and Slade. I don't mean you should add these, but mention it only in case you didn't read those entries. The one on Slade is particularly revealing.

On page 1 you say magicians use the term "shut-eyes." You may have evidence for this, but I have never heard the term before.

On page 3 you list Beloff as being "well informed about the methodology of magic." Incredible! Beloff knows *nothing* about magic, and freely admits it. Take him off this list! I know of no parapsychologist more ignorant of magic.

You suggest that magicians reflect a higher belief in ESP than the public. This could be true. If so, I would attribute it to a sense of wonder. It could contribute both to interest in magic and interest in the paranormal.

On p. 6 you call Francis Carlyle a "stage performing medium." Maybe you know something I don't. I knew Francis well. He was a nightclub performer and a card worker, and an alcoholic, but I never knew him to pose as a medium.

Aside: Did you know that Lulu Hurst wrote a fascinating autobiography?

Among magicians who develop their own tricks independent of tradition, two others come to mind. The late Johnny Paul, of Chicago and Vegas, and Helmut Green, a European card expert of unbelievable skill.

Along with your usual bashing of Randi, I would have liked to see some bashing of Rhine for his failure to allow magicians to observe tests, and his refusal to publish reports of tests that were failures, and his valiant efforts to avoid admitting cases of outright fraud. There should be a mention somewhere of the tendency of many top parapsychologists to fudge data. I would have liked to see some stress of the importance of having a knowledgeable magician *present* during tests of physical psi phenomena, rather than calling him in after the fact for his opinion. Trying to reconstruct a car accident is a lot different from being *there* to witness it! As you know, parapsychologists exclude magicians as observers on the grounds that they influence the results negatively. Witness Eisenbud's fury with having magicians present to observe Ted Serios!

I assume you know that Charlie Reynolds, a very knowledgeable magician, became friendly with Uri and helped him work out a magic show that turned out to be a huge flop. It would be interesting to know if Geller offered to pay Charlie, and if so, did he ever collect?

Best,
Martin

[Chestnut Road]   12 August 1996
Dear Marcello:

When I mentioned a Mr. Green as a magician who fools other magicians because he does only his own tricks, uninfluenced by others, I think I got his first name wrong. He is Lennnart Green, of Sweden. He does incredible close-up work. There are video tapes on sale in which he explains his methods.

Best,
Martin

[Chestnut Road]   22 December 1997
Dear Marcello:

I'm currently working on an SI column about Puthoff and his research on extracting energy from the fluctuations of virtual particles in the vacuum of empty space. I was struck by the fact that in the article about this, in *Scientific American* (December '97 issue), nowhere did Philip Yam, the author, mention Puthoff's background at SRI, or his youthful association with the Scientology church. Most physicists think that machines for extracting this zero-point energy are as hopeless as perpetual motion machines.

Now for my reason for writing. There were rumors that Puthoff and Targ had a falling out before they left SRI to go their separate ways. Do you know if this is true? I don't want to mention this and have it turn out they are still good friends.

All best,
Martin

Dear Martin,

Re your inquiry about the current relationship between Puthoff and Targ, I don't know if "falling out" is an appropriate term. As far as I know, they remain on good terms but are not working together. Targ was let go at SRI, and I think there was some dissatisfaction with his work. Targ has always contended he left SRI because they wanted to work on military and espionage applications for remote viewing, and he wanted to do pure research. Though I suspect the parting was not entirely friendly, I have never heard either T or P say anything hostile about the other.

I note that you expressed surprise that no reference was made to P's past work with psi or Scientology in the *Scientific American* article. It seems to me that is pretty appropriate. After all, why bring up Edison's connection with Theosophy or his spirit contact efforts when discussing his work on the electric chair? Critics like ourselves regularly complain that scientists who work in psi research lose rigor when working outside their fields of expertise. Yet, P's recent work on zero-point energy is clearly within his field of expertise. I would think you would welcome that return (after all, his work on zero-point stuff has been in some top refereed physics journals, and I believe he just gave a talk on it at NASA), and he is far from alone among qualified physicists (though admittedly in a minority) in thinking energy extraction may be possible. Visionaries may wrong, but we need to encourage their existence (especially if their funding is from private sources). I also think it is important that P has been doing some important tests on other people's alleged "free energy" machines which revealed their failures. So, some of his recent work has been significant debunking.

My best wishes to you for a wonderful 1998!

Cordially,
Marcello

Dear Marcello:

There are rumors that you are writing a biography of Uri Geller. Is this true? If so, do you have a publisher and a pub date?

My current project, and a very pleasant one, is writing an article on Hugo Gernsback for *Scientific American*. I have a complete run of his

Science and Invention, on which my piece will focus, with reproductions of many of Paul's great covers.

> Best,
> Martin

Index

451

Printed in the United States
By Bookmasters